Studies in Logic
Volume 6

How To Sell a Contradiction
The Logic and Metaphysics of Inconsistency

Volume 1
Proof Theoretical Coherence
Kosta Dosen and Zoran Petric

Volume 2
Model Based Reasoning in Science and Engineering
Lorenzo Magnani, editor

Volume 3
Foundations of the Formal Sciences IV: The History of the Concept of the Formal Sciences
Benedikt Löwe, Volker Peckhaus and Thoralf Räsch, editors

Volume 4
Algebra, Logic, Set Theory. Festschrift für Ulrich Felgner zum 65. Geburtstag
Benedikt Löwe, editor

Volume 5
Incompleteness in the Land of Sets
Melvin Fitting

Volume 6
How to Sell a Contradiction: The Logic and Metaphysics of Inconsistency
Francesco Berto

Volume 7
Fallacies — Selected Papers 1972-1982
Douglas Walton and John Woods, with a foreword by Dale Jacquette

Studies in Logic Series Editor
Dov Gabbay   dov.gabbay@kcl.ac.uk

How To Sell a Contradiction
The Logic and Metaphysics of Inconsistency

Francesco Berto

© Individual author and King's College 2007. All rights reserved.

ISBN 978-1-904987-43-7
College Publications
Scientific Director: Dov Gabbay
Managing Director: Jane Spurr
Department of Computer Science
Strand, London WC2R 2LS, UK

Original cover design by Richard Fraser
Cover produced by orchid creative www.orchidcreative.co.uk
Printed by Lightning Source, Milton Keynes, UK

All rights reserved. No part of this publication may be reproduced, stored in a retrieval system or transmitted, in any form, or by any means, electronic, mechanical, photocopying, recording or otherwise, without prior permission, in writing, from the publisher.

Table of Contents

Acknowledgments x

Introduction xiii

Part I. The Reasons 1

1 "The Most Certain of All Principles" 3

1.1 A Strange Subject 3
   1.1.1 A πολλαχῶς λεγόμενον 4
   1.1.2 The Law(s) 11
1.2 Getting in Touch with the T-schema 16
1.3 Can We Believe a Contradiction? 19
   1.3.1 The Aristotelian argument 20
   1.3.2 Łukasiewicz's criticisms 22
   1.3.3 Beliefs in the *Tractatus* 24
   1.3.4 Moving On 25
1.4 Challenging the Law 26
   1.4.1 What is a Paradox? 27
   1.4.2 Kantian vs. Hegelian Strategies 31
1.5 Semantic and Set-Theoretic Paradoxes 35

2 True Lies 37

2.1 Liars 37
   2.1.1 Informal Liars 37
   2.1.2 Semantic Closure 40
   2.1.3 Formal Liars 42
2.2 Two Lines of Attack 47
2.3 Parameterisation, Part One 47
   2.3.1 Its Failure 49
   2.3.2 Liar II – the Revenge 51
2.4 Gaps 53
   2.4.1 More Revenge Liars 58
2.5 The Essence of the Liar 60

3 Unbounded Abstraction — 63

3.1 Existence and Objectivity, or the Abstraction Principle — 63
   3.1.1 Notable Paradoxes — 64
3.2 Logical Types — 69
   3.2.1 The Vicious Circle Principle — 69
   3.2.2 Parameterisation, Part Two — 72
3.3 Aristotle, **ZF**, and the Limitation of Size — 76
   3.3.1 The **ZF** strategy — 76
   3.3.2 *Aussonderung* — 78
   3.3.3 Cantor's Domain Principle — 79
3.4 Von Neumann's Proper Classes — 82
   3.4.1 Axiom (IV.2) — 83
   3.4.2 Bernays and Classes Taken (Not) Seriously — 85
3.5 The Cumulative Hierarchy — 87
   3.5.1 Logical and Mathematical Problems — 89

4 The Gödel Paradox — 93

4.1 Peano Arithmetic, **PA** — 93
4.2 Gödel, First Half — 94
4.3 The Argument — 98
   4.3.1 The Naïve Mathematical Theory — 98
   4.3.2 Its Inconsistency — 100

Part II. The Theories — 105

5 "On Detonating" — 107

5.1 The Scotist Conception of the Absurd — 107
   5.1.1 Inconsistent and Trivial — 107
   5.1.2 *Ex falso quodlibet*, Paraconsistency — 108
   5.1.3 "...Understood By Every Thinking Man" — 111
5.2 Disjunctive Syllogism and the Conditional — 114
5.3 "Change of Logic, Change of Subject" — 116
   5.3.1 Three Battlefields for Intuitions — 117
   5.3.2 The Avoid Overload Condition — 120
5.4 The Curry Paradox — 121
5.5 The Classical Recapture — 122
5.6 Weak Paraconsistency, Dialetheism, and Something in the Middle — 124

6 Non-Adjunction and Impossible Worlds — 131

6.1 Discussive Logic — 131
6.2 Impossible Worlds — 133
   6.2.1 Rescher and Brandom's Logic of Inconsistency — 133
   6.2.2 Impossible Worlds in General — 134
   6.2.3 Non-Adjunctive Worlds — 135
6.3 Non-Adjunctive Troubles — 139
6.4 Truth-Functionality, Sub-Valuations, and the Italics Argument — 141
6.5 Which World are You Living In? – Part One — 144
   6.5.1 The Russellian Fallacy — 145
6.6 David Lewis' Fragmentation Strategy, and Preservationism — 148

7 Positive-Plus Systems — 151

7.1 Logics of Formal Inconsistency and C-Systems — 151
   7.1.1 Bits of Positive-Plus Syntax — 152
   7.1.2 Climbing the C-Ladder — 153
7.2 Da Costa Negation and its Troubles — 157
7.3 Adaptive Strategies — 162

8 The Logic of Paradox — 167

8.1 Overview — 167
8.2 True, False, True and False — 168
   8.2.1 The Original **LP** — 168
   8.2.2 Fixing the Conditional — 173
   8.2.3 **LPQ** and Multi-Criterial Terms — 175
8.3 **LPm** and the Classical Recapture in Priest's Approach — 177
   8.3.1 Consistency as the Default Option — 179
8.4 Consistency and the Metalanguage — 181
   8.4.1 Semantic Operators in **LP(Q)** — 181
   8.4.2 Forcing Consistency — 182

9 Relevance Logic and Impossible Worlds Again — 187

9.1 Overview — 187
9.2 Relevant Implication — 189
   9.2.1 The Paradoxes of Material Conditional — 189
   9.2.2 Implication and the Conditional (Boil Down to the Same) — 190
   9.2.3 "Washing Dirty Money Through Mexico" — 191
   9.2.4 The Variable Sharing Property — 192
9.3 Relevance and Contradiction — 194
9.4 Bits of Relevant Syntax — 198

9.4.1 The Mainstream Systems	198
9.4.2 Dialectical Logics by Routley and Meyer, **DL** and **DK**	201
9.5 Relevant Semantics	203
9.5.1 Pure Semantics, Applied Semantics	204
9.5.2 Four Corners and the American Plan	206
9.5.3 The Australian Plan and Impossible Worlds	207
9.5.4 The Routley Star	209
9.6 Ultralogic	211
9.7 Relevant Troubles	214
9.7.1 Strong and Weak Assertions	214
9.8 Which World are You Living in? – Part Two	217
9.8.1 Infons and Beyond	217
9.8.2 The Russellian Fallacy Again	220
9.9 Strong Paraconsistency as an Idealism	222
9.10 Brady's Logic of Entailment, **DJdQ**	224
9.10.1 Bits of Syntax	224
9.10.2 Meaning Containment Semantics	226

Part III. The Uses — 229

10 Semantically Closed Language — 231

10.1 Applications	231
10.2 *Desiderata*	232
10.3 Priest's Semantically Closed Theory	233
10.3.1 The Language of the Theory	234
10.3.2 Axioms and Details	235
10.4 Does it Work?	238

11 Paraconsistent Set Theory and Metalogic — 241

11.1 *Desiderata*	241
11.2 Gödel, Second Half	242
11.2.1 "Consistency… Must Be Taken in Faith"	243
11.2.2 Escape from the Undecidable	244
11.3 Variations on Quine	246
11.4 Routley's Dialectical Set Theory, **DST**	248
11.4.1 The Basic Ideas	248
11.4.2 The Pros and Cons of **DST**	251
11.5 Brady's Paraconsistent Set Theory	253
11.6 The Inclosure Schema	255

12 Impossible Numbers — 259

12.1 Meyer's and Routley's Relevant Arithmetics, **R#** and **DKA** — 259
12.2 $n = n + 1$ — 260
   12.2.1 Natural, Supernatural, and Unnatural Numbers — 261
   12.2.2 The Collapsing Lemma — 263
12.3 Turing vs. Wittgenstein — 266

Part IV. The Problems — 269

13 Hypercontradictions — 271

13.1 "This sentence is true and false, false only, true only, neither true nor false…" — 271
13.2 The Super-Liar — 273
13.3 Relational Semantics — 274

14 The Exclusion Problem — 279

14.1 Overview — 279
14.2 Aristotle's informal remarks on ἔλεγχος — 280
14.3 Looking for the Exclusion-Expressing Device — 284
   14.3.1 Failure at the Level of the "Object Language" — 285
   14.3.2 Failure at the Level of the "Metalanguage" — 286
   14.3.3 Inconsistency All the Way Up — 288
14.4 The Pragmatic Way Out — 292
   14.4.1 Rejection-Consistency and the Rationality Principle — 292
   14.4.2 Some *Inconvenientia* — 297
14.5 The Notion of Material Incompatibility — 303
   14.5.1 The Intuition of Exclusion — 304
   14.5.2 Whither Formalization? — 306
   14.5.3 The Niceties of NOT — 309
   14.5.4 Consistency and Fallibility — 311
14.6 The ἀδύνατον — 315

References — 317

Acknowledgments

To a large extent, this volume expands a book of mine, published in Italian as *Teorie dell'assurdo* (Rome: Carocci, 2006). By the time John Woods and Dov Gabbay kindly proposed me to prepare an English edition, some of my thoughts on the topic had already evolved, and I decided to introduce various adjustments (I have been told that young scholars sometimes change their mind pretty quickly, and I am no exception). This new version occasionally draws on papers that I have recently published, or that I am in the course of publishing, in various journals. The taxonomy of Chapter 1 greatly extends the one proposed in *Meaning, Metaphysics and Contradiction* ("American Philosophical Quarterly", 43, 2006), and the considerations on the "exclusion-expressing device" advanced there appear, with minor modifications, in Chapter 14. I am grateful to Neil Tennant for commenting on the paper before sending it out for anonymous refereeing as the editor of A.P.Q.; and to Greg Restall, Ross Brady, and the referees, for further remarks. The description of "material exclusion negation" in Chapter 14 develops ideas that have been sketched in *Characterizing Negation to Face Dialetheism* ("Logique et Analyse", 195, 2006). Thanks to Graham Priest for his comments on that paper. The presentation of Gödel's Incompleteness Theorems in Chapters 4 and 11 draws on *Il primo teorema di Gödel e l'indeterminabilità del riferimento* ("Epistemologia", 27, 2004), and on a paper presented during a scholarship at the University of Notre Dame, when I proposed my Gödelian issues in a seminar on the philosophy of mind of the 20th Century. I am grateful to all the participants in that debate. Topics concerning idealism and realism discussed in Chapters 6 and 9 have been hinted at in a paper, *Is Dialetheism and Idealism?*, forthcoming in "Dialectica"; thanks to three anonymous referees for their comments on that. References to Hegel and Kant throughout the book come from my volume *Che cos'è la dialettica hegeliana?* (Padua: Il Poligrafo, 2005), and from a paper, *Hegel's Dialectics as a Semantic Theory*, forthcoming in the "European Journal of Philosophy". I am indebted to Vittorio Hösle, Paul Redding, and Diego Marconi, for their comments and encouragements on those Hegelian subjects. I am very grateful to all the editors and publishers, and particularly to Nicholas Rescher, Robert Stern, and Jean-Paul van Bendegem, for their permission to reuse the material.

The book has been submitted to several readers. I am particularly indebted to Achille Varzi, Max Carrara, Luigi Vero Tarca, Luigi Perissinotto and, again, Diego Marconi, who read the original Italian manuscript, for their kindness and their valuable remarks. Achille has helped me all the way through with his encouragements; besides admiring (and envying!) his philosophical skills, I have also come to learn what a nice person he is. My debts towards Marconi's studies on Hegel's dialectics, paraconsistent logics, and the issues of lexical competence, permeate my works.

During my postdoctoral fellowship at the Department of Philosophy of the University of Padua, my collaboration with Franco Chiereghin, Luca Illetterati, Antonio Nunziante, Francesca Menegoni, and Federico Perelda, has provided me a comfortable environment and philosophical stimuli. The lectures on Hegel's dialectics, given at the Doctorate School of Padua, have given me the chance to make up my mind on several issues concerning the Law of Non-Contradiction.

Youngsters help a lot, too. With their intuitive rejection of the idea that everything follows from a contradiction, the students of the University of Venice attending my lectures have instilled in me the curiosity towards paraconsistent logical strategies. Helping Silvia Gaio with her degree thesis on dialetheism helped me to clarify my views on various subjects. Climbing the ladder of wisdom through the logical paradoxes with Smullyan-devotee Matt Plebani added a lot of fun to the whole enterprise.

To Marianna

Né pentere e volere insieme puossi
Per la contradizion che nol consente

Introduction

> I'm a man without conviction,
> I'm a man who doesn't know
> How to sell a contradiction
>
> Culture Club, *Karma Chameleon*

1. Analytic philosophy has been dealing with contradictions since its beginnings. Its foundation myth includes the chronicle of the logicist enterprise: how Frege and Russell ran into the most famous antinomy of contemporary thought – "the contradiction", as Russell simply called it – and tried to overcome it. Frege and Russell considered contradictions as crises, and they were quite right, given the logic they had founded. Following them, all the main authors in the analytic tradition – with the relevant exception of the so-called later Wittgenstein – have tried to push contradictions out of the philosophical market. However, things have recently changed, mainly owing to a handful of men with strong conviction, who know how to sell contradictions at affordable prices. If you want to have elucidations on how and why it happened, I suggest that you read this book.

In 2004, Oxford University Press published a 450-page volume entitled *The Law of Non-Contradiction*.[1] It includes contributions by David Lewis, Patrick Grim, R.M. Sainsbury, Graham Priest, J.C. Beall, Achille Varzi, Stewart Shapiro, Michael Resnik, and several others. Of its five parts, the last two are named "Against" and "For" (the Law of Non-Contradiction). What Łukasiewicz called the "unassailable *dogma*"[2] on behalf of the Law has faded, in favour of a vast and lively discussion. If you look at the *References* at the end of this book, you will find that the debate involves the most important reviews in the fields of logic and philosophy in general. This is mainly due to the recent development of *paraconsistent logics*.

The core thought behind paraconsistency is to provide logics that do not permit to infer anything indiscriminately from inconsistent premises. These may come around in databases, counterfactual-impossible situa-

---

[1] Priest, Beall and Armour-Garb (2004).
[2] See Łukasiewicz (1910b): 62.

tions, inconsistent evidence presented in trials, fiction, etc. In recent years, a couple of very good introductions to the subject (Priest (2002c), Bremer (2005)) have been published. Philosophers, computer scientists, and mathematicians may be intererested in paraconsistency. This book also aims at initiating its reader into it. My focus, though, is the Law of Non-Contradiction: in particular, the philosophical issue whether there might be *true* contradictions or, as they have come to be called, *dialetheias*, in violation of the Law. This means, first, that I have taken into account paraconsistent logics (which are themselves as neutral as any other logic on what is true) mainly, though not exclusively, as tools that can help dealing with that issue. Second, I have coped at length with so-called *strongly* paraconsistent, or dialetheic, uses of the logics in question, *and* their problems. Third, I have explicitly addressed philosophers and, although I have not spared the reader from logical technicalities, philosophical issues have been privileged.

2. The book is divided into four parts. The first one, *Reasons*, presents some evidence that should lead us to dismiss the "unassailable dogma". In Chapter 1, I propose a taxonomy of formulations of the notion of contradiction and of the Law of Non-Contradiction. I take into account, quite rapidly, the question of the so-called "psychological version" of the Law, with the related problem whether we can believe a contradiction. Chapters 2 and 3 present the main proposals to save from harm the Law in the presence of, respectively, the semantic and set-theoretic paradoxes. Here I take into account numerous traditional and fresh arguments to the effect that none of those proposals works. In Chapter 4, I introduce and discuss an argument against the Law, put forth by Graham Priest and Richard Routley via a bold interpretation of Gödel's First Incompleteness Theorem.

In the second part of the book, *Theories*, I introduce the most qualified families of paraconsistent logics, and gesture in the direction of some of their specific applications. The preliminary Chapter 5 collects from the relevant literature a cluster of conditions any logic in which contradictions are allowed should fulfil, thereby providing a methodological reference to evaluate the logical theories presented in the following. Chapter 6 exposes a series of theories labelled as *non-adjunctive*, due to Stanislaw Jaskowski, Nicholas Rescher and Robert Brandom, Achille Varzi, David Lewis and others, and introduces the important subject of impossible worlds. Chapter 7 talks about the so-called *positive-plus* systems due to Newton da Costa and his co-workers, and mentions Diderik Batens' adaptive approach. Chapter 8 is devoted to Graham Priest's *logic of paradox*, and discusses some of his philosophical ideas (together with Routley, Priest is probably the author who has produced the best philosophical works on the topic of contradiction within analytic philosophy). Chapter 9 talks about the systems of relevance logic, and focuses on the vast debate on their semantics.

The third part of the book, *Uses*, looks at some applications of paraconsistent logics in greater detail. Philosophers (including me) are usually much more interested in foundational issues than in investigating the applications of a research program to particular fields – which is why "the reasons" took so much space. Nevertheless, a part of the reasons for accepting a theory has to do with its efficacy. It may therefore be nice to discover the epistemic virtues – such as unity, completeness, simplicity, explanatory power, etc. – displayed by theories lacking the virtue of consistency when one comes to their applications. In Chapter 10, I expose Priest's formal semantics for a semantically closed language (in the Tarskian sense). In Chapter 11, after saying something on the interesting metatheoretical situation produced by some formalized inconsistent theories, I cope with paraconsistent set theories. In Chapter 12, I take into account inconsistent arithmetics – perhaps the strangest outcomes of paraconsistency.

The last part, *Problems*, considers precisely the troubles. Chapter 13 introduces the issue of so-called *hypercontradictions*: particularly infectious contradictions whose treatment, according to some, causes difficulties also within a strongly paraconsistent framework. Chapter 14 examines some expressive and theoretical difficulties facing dialetheism, and particularly linked to the necessity of providing a coherent notion of *exclusion*.

3. As for the symbolism, and the role of formal logic in the book, the reader is supposed to know elementary logic; not much more, though. I have always tried to avoid dismissing intuitive understanding in favour of formal rigidity. This led, among other things, to a certain elasticity on the use/mention distinction. Quine's quasi-quotation marks, as well as other quotational tricks, have been avoided in all cases in which no dangerous confusion could result from their omission, thereby employing (formal as well as informal) expressions as names of themselves. In a word, let the context be your guide.

The notation for the formal language is the standard one – or one of the standard ones – and the metalanguage is usually informal, or semi-formalized (we will learn at length, however, that collapsing the language/metalanguage distinction is somehow a main aim of the champions of inconsistency). I have often used a meta-conditional, $\Rightarrow$, with the corresponding biconditional, $\Leftrightarrow$, to make my life easier; $\vDash$ for logical (semantic) consequence; $\vdash$ for syntactic (proof-theoretic) consequence or theoremhood. As one may expect, non-standard logical notions are around when one talks about paraconsistency. Each time this required the introduction of new symbols, a contextual explanation has been provided. I have translated into the mainstream formalism the symbolisms of works, such as Jaskowski's, that adopted the Polish notation; but I have also preserved some peculiarities of the notations of the original essays, when they appeared to be interesting and/or perspicuous. Again, I am confident in the flexibility of my reader.

The calculus adopted in the (not numerous) formal arguments is Gentzen's well-known natural deduction with introduction/elimination rules for the logical operators. Proofs have a linear presentation. Each assumption gets a numeric index (corresponding to the number of the line in which it is introduced); the index is transmitted each time an inference rule is applied, thus keeping a record of the hypotheses on which a given formula depends (some assumptions being "discharged", in any case, by some rules). Formulas taken as logical or specific principles of a theory are contextually introduced without indexes, not depending on any assumption.

I have followed a common way of numbering Chapters, Sections and Sub-Sections. They are referred to by means of "§", followed by the respective number. So "§ 5" points at Chapter 5, and "§ 9.8.2" points at Sub-Section 2 of Section 8 in Chapter 9. Usually, "if and only if" has become "iff" (to be read, in Lewisian fashion, "iffffff").

4. Wittgenstein claimed that the *Tractatus logico-philosophicus* "is [...] not a text-book".[3] The same holds for this much more modest book, in various senses. Given the amount of literature published on its subject in the last forty years, some things had to be left in the background. Now and then, my lack of competence has been responsible for this. For instance, I have ignored some applications of paraconsistency in computer science and quantum physics, simply because I am ill equipped to talk about these subjects. However, my judgment (therefore, perhaps, my

---

[3] Wittgenstein (1921): 3.

prejudices) has also influenced my choices. For instance, at times I have privileged model-theoretic over proof-theoretic approaches. A key point concerning the plausibility of strong paraconsistency, in my opinion, has to do with models: in which sense the model-theoretic structures for these logics can provide counterexamples to the Law of Non-Contradiction? How plausible is the intuitive reading of such models? What does "intuitive reading" mean here, by the way? The result is what one may call an *opinionated* introduction, with the author taking an explicit stance on many issues. This happens in several Chapters (especially in the second part of the book), but becomes particularly evident in the last one.

The book also includes a large amount of quoting and references. On the one hand, this is because the various authors sometimes made their points in a way I could hardly improve. On the other hand, some references are indirect stimulations for my reader, in the hope that she gets interested in the original texts. Many among them surpass the technicalities of logic, and have the strong flavour that only philosophical inquiries on the most fundamental themes can give.

The book includes topics in the philosophy of logic and language, the philosophy of mathematics, theoretical philosophy in general, and metaphysics. This last is the field towards which, in my opinion, the main debates on paraconsistency are going to turn – witness to this the recent developments of the debate on impossible worlds, whose ontological status will be discussed at length in the following. The very last words of the new 2002 edition of Graham Priest's *Beyond the Limits of Thought* comment on the newly added Chapters along these lines:

> Though this is not a novel development in the book, these two new chapters, which put *what is* at centre stage, certainly mark an ontological turn in focus. The philosophy of language took pride of place in twentieth-century philosophy. Certainly there is no going back to how things were before this. But maybe this century will see a return to the mainstreaming of a more traditional philosophical issue, the nature of reality – and if I am right, a nature that is contradictory.[4]

Aristotle would have disagreed with his main contemporary challenger on almost anything, except this: if metaphysics should be placed (once again) at the very core of philosophy, the debate on the Law of

---

[4] Priest (1995): 295.

Non-Contradiction should be placed at the very core of metaphysics. We are about to enter the core of the core: on with the tale, then.

Part I.
The Reasons

1 "The Most Certain of All Principles"

> Nothing is, and nothing could be, literally both true and false. This we know for certain, and *a priori*, and without any exception for especially perplexing subject matters. […] That may seem dogmatic. And it is: I am affirming the very thesis that [the rivals of the Law of Non-Contradiction] have called into question and – contrary to the rules of debate – I decline to defend it. Further, I concede that it is indefensible against their challenge.
>
> David Lewis, *Logic for Equivocators*

1.1 A Strange Subject

Aristotle calls it βεβαιοτάτη πασῶν ἀρχή, "the most certain of all principles"[1] –*firmissimum omnium principiorum*, as the medieval authors said. It is the Law of Non-Contradiction – henceforth: (LNC). The property of being *firmissimum* expresses the fact that the (LNC) has been taken as *the* most indubitable and incontrovertible law of thought and being, and as the supreme cornerstone of knowledge and science. Such a view on the logical, metaphysical, and psychological status of the (LNC) is due to Aristotle, and a large part of this Chapter will be dedicated to investigate the way he settled the issue in the celebrated Book IV of his *Metaphysics*. Actually, hardly anyone has taken upon herself to *defend* the (LNC) since Aristotle.[2] Thomas Reid put the Law, in the form: "No proposition is both true and false", among the dictates of common sense (together with other alleged self-evident truths, such as that every complete sentence must have a verb, or that those things did really happen which I distinctly remember). Common sense is nameless and spontaneous: it does not bother do defend its views unless someone challenges them.

In the following two §§, I will propose a taxonomy of formulations taken from the relevant literature. Following the catalogue might turn out to

---

[1] *Met*. 1005b24.

[2] Among the few exceptions are the British Idealists: see Bradley (1898): 120, and McTaggart (1922): 8.

be a bit boring, but it will prove useful throughout the rest of the book. Since the (LNC) is (expressed by) a piece of language, exactly what does it *say*? As soon as it appears on the horizon, the Law presents strange features. If it is (expressed by) a sentence, it should make sense to wonder whether it is true or false. However, some doubt it is possible just to *conjecture* that it might be false. Its status of "Principle", or "Law", tells us that something should follow from it. Traditionally, philosophers meant by "Principle" something like: basis or ground of (the truth of) of something else. However, the medieval scholars had already realized that very few things are deduced from it – *nulla demonstratio accipit hoc principium*, said Thomas Aquinas, echoing his master. In the *Posterior Analytics*, Aristotle says: "The impossibility of joint affirmation and denial is presupposed in no proof (syllogism) unless the conclusion itself was also to have demonstrated such".[3] It is plain, though, that the (LNC) aims at somehow forbidding, or excluding, contradictions. Therefore, our next question is: what is a contradiction?

1.1.1 A πολλαχῶς λεγόμενον

To stick to the Aristotelian jargon, "contradiction" is a πολλαχῶς λεγόμενον: it is spoken of in many ways. In a recent study, Patrick Grim has identified so many formulations of the relevant notions in the literature that, by combining them, we get about 240 definitions.[4] I will simplify such a variety into four main groups.

(1) First, we have what we may call *syntactic* formulations. A contradiction is a syntactic object of such and such a *form*:

(C1) $\alpha \wedge \neg \alpha$,

i.e., it is the conjunction of a sentence[5] and its negation. A contradiction is sometimes taken, not as a conjunction, that is, an individual sentence, but as a pair of sentences, one of them negating the other:

---

[3] 77a10-3.
[4] See Grim (2004).
[5] Lower case Greek letters: $\alpha, \beta, \ldots$ (occasionally indexed: $\alpha_1, \ldots, \alpha_n$) in informal contexts are sentential variables. In formal contexts, such as the proof-theoretic presentations of some formal systems I will discuss in the following, they are metavariables for formu-

($C1_{dist}$) $\alpha, \neg\alpha$.

It is customary to call (C1) and ($C1_{dist}$), respectively, *collective* and *distributive* formulations.[6] We will deal mostly with collective versions. The distinction will become relevant, though, when we come to talk of certain theories, called *non-adjunctive*, in which the ordinary treatment of conjunction is altered.[7] Here are some examples of syntactic (collective or distributive) formulations from the literature:

Contradiction
Wff* of the form 'A & --A'; statement of the form 'A and not A'. (Haack (1978): 244)

A contradiction consists of a pair of sentences, one of which is the negation of the other. (Kalish, Montague, and Mar (1980): 18)

The *formal* usage of 'contradiction' has it that contradictions are sentences *of the form* $A \wedge \neg A$, where $\wedge$ is conjunction and $\neg$, as above, is negation. (Beall (2004a): 4)

(2) Then come formulations we may label as *semantic*, since they make use of the semantic notions *par excellence*: truth and falsity. Let us use T and F for the truth and falsity predicates, applying to sentence names,[8] and let us assume that, in general, $\ulcorner \alpha \urcorner$ is the name of α:

---

las of the system (I will usually state the axioms of formal systems and theories schematically).

[6] See e.g. Rescher and Brandom (1980): 7; Varzi (2004): 94.

[7] The distinction between a collective and a distributive reading is transversal to the various kinds of contradictions we shall consider in the following: they can all be expressed conjunctively, or in terms of pairs.

[8] Which is the right sort of thing that can bear truth values? Among the candidates are beliefs, propositions, judgments, statements, assertions, sentence tokens, sentence types, etc. Such a discussion is certainly interesting for those who specialize in theories of truth. But to those who have only a collateral interest in the issue, the matter soon begins to look more as one of "*choice*, not discovery" (Kirkham (1992): 59). For the sake of simplicity, in the following I will usually take *sentences* as the truth-bearers as the default option. In most contexts, one will be free to replace "sentence" with the name of her other favourite truth-bearer.

(C2a) $T(\ulcorner\alpha\urcorner) \wedge F(\ulcorner\alpha\urcorner)$,

whose informal reading is something like: "Sentence α is both true and false". (C2a) is equivalent to:

(C2b) $T(\ulcorner\alpha\urcorner) \wedge T(\ulcorner\neg\alpha\urcorner)$,

("Both sentence α and its negation are true"), if we accept a characterization of falsity as the truth of negation, i.e., if we accept:

(Neg1) $F(\ulcorner\alpha\urcorner) \leftrightarrow T(\ulcorner\neg\alpha\urcorner)$;

informally: "Sentence α is false iff its negation is true". The idea that "false" means just "has a true negation" is generally accepted both by the friends and (with few exceptions) by the foes of the (LNC). I will stick to (Neg1) throughout this book.

Much more controversial is the equivalence between falsity (that is to say, given (Neg1), truth of negation) and *untruth*:

(Neg2) $T(\ulcorner\neg\alpha\urcorner) \leftrightarrow \neg T(\ulcorner\alpha\urcorner)$;

informally: "The negation of α is true iff α is not true". It is usually claimed that (Neg2) expresses the semantics of *classical* negation, or the so-called exclusion condition of classical negation.[9] As we shall see in the

---

[9] (Neg1) and (Neg2) are normally taken as semantic clauses. As the usual story goes, to avoid confusion expressing them one should use metalinguistic symbols (be the metalanguage informal, or itself a formalized or semi-formalized one) of a different sort from those of the object language. For instance:

(Neg2) $T(\ulcorner\neg\alpha\urcorner) \Leftrightarrow \text{Not } T(\ulcorner\alpha\urcorner)$,

where the whole expression, including the truth predicate for the sentences of the object language, belongs to the metalanguage, which contains names for the expressions of the object language, $\Leftrightarrow$ is a meta(bi)conditional, "Not" a meta-negation. To be sure, this is the procedure we have learned from Tarski, who rejected *semantic closure*, i.e., roughly, the idea that the semantics of a language can be given within that very language. I shall characterize the notion in detail in the following Chapter. According to most critics of the (LNC), though, the separation between object language and metalanguage is somewhat following, (Neg2) is rejected in various non-classical logics. If we accept it, on the other hand, (C2a) and (C2b) turn out to be equivalent to:

(C2c) $T(\ulcorner\alpha\urcorner) \wedge \neg T(\ulcorner\alpha\urcorner)$,

("Sentence α is both true and untrue"). Following Priest (1987): Ch. 4, we can call things of the form of (C2a) and (C2b), *internal* contradictions, and things of the form of (C2c), *external* contradictions. Some literature:

> When the going gets tough, and we encounter true sentences whose negations also are true, then the relevant logician gets going. (Lewis (1982): 97)

> Dialethism, the thesis that a single proposition can be both true and false at the same time. (Saka (2001): 6)

> Dialetheism is the view that some contradictions are true: there are sentences [...], α, such that both α and $\neg\alpha$ are true, that is, such that α is both true and false. (Priest (2006): 1)

(3) Then come formulations, which may be labelled as *metaphysical*, since they adopt typically ontological notions, such as *object, individual, property, state of affairs*, etc. Formalization is not so straightforward in this case, but we may express the ontological commitment via quantification in a predicative language; a second order formulation (with quantifiers binding also predicative variables) will do:

(C3) $\exists x \exists P(P(x) \wedge \neg P(x))$,[10]

---

illusory: our ordinary language *is* semantically closed, and a formal theory that aims at capturing the fact has to express the truth conditions of sentences within the very language it gives the semantics of. We will also see how contradictions of the (C2)-kind follow from this. Nevertheless, according to the friends of inconsistency, this is precisely what is supposed to happen.

[10] In the notation of the book, upper case Latin letters, $P, Q, ...$ (occasionally indexed: $P_1, ... , P_n$) sometimes play the role of predicative constants, sometimes of (quantifiable) variables when a higher-order language is needed. The ambiguity is resolved contextually.

"Some object x both has and lacks some property P". (C3) is not the most general metaphysical formulation available, since it directly concerns only properties, not n-ary relations; this is a minor point, though, and many authors belonging to the tradition (from Aristotle to Leibniz) have taken the general case as reducible to this one. Some formulations one could qualify as metaphysical:

> A contradictory situation is one where both B and ~B (it is not the case that B) hold for some B. (Routley and Routley (1985): 204).

> A contradictory state of affairs would be one in which something had a particular property and also an incompatible property, or in which something both had a particular property and lacked that property [...]. A contradictory situation would be one in which it is the case that something is P and also the case that that thing is not P. (Grim (2004): 53 and 67)

> A tru[e contradiction] like $F(a) \wedge \neg F(a)$ means at first sight that the object a has property F and does not have property F. (Bremer (2005): 199)

It is sometimes claimed that it does not make sense to talk of inconsistent objects, situations, or states of affairs. The world is all there, all together: how could some pieces of it *contradict* some other pieces? Consistency and inconsistency are properties of sentences, or theories (sets of sentences closed under logical consequence), or propositions (what sentences express), or maybe thoughts, or (sets of) beliefs, etc. Contradiction (*Widerspruch*, the Latin *contradictio*) has to do with discourse (diction, *sprechen*, *dicere*). The world, with its non-mental and non-linguistic inhabitants – houses, rivers, people – is not the right *kind* of thing that can be consistent or inconsistent, and ascribing such properties to a part of the world is, to use Gilbert Ryle's jargon, a category mistake.[11]

For the time being, one may simply stress that consistency and inconsistency can be ascribed to (pieces of) the world in a *derived* sense. As Priest claims in *In Contradiction*, "to say that the world is consistent is just to say that any true purely descriptive sentence about the world is consistent".[12] One may take an intermediate stance and claim that, al-

---

[11] See e.g. Bobenrieth (1998): 29: "There is not much sense in saying that facts 'contradict' each other [...] since facts do not 'say' anything, they do not 'speak' of each other or of themselves. [...] Therefore, the problem of contradictions does not lie in reality".

[12] Priest (1987): 200.

though talk of inconsistent ("non-propositional", as it were) *objects* is a category mistake,[13] it does makes sense to talk of inconsistent states of affairs or facts, e.g., of the world being occupied both by the fact that $P(a)$, and by the fact that $\neg P(a)$. This may entail a commitment to negative facts. I will return to all these issues in the following. Notice, however, that this may appear, again, as a definitional matter: one may well call an object, a, inconsistent, or (self-)contradictory, precisely when such facts involving it as its being P and its not being P (or its being non-P, which, as is well known, may not amount to the same) hold simultaneously. Consequently, and not accidentally, it is quite common in the current literature on (pro and against) the (LNC) to straightforwardly speak of inconsistent objects, states of affairs, and entire inconsistent *worlds*. And I will stick to the routine.

(4) Finally, a fourth cluster is constituted by a variety of formulations I will group under the label of *pragmatic*, for they adopt notions variously connected to pragmatics, more than to syntax, ontology, or semantics strictly conceived. "Pragmatics" will be understood in a broad sense, though, as concerning not only linguistic behaviour but also beliefs, belief management, and overall rational activity. To clarify things, let us use the following terminology: by *acceptance* I shall mean the cognitive, mental state a subject x has towards a sentence α (it is usual to say: towards the *proposition*, or the content, expressed by a sentence, but the distinction is of lesser importance here). Accepting something will be regarded as equivalent to believing it: x accepts α iff x believes (that) α. The polar opposite of acceptance is *rejection*. To reject something is to definitely refuse to believe it. By *assertion* and *denial*, on the other hand, I shall mean (typically) linguistic acts or, equivalently, illocutionary forces attached to utterances. Roughly, assertion and denial are the linguistic counterparts of acceptance and rejection: when x asserts/denies α, supposing x is sincere, x aims at expressing that she accepts/rejects α and, secondarily, x may also aim at getting those who listen to accept/reject it.[14] Philosophers often conflate acceptance and assertion, and, respectively, rejection and denial. As we shall see towards the end of the book, the two couples can come apart in one important respect (important, that is, with respect to the (LNC) issue), so they should be kept conceptually distinct.

---

[13] Which does Priest sometimes declare: see Priest (1995): 295, (2006): 51.
[14] See e.g. Sainsbury (1997): 218.

Nevertheless, for most of our purposes linguistic acts and the corresponding mental states can match.

I shall use two sentential operators, "$\vdash_x$" and "$\dashv_x$", whose intuitive reading is, respectively, something like "rational agent x accepts/asserts (that)" and "rational agent x rejects/denies (that)".[15] We have, then, our pragmatic versions:

(C4a) $\vdash_x \alpha \land \vdash_x \neg \alpha$,

to be read as: "(Rational agent) x accepts/asserts both α and $\neg\alpha$";

(C4b) $\vdash_x \alpha \land \dashv_x \alpha$,

"(Rational agent) x both accepts/asserts and rejects/denies α". (C4a) and (C4b) turn out to be equivalent, if we accept the following principle:

(Acc) $\dashv_x \alpha \leftrightarrow \vdash_x \neg \alpha$.

If one understands it in terms of linguistic acts, (Acc) is the claim, famously held by Frege and Peter Geach, according to which to deny something is just to assert its negation.[16] The equivalence is controversial, also independently of issues concerning the (LNC). I will come to this subject in due course, though. Meanwhile, here are some pragmatic examples:

Contradiction: the joint assertion of a proposition and its denial. (Brody (1967): 61)

One can certainly believe something and believe its negation. (Priest (1987): 122)

---

[15] The notation goes back to Łukasiewicz (1957), but the version with subscripts is credited to Richard Routley. $\vdash_x$ should not be confused with the normal symbol for (syntactic) consequence or theoremhood I will use in the following. This either will always appear with no subscripts, or followed by upper case letters as subscripts when a specification is due of the theory or formal system within which a formula is a theorem: for instance, the notation $\vdash_{PA}$ says that the formula following it is a theorem of Peano arithmetics, labelled as **PA**.

[16] See Geach (1965).

A contradiction both makes a claim and denies that very claim. (Kahane (1995): 308)

1.1.2 The Law(s)

If the essence of the (LNC) consists in barring contradictions, since there are different forms of contradiction, we will have different forms of barring. One may claim that contradictions cannot be true, or that they are necessarily false (semantic exclusion); that they cannot be sensibly asserted or, maybe, believed (a psychological exclusion, on which something shall be said soon); or that contradictions cannot possibly *exist*, in the sense that there cannot be contradictory objects, or states of affairs (metaphysical exclusion). Corresponding to the fourfold classification given above, we can have, in particular, four kinds of formulations of the (LNC) – of course, the fact that all the following formulations are taken as *laws* means that they are supposed to hold for *any* α, that they are put forth as necessary truths, etc.

(1) Syntactic versions:

$(LNC1)\ \neg(\alpha \wedge \neg\alpha)$,

for instance:

The law of non-contradiction, $\neg(\alpha \wedge \neg\alpha)$ (Priest (1987): 76)

$\sim(A \wedge \sim A)$ […] the law of non-contradiction has traditionally been seen as a central property, if not a defining characteristic, of negation. (Priest and Routley (1989c): 164-5)

(2) Then come semantic formulations corresponding to contradictions of the (C2a-c)-kind:

$(LNC2a)\ \neg(T(\ulcorner\alpha\urcorner) \wedge F(\ulcorner\alpha\urcorner))$,

"The same sentence cannot be both true and false";

$(LNC2b)\ \neg(T(\ulcorner\alpha\urcorner) \wedge T(\ulcorner\neg\alpha\urcorner))$,

"A sentence and its negation cannot both be true"; and:

(LNC2c) $\neg(T(\ulcorner\alpha\urcorner) \wedge \neg T(\ulcorner\alpha\urcorner))$,

"The same sentence cannot be both true and untrue". The first case to mention is the Aristotelian one:

> The most indisputable of all beliefs is that contradictory statements are not at the same time true. (*Met.* 1011b13-4)

To appreciate the point, we should keep in mind that the notions of truth and falsity are also used to characterize the relation of *contradictoriness* between sentences. Following a schema codified by the famous "square of opposition" of traditional logic, it is claimed that two sentences α and β are contraries iff they cannot be true together (i.e., their conjunction is a logical falsity); that α and β are sub-contraries iff they cannot be false together (i.e., their disjunction is a logical truth); finally, that they are contradictory iff they are both contraries and sub-contraries. Some examples:

> Contradictories, or propositions one of which must be true and the other false… (DeMorgan (1846): 4)

> Contradictory
> The contradictory of a wff* (statement) *A* is a wff* (statement) which must be false if *A* is true and true if *A* is false. (Haack (1978): 244)

> Contradictories: two propositions are contradictories if and only if it is logically impossible for both to be true and logically impossible for both to be false (Sainsbury (1991): 369).

Now, contradictory statements (ἀντικειμένας φάσεις) according to Aristotle are a sentence (or its avowal, κατάφασις) and its negation (ἀπόφασις). Therefore, to say that contradictory statements cannot both be true amounts to a claim to the effect that a sentence and its negation cannot both be true, as in (LNC2b). Especially since Łukasiewicz's work, formulations of this kind have also been qualified as *logical*:

> Logical formulation: *Two contradictory sentences cannot be true at the same time*. By 'sentence' I mean a sequence of words or other perceptible symbols

which has a meaning insofar as it predicates or denies some property of some object. (Łukasiewicz (1910b): 51)

Further examples:

The law of contradiction asserts that a statement and it direct denial cannot be true together. (Prior (1967): 481)

The Law of Contradiction [...]: "no statement is both true and false". (van Benthem (1979): 335)

The *law of noncontradiction*: nothing is both true and false. (Priest (1998): 416)

(3) A metaphysical way of stating the (LNC) would be, again, a second-order one:

(LNC3) $\forall x \forall P \neg (P(x) \wedge \neg P(x))$,

"The same object cannot both have and not have the same property" – which is quite close to the first Aristotelian formulation of the *Metaphysics*:

The same attribute cannot at the same time belong and not belong to the same subject in the same respect; we must presuppose, in the face of dialectical objections, any further qualifications which might be added. (1005b19-22)

Let us say something straight away on the famous "further qualifications", to be added in order to avoid "dialectical objections". In the *De interpretatione*, Aristotle explains that he speaks of "statements as opposite", in the sense that they are really contradictory, only "when they affirm and deny the same thing of the same thing – not homonymously".[17] Only if the meaning of the subject, and that of the predicate, is the same in both sentences, we have a genuine contradiction. Such remarks have to do with the ancient technique of *parameterisation*, or *distinction of respects*. This is a variation on a typical move made in philosophical argumentation, that is, a plea of ambiguity. A common strategy we will often meet in the following, when one collides with a contradiction such as $\alpha \wedge \neg \alpha$, is to treat α, or some of its parts, as having different meanings, that is to say, as ambiguous

---

[17] *De int.* 17a34-6.

(maybe just *contextually* ambiguous). For instance, if it seems that $P(a) \wedge \neg P(a)$, one claims that, actually, a is P and is not P under different parameters or respects – say, r_1 and r_2. This discrimination not emerging, contradiction jumps in. But it can be resolved by clarifying that $P_{r1}(a) \wedge \neg P_{r2}(a)$ (Juliette Binoche is and is not a star, but she is a star in the sense that she is a great actress, not a star in the sense of Alpha Centauri). Therefore, in the *Metaphysics* Aristotle also stresses that the rival of the (LNC) does not get the point insofar as he plays with the equivocal meanings of some words: "for to each formula there might be assigned a different word".[18] Aristotle has plenty of followers on this way for sure. For instance, according to van Benthem:

> When a contradiction arises, this may be interpreted as a symptom of poverty: a relevant aspect has been neglected. Addition of suitable "parameters" will usually restore consistency, i.e., freedom from contradictions. [...] When viewed in this way, contradictions play a role in the refinement of our theories, not because of their acceptance, but precisely because of their rejection![19]

Parameterisation is usually a win-win game, in the sense that apparently conflicting beliefs may be both true, because the conflict itself is apparent. Notice, nevertheless, that this is hardly an argument against the rival of the (LNC). That parameterisation works in all cases *presupposes* precisely that the Law is considered as non-negotiable.

The most concise version of the (LNC) in the *Metaphysics* is probably the following ontological variation:

> A thing cannot at the same time be and not be. (*Met.* 996b29-30)

The usage of calling "ontological" such formulations comes, again, from Łukasiewicz:

> Ontological formulation: *The same property cannot belong and not belong to a single object at the same time.* By 'object' I understand, with Meinong, anything that is 'something' and not 'nothing'; by 'property' I mean anything that can be predicated of an object. (Łukasiewicz (1910b): 51)

---

[18] 1006b1-2.
[19] Van Benthem (1979): 336-7.

It is uncontroversial that the ontological or metaphysical variant was the one to which Aristotle attributed the greatest importance. The issue of the incontrovertible validity of the (LNC) is not taken up in the *Organon*, i.e., in the logical treatises (albeit we find formulations of the Law there), but precisely in the *Metaphysics*. Here Aristotle affirms that the discussion of the "axioms" (and the axiom *par excellence* is the (LNC)) is a business reserved to the metaphysician, since "these truths hold good for everything that is, and not for some special genus apart from others"; therefore, "he who studies being qua being will inquire into them too".[20] The (LNC) is a feature that all entities have merely because of their being entities. Further examples from the literature:

> The Law of Contradiction [...]: "nothing can possess a property as well as its complementary property". (van Benthem (1979): 335)

> Something cannot both be and not be so unequivocally and in one and the selfsame respect. (Rescher and Brandom (1980): 2).

> ONTOLOGICAL (NON-)CONTRADICTION: No 'being' can instantiate contradictory properties. (Beall (2004a): 3)

> It is not possible that there is an object, *a* and property, *F*, such that $Fa \land \neg Fa$. (Priest (2006): 8)

(4) Finally, we have pragmatic versions corresponding to (C4a) and (C4b):

(LNC4a) $\neg(\vdash_x \alpha \land \vdash_x \neg \alpha)$,

(LNC4b) $\neg(\vdash_x \alpha \land \dashv_x \alpha)$.

In Aristotle's words:

> It is impossible for any one to believe the same thing to be and not to be, as some think Heraclitus says. (*Met.* 1005b23-5)

---

[20] 1005a22-9.

Some doubt it is accurate to call such principles "Law of Non-Contradiction", too.[21] Łukasiewicz, though, talked of a "psychological formulation":

> Psychological formulation: *Two beliefs which answer to two contradictory sentences cannot exist at the same time in a single consciousness.* By 'belief' I mean a mental act which is sui generis: it is also designated by the terms 'conviction', 'recognition' etc. (Łukasiewicz (1910b): 51).

Some other pragmatic cases:

> Someone who rejects A cannot simultaneously accept it, any more that a person can simultaneously catch a bus and miss it, or win a game of chess and lose it. (Priest (1989): 618)

> There seems to be a role in dialogue for an expression whose significance is captured by the law of non-contradiction: by the principle that a proposition and its negation cannot both be accepted. (Price (1990): 224)

> It is a necessary condition of a sentence's being rejected that it also not be accepted. (Mares (2000): 504)

1.2 Getting in Touch with the T-schema

As for the interactions between the various formulations of the (LNC), Łukasiewicz thought that the semantical and metaphysical Laws were equivalent.[22] This is easily obtained by accepting the famous Tarskian schema for the characterization of truth:

(T) $T(\ulcorner \alpha \urcorner) \leftrightarrow \alpha$.[23]

---

[21] It is also doubtful that a *negation* has to be used as the main operator of (LNC4a-b), since according to some the situation is better expressed by a *denial* of what lies inside the parentheses. More on this in § 14.

[22] See Łukasiewicz (1910b): 52.

[23] If Tarski is right on the non-definability of the truth predicate for a language within that very language, the T-schema is constitutively metalinguistic (but you may recall what I have anticipated above on the language/metalanguage distinction). (T) corresponds to what Łukasiewicz, who, of course, wrote before Tarski, called "definition of true judgment": "I call an affirmative sentence true if it ascribes to an object a property which belongs to it" (Łukasiewicz (1910b): 54).

Now, a metaphysical formulation of the (LNC), such as: $\forall x \forall P \neg(P(x) \land \neg P(x))$, once quantifiers have been removed, has the form of the syntactic (LNC1), i.e., $\neg(\alpha \land \neg\alpha)$. Given (T), we can derive the semantic version, (LNC2b), i.e., $\neg(T(\ulcorner\alpha\urcorner) \land T(\ulcorner\neg\alpha\urcorner))$, and *vice versa*, by replacing equivalents. The equivalence extends to (LNC2a), since this last is itself equivalent to (LNC2b), given (Neg1).

Concerning the Tarskian idea, it is common to talk about the "semantic theory of truth".[24] Which notion of truth underlies the T-schema? Scholars have been discussing for decades on the issue whether Tarski's is a *correspondence* conception of truth in which, roughly, a sentence is true iff it corresponds to facts, to how things actually are in the world. In the opening notes of *The Concept of Truth in Formalized Languages*, Tarski maintains that the idea according to which "a true sentence is one which says that the state of affairs is so and so, and the state of affairs indeed is so and so" expresses "the classical view of truth".[25] He quotes, then, a famous passage from Aristotle's *Metaphysics* that we shall meet soon. The famous Tarskian convention for (good) theories of truth,[26] and the very notion of satisfaction of an atomic formula $P(t_1, \ldots, t_n)$ by a sequence of objects in a domain, according to some, stem from a conception admitting an objective reality, irreducible to thought and language, with respect to which interpreted sentences are evaluated. According to others, and regardless of what Tarski may have thought of his own theory, it is not at all mandatory that the T-schema be tied to a correspondence conception.[27] On the contrary, it is claimed that the schema has precisely the advantage of providing a guiding principle for the specification of the truth conditions without employing heavy theoretical or metaphysical commitments. Bearing in mind the Austin-Strawson controversy, we may want to avoid admitting facts as the truth-makers of sen-

---

[24] See e.g. Kirkham (1992): Ch. 5. Tarski talked about a semantic *conception* of truth (Tarski (1944): 120).

[25] Tarski (1956): 155.

[26] According to Tarski a good definition of truth for a given language is properly given via a series of clauses, from which it is possible to derive by purely logical means any instance of the T-schema, i.e., for any sentence α of the language in question, the corresponding biconditional. I will come back to this point in the following Chapter.

[27] Susan Haack believes Tarski did not think of his theory as a correspondence theory (Haack (1978): 112ff), being partly criticized for this by Kirkham (1992): 170. It seems to be Tarski's fate that people believe he was confused on how to interpret the outcomes his own work.

tences,[28] but it is doubtful that the T-schema commits us to some metaphysics of facts. We may consider (T) as just a disquotational schema, following the current deflationist theories of truth.[29] No matter what our ontological preferences are, however, it seems most of us agree on the point that "Snow is white" is true iff snow is white.

Some semantic theories, as a matter of fact, reject the T-schema, or at least one of the two conditionals composing it. The most famous case it supervaluationism. The first supervaluational semantics is due to van Fraassen, who used it to provide a treatment of presuppositional phenomena – and semantic paradoxes, too.[30] The Law of Excluded Middle is accepted by the theory:

(LEM) $\alpha \vee \neg\alpha$;

but the Principle of Bivalence, according to which all sentences are either true or false, is rejected. Given (Neg1), we can express Bivalence as follows:

(PB) $T(\ulcorner\alpha\urcorner) \vee T(\ulcorner\neg\alpha\urcorner)$.

In order to avoid the entailment from the (LEM) to (PB), supervaluationism has to reject the right-to-half direction of the T-schema.[31] I shall argue later that the T-schema gives us a minimal condition for a predicate to be a *truth* predicate. Meanwhile, it will suffice to observe that Aristotle not only endorsed (T), but also interpreted it in a realist direction. It is true that oscillations on this subject are found in his works (for instance, according to some interpreters, in Ch. 9 of *De interpretatione* Aristotle shows some anti-realist tendencies when dealing with the famous problem of con-

---

[28] See Pitcher (1964).

[29] See Horwich (1990); Kirkham (1992): Ch. 10.

[30] See van Fraassen (1966). Supervaluations have later been extended to the treatment of vagueness (the canonical work is Fine (1975)). For a short presentation, see Beall and van Fraassen (2003): 133ff.

[31] The rejection of (T) within the supervaluationary approach, actually, is more articulated. Van Fraassen (1966) has pointed out that supervaluationists can retain an alternative version of (T) phrased as an inference rule.

tingent sentences about the future).[32] Nevertheless, he certainly claimed that:

> To say of what is that it is not, or of what is not that it is, is false, while to say of what it is that it is, and of what is not that it is not, is true. (*Met.* 1011b26-7)

> He who thinks the separated to be separated and the combined to be combined has the truth, while he whose thought is in a state contrary to that of the object is in error. (*Met.* 1051b3-6)

Most interpreters[33] have seen in these claims the historical stronghold of a realist conception:

> It is not because we think that you are white, that you *are* white, but because you are white we who say this have the truth. (*Met.* 1051b7-9)

1.3 Can We Believe a Contradiction?

The relationships between the formulations of the (LNC) of kinds no. 1, 2 and 3 are, therefore, reasonably uncontroversial. What about the connection between these and kind no. 4? Things get complicated here. In this and the subsequent §§ I shall examine the subject, albeit briefly: as we shall see, a couple of reasons suggest not do dedicate too much space to the issue in this book. The general issue here is whether we can *accept*, or *believe* an impossibility (contradiction being usually taken as the pinnacle of impossibility). A venerable philosophical tradition rejects the idea. One of the Humean slogans inherited by the empiricist tradition, according to which what is thinkable is possible, entails that the impossible, the absurd, far from being believable, cannot even be *conceived*. *Positivismus und Realismus*, Moritz Schlick claimed that, while the merely practically impossible is still conceivable, the logically impossible, such as contradictions, is simply unthinkable.[34] Nowadays many authors, from Dennett[35] to Ruth Barcan-Marcus, hold the impossibility of believing the impossible. Marcus' point is based upon the observation that belief ascriptions are al-

---

[32] Anscombe (1956) has famously challenged this traditional reading.
[33] See Kirkham (1992): 119ff.
[34] See Schlick (1932): 42.
[35] See Dennett (1987).

ways defeated when it turns out that no state of affairs could possibly make the belief true: just as knowledge requires truth, so belief requires possibility.[36] Foley has exploited a fundamental idea of truth-conditional semantics, according to which understanding amounts to knowing truth conditions: "To understand a proposition means to know what is the case, if it is true", as the Tractarian maxim goes. Now, one can believe only what one understands. Since a contradiction is true under no condition, nobody can understand a contradiction *and* believe it to be true.[37]

A parallel tradition, nevertheless, has it that contradictions are thinkable, and possibly believable, too. Hegel complained against "one of the fundamental prejudices of logic as hitherto understood", namely that "the *contradictory* cannot be *imagined* or *thought*".[38] More recently, Roy Sorensen has even produced, in *Vagueness and Contradiction*, a "transcendental argument" in support of the possibility of believing the impossible.[39]

1.3.1 The Aristotelian argument

To understand the origin of the problem, and to come closer to our main theme, we have to consider the following passage of Book IV of the *Metaphysics*; in this case, too, Aristotle set up the framework of the dispute. We already know the beginning:

[(LNC4a):] It is impossible for any one to believe the same thing to be and not to be, as some think Heraclitus says; for what a man says he does not necessarily believe. If [(P1):] it is impossible that contrary attributes should belong at the same time to the same subject (the usual qualifications must be presupposed in this proposition, too), and if [(P2):] an opinion which contradicts another is contrary to it, obviously [(LNC4a):] it is impossible for the same man at the same time to believe the same thing to be and not to be; for

---

[36] See Barcan-Marcus (1983).

[37] See Foley (1986): 350.

[38] Hegel (1831): 439-40.

[39] The informal version of the argument is the following: "I argue that it is possible to [...] believe the impossible, say, that there is a largest prime number. The impossibilist objects that I am mistaken. Wrong move! By trying to correct me, the impossibilist concedes that I believe a false proposition. The proposition in question (i.e. that impossibilities can be believed, if false, is necessarily false. Thus, the impossibilist would be conceding that an impossibility can be believed. [...] Belief in 'Some impossibilities are believable' guarantees its own truth" (Sorensen (2001): 124-5).

if a man were mistaken in this point he would have contrary opinions at the same time. (*Met.* 1005b23-32)

Prima facie, Aristotle's strategy is quite strange. Recall that we are in a context where he is trying to produce a defence of the (LNC) in front of possible, and sometimes actual, deniers of the Law – Heracliteans and Protagoreans are mentioned. Such opponents would be made impossible by the impossibility of believing a contradiction. Nevertheless, as Łukasiewicz has pointed out, Aristotle is trying to *infer* the pragmatic-psychological version of the (LNC) directly from the metaphysical one. He tries to deduce the principle that "It is impossible for anyone to believe the same thing to be and not to be", or "It is impossible for the same man at the same time to believe the same thing to be and not to be" (which are versions of (LNC4a)). I have marked the two premises as (P1) and (P2). Now, (P1) is a variation of the ontological (LNC), phrased in terms of *contraries*. Here is a passage in which Aristotle gets the formulation in terms of contraries out of a semantic version of the (LNC) – precisely, a (2b)-kind:

> Now since [(LNC2b):] it is impossible that contradictories should be at the same time true of the same thing, obviously [(P1):] contraries also cannot belong at the same time to the same thing. For of the contraries, no less than of the contradictories, one is a privation – and a privation of substance; and privation is the denial of a predicate to a determinate genus. If, then, it is impossible to affirm and deny truly at the same time, it is also impossible that contraries should belong to a subject at the same time. (*Met.* 1011b16-21)

According to the Aristotelian theory, two *contraries* are a pair of incompatible properties (sometimes concepts, or notions), maximally opposed within the same common genus (like *white* and *black* within the genus *coloured*). One of the two contraries is understood as the privation of the other, which means that an object having one of the incompatible properties necessarily is deprived of, i.e., it does *not* have, the other. Therefore, if an object could be both black and white, that is, if it had both the contrary properties, since to be black is to be deprived of white, i.e., *not* to be white, the object would be white and not white, which would violate the metaphysical (LNC). Fair enough, at least for the supporters of the (LNC). The problem now is with premise (P2). Here Aristotle considers "opinions", i.e., beliefs, as properties or states of the mind, and tries to argue that beliefs having contradictory contents (or contents expressed by contradictory sentences) *are* themselves contrary, that is to say, incompatible properties or states of the mind. If we now read our operator for acceptance/assertion in a psychological fashion, we can claim that $\vdash_x \alpha$ and $\vdash_x \neg\alpha$ express contrary or incompatible properties of believer x (or of believer x's mind). If the very same x, then, could accept or believe (that) α, and accept or believe (that) $\neg\alpha$, then we would face a situation in which the very same thing, x (or x's mind) has two incompatible properties – which is precisely what the metaphysical (LNC) forbids. Therefore, Aristotle is trying to prove that it is *impossible to believe in a contradiction*, by means of an argument that has the very (LNC) as its premise (P1).

1.3.2 Łukasiewicz's Criticisms

Łukasiewicz objected both to premise (P2) of the argument, and directly to the "psychological" (LNC) itself. As for (P2), Łukasiewicz claimed that is it illegitimate to attribute to beliefs or opinions the same kind of logical relations holding between their contents; and he saw this as a case of confusion between logical and psychological issues. He blamed Aristotle for mistakenly attributing to beliefs such features as truth and falsity, which can be properly ascribed just to sentences, or to the thoughts expressed by sentences – with "thought" understood, though, in an objective, Fregean sense.[40]

As for the psychological (LNC), Łukasiewicz maintained that the incompatibility between beliefs, being a merely psychological assumption, could not be secured *a priori*: therefore, formulations of the (4a-b)-kind of the (LNC) are mere empirical hypotheses, subject to inductive confirmations, and having at most a certain degree of probability.[41] The same idea was expressed in Husserl's *Logische Untersuchungen*:

> In the same individual, or rather in the same consciousness, contradictory acts of belief are incapable of lasting for any time, however short. But is this really a *law*? Can we really utter it with such boundless generality? What are the psychological inductions which justify its acceptance? May there not have been people, and may there not still be people who, deceived by fallacies, contrive at times to believe contradictories together?[42]

---

[40] "Aristotle characterises beliefs as true and false; but beliefs, considered as mental acts, can no more be true or false in the primary sense than can sensations, feelings, and the like" (Łukasiewicz (1910b): 53).
[41] See Łukasiewicz (1910a): Chs. IV and V.
[42] Husserl (1900): 114.

Some of these considerations do not appear to be irresistible. For instance, Aristotle's shift in the attribution of truth and falsity is amply justified by commonsensical and philosophical practice. It is quite common to talk about true and false beliefs, or convictions, at least in a derived sense: thinker x has a true belief iff $\vdash_x \alpha$, and α is true: a true belief is a belief in a true sentence (or proposition, etc.). The core point at issue is a different one, however. We have seen that it is common in the literature to call a "contradiction" something of the form:

(C4a) $\vdash_x \alpha \wedge \vdash_x \neg \alpha$,

and (psychological) "Law of Non-Contradiction" a negation of (C4a). As Łukasiewicz pointed out, however, (C4a) is not, as it were, a *bare* contradiction yet, that is to say, something of the form of (C1): $\alpha \wedge \neg \alpha$. We do not have an exposed inconsistency yet when someone believes (that) α and believes (that) $\neg \alpha$. A bare contradiction would be one in which the very same belief both exists and does not exist in the very same mind,[43] i.e.:

(C4c) $\vdash_x \alpha \wedge \neg \vdash_x \alpha$.

In order to move from (C4a) to (C4c) we need the following entailment, which may be called Psychological Incompatibility, since it captures the incompatibility Aristotle was trying to express via his premise (P2):

(PI) $\vdash_x \neg \alpha \rightarrow \neg \vdash_x \alpha$.

"To believe in the antithesis is not to believe in the thesis", as one may say. To accept (PI) – which is also a principle of many doxastic logics – and the shift of the negation outside of the scope of the epistemic operator, is to turn any covered contradiction of the (C4a)-kind into a bare contradiction.

---

[43] See Łukasiewicz (1910a): Ch. IV.

1.3.3 Beliefs in the Tractatus

Why should we accept (PI)?[44] I think a convincing reply would involve a *comprehensive* account of intentionality. It is because Łukasiewicz sticks to a certain (largely empirical) conception of intentional states that he refuses to attribute to beliefs the same kind of properties and relations holding between their contents. A belief "cannot be more closely analysed but has to be experienced":[45] as psychological phenomena, beliefs do not *signify* that something is such and such. They are close to sensations: we can experience them but, strictly speaking, they do not *depict* any fact.

A different view may give more plausibility to (PI).[46] For instance, let us consider Wittgenstein's treatment of propositional attitudes in the *Tractatus*. According to Wittgenstein, beliefs have to be analyzed exactly as psychological situations, which, contrary to what Łukasiewicz assumed, *are* pictures of facts.

In 5.541, Wittgenstein begins to talk of "certain propositional forms of psychology, like 'A thinks, that *p* is the case', or 'A thinks *p*', etc.". And he criticizes both the "contemporary superficial psychology", and Russell's and Moore's treatments of propositional attitudes, according to which, in the Tractarian reconstruction, they should be explained as relations between an object, the believer, and (the fact, or perhaps the state of affairs[47] expressed by) a sentence. On the contrary, as is claimed in 5.542, "here we have no co-ordination of a fact and an object, but a co-ordination of facts by means of a co-ordination of their objects". Which

---

[44] Priest and Routley (1989b): 384-5, argue against it. Bremer (2005): 181, claims that "Giving up [(PI)] for reasons of psychological adequacy, however, does not leave much of a [doxastic] logic worth its name!".

[45] Łukasiewicz (1910b): 51.

[46] Among those who accept (PI) is Strawson, according to whom beliefs with contradictory contents simply cancel each other (see Strawson (1952): 21).

[47] The *Sachverhalt* of the *Tractatus* is a state of affairs, which may be taken as a possible situation. A sentence is the picture of a state of affairs. A fact is "what is the case" (2), i.e., one may argue, not only a possible, but an actual situation: it is the subsisting of a state of affairs. A sentence "*shows* how things stand, *if* it is true", and "it *says*, that they do so stand" (4.022). A more popular version has it that "*Sachverhalt*" names an atomic situation, as opposed to *Tatsache*, a structured, molecular one. This was Russell's view: "facts which are not compounded of other facts are what Mr Wittgenstein calls *Sachverhalte*, whereas a fact which may consist of two or more facts is called a *Tatsache*" (Wittgenstein (1921): xix).

means that the existence of a belief in x's mind is the existence of a state of affairs: a configuration of psychic elements which pictures another state of affairs. For instance, "Pierre Believes that London is beautiful" should be analyzed as: there is a connexion of psychic elements (the belief in Pierre's mind), and such a belief pictures London as being beautiful. Beliefs, as psychological facts, have precisely the role of picturing states of affairs, in line with the mechanisms codified by the more general Tractarian theory of picturing.

One of the upshots of all this is precisely that, according to Wittgenstein, "it is impossible to judge a nonsense" (5.5422). This happens because "the picture contains the possibility of the state of affairs which it represents" (2.203). Thought, as the logical picture of the world, "contains the possibility of the state of affairs which it thinks". "What is thinkable is also possible" (3.02) – which is the Tractarian edition of the Humean motto – and "we cannot think anything unlogical, for otherwise we should have to think unlogically" (3.03). It is doubtful that Wittgenstein's "thought" here should be understood in a psychological sense (as the so-called psychological interpretation of the *Tractatus* maintains). However, the point is that what holds for thought as a logical picture, holds for beliefs as psychological states, too, since the latter *are* portraits themselves. Therefore, the possibility of the believed situation constrains the possibility of belief itself. For Wittgenstein, just as for Aristotle, if a situation is impossible, then it is impossible to believe it, since such a belief would itself be an impossible situation.

1.3.4 Moving On

Then again, we have a couple of good reasons to leave out issues of this kind in (most of) this book. The first one is that their management has to do, if not with empirical psychology, with philosophy of mind, or perhaps cognitive science. It goes beyond the kind of questions we will be for the most part concerned with, chiefly involving logic, semantics, and ontology.

The second and more significant reason is the following. Suppose we accept (PI), or premise (P2) in the Aristotelian argumentation of the *Metaphysics*: beliefs having contradictory contents are incompatible. We would still have to discuss premise (P1), which is a variation on the theme of the *ontological* (LNC): "Contraries (i.e., incompatible properties) cannot belong to the very same thing". Even though (4c) (a "bare" contradiction) follows from (4a) via (PI), the argument is sound on the presupposition that the (LNC) holds in his semantical and ontological versions. If this is the case, then, something of the form of (C4c) and, backwards, (C4a), cannot occur. Therefore, the validity of the semantic and ontological (LNC) is a more *basic* issue than that of the validity of the psychological and pragmatic (LNC), in the following sense: if a contradiction can be true, or if there can be contradictory objects or facts, then believing this should be not only *possible*, but also rationally *required*. Truth is the *telos* of belief (what we aim at when we accept, and assert) for sure. Conversely, only on the presupposition that that no contradiction can be true, or that contradictions cannot obtain, one can (possibly) also defend the thesis that it is not possible to believe it. Talking truth-conditionally: only on the presupposition that a contradiction is true under no circumstance, one is allowed to claim that understanding a contradiction, that is, knowing its truth conditions, entails that it cannot be believed. Now, the difficulties with the (LNC) we shall deal with soon are precisely of such a "more basic" kind: they seem to testify that contradictions *can* be true, albeit in very special circumstances. This does not entail that questions of pragmatics, belief, and rational commitment, shall henceforth be completely abandoned. On the contrary, pragmatic variations on the (LNC) will resurface later – and they will play a strategic role within the very theories that admit true contradictions.

1.4 Challenging the Law

Despite its being entrenched in the orthodoxy, the (LNC) has been challenged every now and then since the ancient times. Heraclitus was considered, mainly because of the Aristotelian references, as the beginner of a procession of philosophers who saw in change and movement the phenomenal evidence of a violation of the (LNC). We go from Hegel, according to whom "something moves, not because at one moment is here and at another there, but because at one and the same moment it is here and not here",[48] to Graham Priest's metaphysics of change.[49] Even before Aristotle, sophists like Protagoras and Gorgias resisted – or were understood as resisting – the early Parmenidean formulation of the Law and its paradoxical defence by Zeno. So-called Gorgian nihilism was a direct overthrowing of the Eleatic theses: nothing is, or being is non-being; even if it were, it would not be thinkable; and even if it were thinkable, it would not be expressible. And the so-called Protagorean relativism, ex-

---

[48] Hegel (1831): 440.
[49] See Priest (1987): Chs. 11, 12 and 15.

pressed by the view that "man is the measure of all things: of those which are, that they are, and of those which are not, that they are not",[50] was seen by Aristotle as entailing a denial of the (LNC). "Many men hold beliefs in which they conflict with one another", and from this and the Protagorean principle follows that "the same thing must be and not be".[51]

During the Middle Ages, the problem resurfaced in connection to the question of the divine omnipotence: can God make a stone too heavy for Him to raise? We find St. Pier Damiani getting close to the point in the *De divina omnipotentia*, by blaming St. Girolamus for having claimed that God cannot overturn the past and twist what happened into something that didn't happen. Since God lives an eternal present, denying Him power on the past equals to denying Him power on current and future events, which is blasphemous. Later on, Cusanus placed at the core of his book *De docta ignorantia* the idea that God is *coincidentia oppositorum*: as an infinite being, He includes all the opposite and incompatible properties, therefore being all things, and none of them. Medieval philosophers, though, also found troublemakers for the (LNC) largely independently from theology. These had to do with what Ockham and others called *insolubilia*, and we nowadays call *logical paradoxes* or, sometimes, paradoxes of self-reference. Since these appear to provide the strongest case against the (LNC), we will have to deal with them at great length.

1.4.1 What is a Paradox?

On page one of most books on paradoxes, we find a definition. However, not unexpectedly, definitions diverge, so that also "paradox" is spoken of in many ways. I will use the term ambiguously throughout this book, too, and three main meanings between which I will alternate are the following.

(a) Firstly, a paradox is characterized as the absurd or blatantly counter-intuitive *conclusion* of an argument, which starts with intuitively plausible premises and advances via seemingly acceptable inferences. In *The ways of Paradox*, Quine claims that "a paradox is just any conclusion that at first sounds absurd but that has an argument to sustain it".[52] We

---

[50] Plat. *Theaet.* 152a2-4.
[51] Arst. *Met.* 1009a10-12.
[52] Quine (1976): 1. Sorensen's (1995): 1 definition is: "an apparently unacceptable conclusion derived by apparently acceptable reasoning from apparently acceptable premises".

shall be particularly concerned not just with sentences that are paradoxical, in the sense that they are implausible, or contrary to common sense ("paradox" intended as: opposed to the dòxa, or to what is œndoxon, entrenched in pervasive and/or authoritative opinions); but with sentences that constitute violations of the (LNC) in one or another of the formulations listed above. A paradox in this strict sense is also called an *antinomy*. Some paradoxes are not, therefore, properly antinomies, but counter-intuitive sentences which do not constitute, nevertheless, apparent violations of the (LNC). A classic example is provided by the paradoxes of material conditional, of which I shall speak at length in the following.[53]

(b) Sometimes the whole *argument* is also called a paradox.[54] So we have Priest retaining that "[logical] paradoxes are all arguments starting with apparently analytic principles […], and proceeding via apparently valid reasoning to a conclusion of the form 'α and not-α'".[55]

(c) Thirdly, at times a paradox is considered as a *set* of jointly inconsistent sentences, which are nevertheless credible when taken individually. This definition is taken as having some advantages over the previous ones by Sorensen;[56] and a generous set-theoretic account of paradoxes is developed, e.g., by Nicholas Rescher.[57]

The debate on logical paradoxes nowadays involves technicalities that get closer and closer to esotericism. Nevertheless, the philosophical importance of paradoxes can hardly be overestimated. This is due to the following simple fact: they involve the most decisive categories of thought – such notions as *demonstration, negation, predication, totality, sethood, membership* and, of course, *truth*. Our ordinary language (taken as inclusive of the jargon of philosophy, science, mathematics, etc.) includes some quite general expressions – so general that we cannot even begin to *discuss* them without *using* them – such as "not", "member of", and

---

[53] See e.g. Susan Haack's terminology: "Paradoxes: (i) (Also known as 'antinomies'.) Contradictions derivable in semantics* and set* theory; […] (ii) The 'paradoxes' of material and strict implication* are theorems* of classical, 2-valued and modal logic […]. I use scare quotes because these 'paradoxes' involve no contradiction" (Haack (1978): 249).

[54] See e.g. Mackie (1973); Beall and van Fraassen (2003): 119 claim that "a *paradox* […] is an argument with apparently true premises, apparently valid reasoning, and an apparently false (or untrue) conclusion".

[55] Priest (1987): 9.

[56] See Sorensen (2003): 364

[57] See Rescher (2001).

"true". Intuitive principles govern their application. Some of them we have already met – the T-schema:

(T) $T(\ulcorner \alpha \urcorner) \leftrightarrow \alpha$,

and our characterization of falsity as the truth of negation:

(Neg1) $F(\ulcorner \alpha \urcorner) \leftrightarrow T(\ulcorner \neg \alpha \urcorner)$.

We have, then, what is usually called the *Comprehension* or *Abstraction* Principle, which, having set abstracts, may be formulated for the time being as follows:

(Abs) $x \in \{y \mid P(y)\} \leftrightarrow P(x)$,

"*x* is a member of the set of the *P*s iff *x* is (a) *P*" (Jeffery Deaver is a member of the set of writers iff Jeffery Deaver is a writer). They look so obvious that some take them as simply (partly) defining the terms in question. As we shall see in §§ 2 and 3, they also apparently produce violations of the *principium firmissimum* when we consider some particular property *P*, or sentence α. Before we enter the strange realm of logical paradoxes, and of the strategies proposed by logicians, philosophers and mathematicians in order to solve them, let us wonder: what, exactly, does "solving" a paradox mean? If a paradox is an argument of the kind characterized in (b) above, we can list, in order of increasing difficulty, three requirements found in the literature which a theory that aims at solving a paradox should fulfil.

(1) Clearly, the theory should indicate precisely which premise is false, or which inference is invalid. This is what Haack calls "*formal solution*",[58] and Kirkham "criterion of specificity".[59]

(2) But the theory should also explain independently *why* this is so. As John Woods has said, a defender of the (LNC):

Must non-question-beggingly *identify* the defective components: that is, he must try to find some theorems that he is able to discredit quite independently

---

[58] See Haack (1978): 138-9.
[59] Kirkham (1992): 273.

of their contribution to the paradox. He must *replace* those theorems with paradox-resistant counterparts...[60]

Anyone can choose a premise or inference rule at random and reject it. However, the mere intention of avoiding the paradoxical conclusion is not sufficient to motivate the choice of some or some other premise. Such a merely formal remedy would look like the prescription of Groucho Marx's doctor:

Groucho: Doctor, my shoulder hurts when I lift my arm like this.
Doctor: Don't lift it like that.

In other words, the choice should be independently motivated, not *ad hoc*. Haack dubs such a condition "philosophical solution",[61] and the denomination is quite appropriate. The philosophical point is a Socratic one: you might not be ready to follow the argument wherever it leads, but you cannot dismiss it *only* because of the unacceptability of its conclusion. If you do, you undermine the key function of arguments as tools of *discovery* – and discoveries can be surprising by definition; or even unpalatable.

(3) Finally, the theory may explain why the imputed (and amputated) premise or rule *appeared* to us so plausible that only the derivation of an antinomic conclusion made it look suspect. This last requirement has been put forth, e.g., in Priest (1979) and Priest and Routley (1989d).[62] It may look too strong to be fair, presupposing as it does a difficult analysis of the intuitions deposited in our common sense. For instance, Quine has conjectured that there is nothing wrong with paradoxes from the point of view of our intuitions: intuitions are doomed to death *because they* lead to paradoxes, and we cannot but postulate a counter-intuitive alteration of

---

[60] Woods (2003): 172.

[61] "What is intended is that it should be shown that the rejected premise or principle is of a kind to which there are independent objections – objections independent of its leading to paradoxes, that is. It is impossible, though difficult, to avoid 'solutions' which simply *label* the offending sentences in a way that seems, but isn't really, explanatory" (Haack (1978): 139).

[62] "Someone who wants to [solve the paradoxes] must not only locate the source of the unsoundness but must explain, in a non-question-begging and coherent way, what is wrong with it. In fact, even more than this is required. An explanation of why the incorrect principle was found plausible in the first place is also required. [...] These are tall orders, which have never been met..." (505).

our "natural and established usage".[63] This seems incorrect to Kirkham, since it threatens condition no. 2.[64] Nevertheless, we will see that standard solutions of the logical paradoxes usually have difficulties even in reaching level 2.

1.4.2 Kantian vs. Hegelian Strategies

Even when they reach it, other problems creep around. Let us begin with a schema we shall find instances of in the following Chapters, and which has been informally introduced as follows by Priest, at the beginning of his *Beyond the Limits of Thought*. Priest has conjectured that the particular notions generating the paradoxes constitute some limit cases of what can be expressed, described, conceived, or of the iteration of certain recursive operations of thought:

> Limits of this kind provide boundaries beyond which certain conceptual processes (describing, knowing, iterating, etc.) cannot go; a sort of conceptual *ne plus ultra*. [...] The contradiction, in each case, is simply to the effect that the conceptual processes in question *do* cross these boundaries. Thus, the limits of thought are boundaries which cannot be crossed, but yet which are crossed.
> In each of the cases, there is a totality (of all things expressible, describable, etc.) and an appropriate operation that generates an object that is both within and without the totality. I will call these situations *Closure* and *Transcendence*, respectively. In general, the arguments both for Closure and for Transcendence use some form of self-reference, a method that is both venerable and powerful. [...] Closure is usually established by reflecting on the conceptual practice in question. [...] Arguments for Transcendence are of more varied kinds; often they involve applying a theory to itself. Some of them are more technical: a paradigm of these is diagonalisation, a technique familiar from the logical paradoxes.[65]

I will come to technical details and diagonalisation later. Meanwhile, one could conjecture that an initial, but comprehensive, grasp of the situa-

---

[63] See Quine (1976): 4-9

[64] So that "Quine leaves no room for any further adjudication between proposed revisions" (Kirkham (1992): 275).

[65] Priest (1995): 3-4.

tion involving the schema[66] is not due to some mathematical logician, but to two classical philosophers, namely, Kant and Hegel. Several authors, including Zermelo and Fraenkel,[67] have noticed the remarkable similarities between the Kantian antinomies and some logical paradoxes – precisely, the paradoxes of the Cantorian infinite.[68] With a little bit of rubdown, the whole debate on logical paradoxes may be viewed as a ramification and formal specification of the Kant-Hegel dialectics.

Kant believed that rational antinomies were produced by an illicit use of pure concepts; he also held, though, that such an illicit use was a "*natural* and inevitable *illusion*"[69] – a side effect of the reason's pursuit of completeness in knowledge. Given some phenomenon, we can be curious on its "condition", as Kant says. This condition being another phenomenon, now we can be curious on *its* condition. *Et cetera*. Reason asks us to inquire further, but it also gives us some idea of an unconditioned *totality* of all conditions of a certain realm. The antinomies of pure reason, in particular, are originated by such basic concepts as *before*, *part of*, *cause*, *depends on*. As soon as children begin to make use of reason, they start asking, what is beyond that? What was before this? Again, the question can be iterated – What is beyond *that*, then? Curiosity is good – makes us human. The "transcendental illusion" begins when we turn what should just be a regulative ideal into a limit-*object*. *Legitimate* inferences on the world as a whole (a totality which is never given to us as such) can lead us to conclude both that it has a beginning in time and a limit in space, and that it has no beginning nor limits in space, that it is infinite in space and time.[70] Both horns assume the opposite thesis and seemingly perform a *reductio*. The fallacy lies in treating *the world as a whole* as an object – in mistaking a subjective condition for an objective reality.[71]

---

[66] Which can also be found in Russell's works, and can be made formally precise within an inconsistent set theory – as we shall see in § 11.

[67] See Hallett (1984): sect. 6.2.

[68] According to G. Martin, "[the] conflict between concluding and beginning anew, between forming a totality and using this totality as a new element, is the actual ground of the [Cantorian] antinomy. It is this conflict that gives the connection with the Kantian antinomies. Kant saw quite clearly that the antinomies rest on this antithesis between making a conclusion and going beyond the conclusion" (Martin (1955): 55).

[69] Kant (1781): 300.

[70] See *Ibid*: 396ff.

[71] This is Kant's first solution. His second solution works only for the third (there are/are not uncaused causes) and fourth (there is/is not a necessary being) antinomy, and

Now, according to Hegel such a conception has a *pro* and a *contra*. Kant has a point in showing via the antinomies that dialectics is "a necessary function of reason"; in defending the *"objectivity of the illusion* and the *necessity of the contradiction* which belongs to the nature of thought determinations".[72] However, Kant mistakenly imputes objectification, as an error, to reason: "the result is only the familiar one that reason is incapable of knowing the infinite".[73] On the contrary, we should abandon such "tenderness for the things of this world", and the idea that "the stain of contradiction ought not to be in the essence of what is in the world; it has to belong *only* to thinking reason".[74] Contrary to what Kant held, the Kantian antinomies are not a *reductio* of the illusions of reason. They are perfectly sound arguments, deducing the contradictory nature of the world. Or, at least, this is the ordinary interpretation of Hegel's position.[75]

The standard strategy in order to overcome the logical paradoxes is a strictly Kantian one: once a contradiction is detected, restrictive measures are adopted in order to rule out such dangerous totalities, and/or our capacity of grasping and describing them. The cases at the limits of thought produce *absurdities*, and the rejection of absurdity is a minimal constraint on rationality. This is, for instance, the central thesis of Patrick Grim's book *The Incomplete Universe*.[76] This book, on the other hand, aims at exploring the destiny of those who followed the Hegelian way:

> It is the supreme inconsistency to admit, on the one hand, that the understanding is cognizant only of appearances, and to assert, on the other, that this cognition is *something absolute* – by saying: cognition *cannot* go any further, this is the *natural*, absolute *restriction* of human knowing. [...] Something is

---

rests on the distinction between phenomena and noumena: determinism, for instance, holds in the phenomenal world, but freedom may have a place in the realm of noumena.

[72] Hegel (1831): 56.

[73] *Ibid*: 39.

[74] Hegel (1830): 92.

[75] In Berto (2005) and (2007) I argued that attributing such a violation of the (LNC) to Hegel may be a bad way to reconstruct his dialectics.

[76] "This is an explanation of a cluster of related logical results. Taken together, these seem to have something philosophically important to teach us: something about knowledge and truth and something about the logical impossibility of a *totality* of knowledge and truth" (Grim (1991): 1).

> only known, or even felt, to be a restriction, or a defect, if one is at the same time *beyond* it.[77]

> With respect to the form of the *limitation* [...] great stress is laid on the limitations of thought, of reason, and so on, and it is asserted that the limitation *cannot* be transcended. To make such an assertion is to be unaware that the very fact that something is determined as a limitation implies that the limitation is already transcended.[78]

Kantian strategies trying to save the (LNC) via restrictive measures seem to fall foul of one of the deepest demands of philosophical thought: the accomplishment of *universality*. Arguably, aspiring to grasp very comprehensive totalities seems to be a feature of Western philosophy (not to speak about *Eastern* philosophy) since its Greek origins. Philosophers were those who began to speculate on the principle of *all* things, thereby assuming as possible, if not mandatory, to talk about the world as a whole. Later on, they dealt with another very general concept, namely, *set*: and for a brief golden age – i.e., until set-theoretic antinomies emerged – they became inclined to assume there is a universal set, which is just the extension of "is a set". Philosophers concerned with the nature of language and thought have aimed at a general characterization of *truth* – not just truth in a (formal) language and in a given structure. And it seems that the full power of ordinary language is required to say what we aim at saying in philosophy. The Kantian strategy would lead us to treat such ambitions as "natural and inevitable illusions". However, the following Chapters will show that the limitative strategies face some problems and failures. These will be of different kinds. We shall find theories trying to solve the logical paradoxes and to protect the (LNC) facing new paradoxes, formulated using *the very notions* originally introduced to overcome the initial trouble. The defensive players systematically score own goals. These new paradoxes will be addressed mostly by refusing that such notions can be expressed within the theory itself. The theory, then, will appear as vacillating between an upsetting self-refutation and a failure to fulfil its initial promises. For instance: the promise was that of providing a semantic theory for ordinary language; but the theory ends up as capable to treat only artificial and expressively weaker ones. Such a

---

[77] Hegel (1830): 105.
[78] Hegel (1831): 134.

recurring fact has even led Roy Sorensen and Manuel Bremer to formulate general maxims about it:

> The phenomenon is general; introducing almost any apparatus to resolve paradoxes makes that tool the subject of other paradoxes.[79]

> A linguistic framework rich enough to avoid some [of] the antinomies, generates its own versions of them.[80]

1.5 Semantic and Set-Theoretic Paradoxes

It is customary to distinguish between two families of paradoxes: the *semantic* and *set-theoretic* ones. The former family typically involves such notions as *truth*, *denotation*, *definability*, etc. The latter, such notions as *membership*, *cardinality*, etc. The distinction, anticipated by Giuseppe Peano, is credited to Frank Ramsey, who formulated it with reference to the list of logical paradoxes examined in Russell and Whitehead's *Principia mathematica*:

> Group A [i.e., antinomies no. 2, 3 and 4 of the original list of the *Principia*: among them, the Russell and Burali-Forti paradoxes, which will introduce themselves in § 3.1.1] consists of contradictions which, were no provision made against them, would occur in a logical or mathematical system itself. They involve only logical or mathematical terms such as class and number, and show that there must be something wrong with our logic and mathematics. But the contradictions in Group B [i.e., antinomies no. 1, 5, 6, 7 of the *Principia*: among them, the Liar paradox, which we shall meet in § 2.1] are not purely logical, and cannot be stated in logical terms alone; for they all contain some reference to thought, language, or symbolism, which are not formal but empirical terms.[81]

The idea that semantic paradoxes contain "empirical" references has vanished today, i.e., after Gödel's and Tarski's formal procedures to obtain self-reference (on which we will concentrate in the following Chapters). Those very procedures make it difficult to draw a sharp line between the two families (for instance, because of the fact that Tarskian

---

[79] Sorensen (2003): 173.
[80] Bremer (2005): 27.
[81] Ramsey (1931): 36-7.

semantics is framed in set-theoretic terms). Nevertheless, the distinction is commonly accepted within the relevant literature, and Priest himself discusses semantic and set-theoretic paradoxes in different Chapters of *In Contradiction*.[82] I shall follow him closely.

---

[82] See Priest (1987): Chs. 1 and 2.

2 True Lies

> Talk is cheap and lies are expensive.
>
> Green Day, *Walking Contradiction*

2.1 Liars

Semantic paradoxes can involve different semantic notions, such as *denotation*, *definability*, etc. I will consider only those who employ the notions of truth and falsity, which are usually grouped under the label of the *liar*. These are the most discussed in the literature – those for which most tentative solutions have been proposed. They also are the most classical ones, having been on the philosophical market for more than two thousand years. The ancient Greek grammarian Philetas of Cos is believed to have lost sleep and health trying to solve the liar paradox, his epitaph claiming: "It was the Liar who made me die/ And the bad nights caused thereby".

2.1.1 Informal Liars

One of the most ancient versions of semantic paradox appears in St Paul's *Epistle to Titus*. Paul blames a "Cretan prophet", who was to be identified as the poet and philosopher Epimenides, and who was supposed to have at one time said:

(1) All Cretans always lie.

Actually, (1) is not a real paradox in the sense of a sentence which, on the basis of our *bona fide* intuitions, would entail a violation of the (LNC). It is just a sentence that, on the basis of those intuitions, cannot be *true*. It is self-defeating for a *Cretan* to say that Cretans always lie: if this were true – that is, if it were the case that all sentences uttered by any Cretan are false – then (1), being uttered by the Cretan Epimenides, would have to be false itself, against the initial hypothesis. However, there is no contradiction yet: (1) can be just false under the (quite plausible) hypothesis that some Cretan sometimes said something true.

We are dealing with a full-fledged liar paradox (also attributed to a Greek philosopher, and probably the greatest paradoxer of Antiquity: Eubulides) when we consider the following sentence:

(2) (2) is false.

As we can see, (2) refers to itself, because it is no. 2 of the sentences highlighted in this Chapter, and tells something of the very sentence no. 2. Also (1) refers to itself, but does it in a different way from (2). This is what makes (1) not strictly paradoxical. (1) claims that all the members of a set of sentences (those uttered by Cretans) are false. Besides, it belongs to that very set, due to the fact of being uttered by a Cretan. Therefore, (1) can be simply false, under the empirical hypothesis that some sentence uttered by a Cretan, and different from (1), is true. This is also what makes it look so odd. As Stephen Kleene observed, it is unsatisfactory that a logical paradox is avoided only via the empirical fact that some Cretan sometimes said something true.[1]

Some form of self-reference can be detected in all paradoxes, so that the phenomenon of self-referentiality (or the related one of *impredicative definitions*, of which I shall talk later) has been considered directly liable of producing the antinomies.[2] Nevertheless, many self-referential sentences are harmless, in that we seem to be able to ascertain their truth value in an unproblematic way. For instance, it is easily observed that, among the following, (3) and (4) are true, whereas (5) is false:

(3) (3) is a grammatically well-formed sentence;

(4) (4) is a sentence contained in *How to Sell a Contradiction*;

---

[1] See Kleene (1952): 39.

[2] One recent exception might be constituted by Stephen Yablo's "infinitary" chains, e.g.:

(Y1) For all n greater than 1, Yn is false
(Y2) For all n greater than 2, Yn is false
(Y3) For all n greater than 3, Yn is false
...

(see Yablo (1993)). But these might be considered as also involving some essential circularity – as claimed, e.g., in Priest (1997b). On this issue, see also Beall (2001c).

(5) (5) is a sentence printed with yellow ink.

On the contrary, (2) is not harmless at all. Let us reason by cases. Suppose it is true: then what it says is the case, so it is false. Suppose it is false: this is what it claims to be, so it is true. If we accept the Principle of Bivalence, (PB), that is, the principle according to which all sentences are either true or false, both alternatives lead to contradiction: (2) is true *and* false, against (LNC2a).

A sentence can refer to itself in various ways, so we can have various versions of (2). For instance:

(2a) This sentence is false;

(2b) I am false;

(2c) The sentence you are reading is false.

The paradox can also be produced without any direct self-reference, but via a short-circuit of sentences. For instance:

(2d) (2e) is true

(2e) (2d) is false.

This is as old as Buridan (his Sophism no. 9: Plato saying "What Socrates says is true"; Socrates replying "What Plato says is false"). If what (2d) says is true, then (2e) is true. However, (2e) says that (2d) is false... And so on: we are in a paradoxical loop. Other versions of the liar are called *strengthened liars*,[3] or also *revenge liars*.

(6) (6) is not true.

(7) (7) is false or neither true nor false.

We shall understand later why these versions deserve such names (whereas (2) may also be called "standard liar"). I will anticipate, nevertheless, one of their peculiarities. Given a line of reasoning analogue to

---

[3] As far as I know, the terminology may be due to a van Fraassen (1968).

that of the standard liar, (6) turns out to be true and *untrue*: which means that it contravenes (LNC2b), even though we reject the principle (Neg2) equating falsity with untruth (therefore, equating "internal" contradictions of the (C2a-b)-kind and "external" ones of the (C2c)-kind).

A different sort of semantic paradox we shall deal with in the following is Curry's (also called the Curry-Geach-Löb[4] paradox). The following sentence produces it:

(8) If (8) is true, then β,

β being any sentence. Intuitively, (8) says something like: "Anything follows from my truth". I will come later to how to derive unwelcome results from (8). A peculiarity of (8) consists in its employing neither the falsity predicate (like the standard liar (2)), nor negation (like the strengthened liar (6)). According to some, (8) shows that neither of these notions plays an essential role in the derivation of the semantic paradoxes. We shall see that its treatment can be tricky also for those who admit true contradictions.

2.1.2 Semantic Closure

Tarski ascribed the presence of the semantic paradoxes in a language to certain features, grouped under the label of "semantic closure conditions". Intuitively, a semantically closed language is a language capable of talking of its own semantics, of the meanings of the expressions of the language itself. This means that it can mention its own expressions and attribute semantic properties to them.[5]

Given our purposes, it will be useful to formulate the closure conditions for a given language as follows.

1. The language has a name for each expression of the language itself.

2. It is possible to express the notion of truth for the language within the language.

---

[4] This name might be due to van Benthem (1978).

[5] Kirkham's characterization is: "a semantically closed language is one with semantic predicates, like 'true', 'false', and 'satisfies', that can be applied to the language's own sentences" (Kirkham (1992): 278).

3. All sentences of the language are either true or false.[6]

It is out of question that English satisfies condition no. 1. Written English has a name for any of its expressions – just put it in here: " ". Apart from the graphic device of quotation marks, ordinary English always allows its expressions to contextually denote themselves, instead of their ordinary denotation (*suppositio materialis*, as Ockham used to say). If we want to dismiss the idea that ordinary English is semantically closed, it seems we have to attack condition no. 2 or no. 3 (or maybe both, but this may sound a bit drastic). Condition no. 3 is the Principle of Bivalence. As for no. 2, the point is that "to express" is loose (some just say: "the language *contains* a truth predicate…"[7] etc.). We may begin to make it more precise by wondering what we should consider, in general, as a good *definition* of truth for a language. Tarski's reply is well known. According to his famous "Convention T", a theory provides a *materially adequate* definition of truth for a given language if, for any sentence of the language, it is possible to infer from the principles of the theory the corresponding instance of the T-schema.[8]

Now, the T-schema is considered by many (including me) a minimal – at the very least, extensional – characterization of the truth predicate, even though according to some it does not exhaust its sense. As Timothy Williamson said, even though "it is not claimed that a Tarskian theory tells the whole truth about truth", nevertheless "it tells an essential part of the truth". As we saw in § 1.2, one may refuse to subscribe to a robust or real-

---

[6] See e.g. Woods (2003): 170-1. Woods uses the notion of *satisfaction*, not that of truth, but in the Tarskian framework the latter is notoriously definable via the former given a few intuitive assumptions. Tarski phrased condition no. 2 in terms of *truth* in Tarski (1944): 124, which led him to add the liar as an "extra" premise. Sometimes condition no. 3 is expressed as the inference rule:

$$\frac{\alpha \leftrightarrow \neg\alpha}{\alpha \wedge \neg\alpha}$$

but this one is equivalent to the Law of Excluded Middle under very weak assumptions, and one can take the failure of the (LEM) as equivalent, in its turn, to a failure of bivalence, given the T-schema. Sometimes condition 3 is not listed among the closure conditions, and languages fulfilling only the first two conditions are called "semantically closed".

[7] See e.g. Armour-Garb (2004): 119.

[8] See Tarski (1944): 120.

ist interpretation; but "without a disquotational schema, it is doubtful that one has a truth predicate at all".[9] Some of our linguistic businesses seem to presuppose the T-schema, and would be unintelligible without it. We sometimes want to approve someone else's words, but simply repeating them may be impossible – because it was too long a speech; or because we do not even know exactly what she said. The T-schema guarantees, so to speak, an operator that reverses quotation.[10] As Quine famously stressed:

> Where the truth predicate has its utility is in just those places where, though still concerned with reality, we are impelled by certain technical complexities to mention sentences. Here the truth predicate serves, as it were, to point through the sentence to the reality; it serves as a reminder that though sentences are mentioned, reality is still the whole point. [...] The truth predicate is a reminder that, despite a technical ascent to talk of sentences, our eye is on the world. This cancellatory force of the truth predicate is explicit in Tarski's paradigm:
>
> 'Snow is white' is true if and only if snow is white.[11]

The T-schema is also at the core of the most traditional theory of meaning, the truth-conditional one, as based on the aforementioned crucial principle from Section 4.024 of the *Tractatus*: "To understand a proposition means to know what is the case, if it is true". According to those who understand this strictly, meaning is given by truth conditions[12] (or, in the generalized version, truth-in-a-possible-world conditions). However, even if we reject the idea that truth conditions are sufficient for meaning, there is wide agreement on their being necessary. So the T-schema should hold for any sentence, as a (partial) specification of its meaning.

2.1.3 Formal Liars

We have listened to Ramsey claiming that the semantic paradoxes contain an unavoidable "empirical", i.e., contextual or indexical, component.

---

[9] Williamson (1992): 268.

[10] "So to speak" because, of course, any linguistic expression can be quoted, whereas what gets disquoted or, more generally, "unmentioned" here are just declarative sentences.

[11] Quine (1970): 10-3.

[12] See for instance Davidson (1984).

In fact, in the paradoxical sentences we have examined so far self-reference are obtained via the numbering device, or via indexical expressions such as "I", "this sentence", and so on. Only factual and contextual information tells us that the denotation of (those tokens of) such expressions is the very sentence in which they appear as the grammatical subjects. This holds also for the "looped liar": suppose (2d) is as above, but (2e) now is "Perth is in Australia". Then (2d) is just true, and no paradox is expected.

Nevertheless, a few years after Ramsey's work, Kurt Gödel showed that a sufficiently powerful theory, expressed in a fully formalized language, could talk about some of its own syntactic and, what interests us more, semantic features, by means of sentences that are as empirical as "2 + 2 = 4". To begin with, such a theory can "talk" of its syntax without contradiction, by having the expressive resources to name terms and expressions of its own language, and to represent some syntactic properties: for instance, the property of *being provable in theory* can be mirrored within the theory.[13]

In order to obtain this, we need a so-called *gödelization*. The basic idea is simple: when we study any formal language, L, and a theory, T, built on it, we are dealing with a denumerable set of objects. Therefore, we can represent such objects by univocally pairing them with natural numbers, via a suitable codification.[14] A gödelization is based upon an (injective) function g, which assigns to each symbol, formula, and finite sequence of formulas of L a natural number. It is so construed that (a) given any expression of L, say, α, we can always establish which natural number (called its *Gödel number*) $g(\alpha)$ corresponds to it; and (b) given a

---

[13] I am talking of "expressing", "representing", and of sufficiently powerful theories in an approximate way here. Logicians usually call *sufficiently powerful* a formal theory capable of representing certain functions, sorted out in Gödel's seminal paper on the incompleteness of arithmetic (Gödel (1931)). A typical example of such a theory is Robinson's arithmetic, **Q**; another is Peano's, **PA**. I shall come back to these issues in a more refined way in § 4.2, where we will meet Gödel's First Incompleteness Theorem. An intuitive idea of how non-contextual semantic paradoxes can be produced in a suitable formal theory shall be sufficient at present.

[14] "The formulas of a formal system [...] in outward appearance are finite sequences of primitive signs (variables, logical constants, and parentheses or punctuation dots), and it is easy to state with complete precision which sequences of primitive signs are meaning formulas and which are not. Similarly, proofs, from a formal point of view, are nothing but finite sequences of formulas (with certain specifiable properties)." (Gödel (1931): 597).

natural number, we can always establish whether it is the Gödel number of some expression of L and, if it is, of which.[15]

Now, if the theory T on L – roughly – includes elementary arithmetics, sentences of T may "talk" about (properties of, or relations between) sentences of T, in the following sense. Natural numbers have a twofold role: besides being what the theory "officially" talks about, they work as codes, univocally associated to expressions of the theory itself. Syntactic statements on T can be mirrored within T as statements on (arithmetical operation and relations between) numbers. As Hofstadter puts it suggestively in *Gödel, Escher, Bach*, T is capable of some "introspection", or self-analysis. Within this framework, self-referential sentences can be produced: sentences "talking about themselves", in the sense that they refer to their own Gödel number. The standard procedure to achieve self-reference is called *diagonalisation*. A diagonalisation is an operation associating to a formula $\alpha[x]$ of L, containing free only the variable x, the sentence β one obtains by substituting the free variable with the name of the sentence itself. β is then called a *fixed point* of $\alpha[x]$. A theory satisfies the *diagonal requirement* if, for every $\alpha[x]$ of its L, the corresponding instance of the Fixed Point Schema:

(FP) $\beta \leftrightarrow \alpha[x/\ulcorner \beta \urcorner]$[16]

is a theorem of T (this is usually called the Fixed Point Lemma, or the Diagonal Lemma). Now, let us assume that T and F are, as usual, the truth and falsity predicates for our language L, and that they are adequately expressible within the same language. Given (FP), it is easy to have a sentence, λ (the standard liar),[17] such that:

(FP$_\lambda$) $\lambda \leftrightarrow F(\ulcorner \lambda \urcorner)$.

---

[15] The function need not be an injection, although Gödel's original one was: in some gödelizations, every number is the Gödel number of an expression of L (see e.g. Smullyan (1992): 6).

[16] In a formal context, therefore, the name of a formula α can be characterized more precisely as the numeral of the Gödel number of α.

[17] Following the current usage, in some contexts I shall use lower case Greek letters also for some famous sentences – for instance, λ, λ_1, λ_2, ... are various formalized liars; later on we shall meet γ, the undecidable Gödel sentence, etc. In this case, too, context should disambiguate.

λ is a fixed point of the falsity predicate. Intuitively, λ says exactly: "I am false", or "This sentence is false". Via (Neg1) (and replacement), we also have $\lambda \leftrightarrow T(\ulcorner \neg \lambda \urcorner)$. We can have a strengthened liar, λ_1:

(FP$_{\lambda_1}$) $\lambda_1 \leftrightarrow \neg T(\ulcorner \lambda_1 \urcorner)$.

λ_1 intuitively says: "This sentence is not true". The Curry paradox, then, will be produced by a fixed point of the schematic $T(x) \rightarrow \beta$ (where β is any sentence), i.e., a sentence λ_2 such that:

(FP$_{\lambda_2}$) $\lambda_2 \leftrightarrow (T(\ulcorner \lambda_2 \urcorner) \rightarrow \beta)$.

The following is a sample formal derivation, showing how a contradiction follows from the standard liar. Keep your eye on the principles employed:

(1)		$\lambda \leftrightarrow F(\ulcorner \lambda \urcorner)$	(FP$_\lambda$)
(2)		$T(\ulcorner \lambda \urcorner) \leftrightarrow \lambda$	(T) (with $\alpha = \lambda$)
(3)		$T(\ulcorner \lambda \urcorner) \leftrightarrow F(\ulcorner \lambda \urcorner)$	1, 2, Repl
(4)		$(T(\ulcorner \lambda \urcorner) \rightarrow F(\ulcorner \lambda \urcorner)) \wedge (F(\ulcorner \lambda \urcorner) \rightarrow T(\ulcorner \lambda \urcorner))$	3, Df$\leftrightarrow$
(5)		$T(\ulcorner \lambda \urcorner) \vee F(\ulcorner \lambda \urcorner)$	(PB) (with $\alpha = \lambda$)
(6)	6	$T(\ulcorner \lambda \urcorner)$	Ass
(7)		$T(\ulcorner \lambda \urcorner) \rightarrow F(\ulcorner \lambda \urcorner)$	4, $\wedge$E
(8)	6	$F(\ulcorner \lambda \urcorner)$	6, 7, $\rightarrow$E
(9)	6	$T(\ulcorner \lambda \urcorner) \wedge F(\ulcorner \lambda \urcorner)$	6, 8, $\wedge$I
(10)	10	$F(\ulcorner \lambda \urcorner)$	Ass
(11)		$F(\ulcorner \lambda \urcorner) \rightarrow T(\ulcorner \lambda \urcorner)$	4, $\wedge$E
(12)	10	$T(\ulcorner \lambda \urcorner)$	10, 11, $\rightarrow$E
(13)	10	$T(\ulcorner \lambda \urcorner) \wedge F(\ulcorner \lambda \urcorner)$	10, 12, $\wedge$I
(14)		$T(\ulcorner \lambda \urcorner) \wedge F(\ulcorner \lambda \urcorner)$	5, 6, 9, 10, 13, $\vee$E

The standard liar is both true and false, against (LNC2a).[18] Given (Neg1) again, one can say equivalently that λ is a true sentence with a true negation, $T(\ulcorner\lambda\urcorner) \wedge T(\ulcorner\neg\lambda\urcorner)$, against (LNC2b).

---

[18] In this kind of linear presentation, the first column just numbers the passages of the proof. The second column spells the lines on which the formula appearing at that line depends (for instance, in this case the formula of line 6 depends on itself, and the one of line 12 depends on that of line 10). This column can be empty at some lines, when the corresponding formula is a logical or specific principle, an axiom or a theorem of the theory at issue (for instance, at lines 2 and 5 we have an instance of the T-schema and one of the Principle of Bivalence, concerning the standard liar λ). The last column has the abbreviation for the name of the principle, or of the inference rule used in order to obtain that formula at that step (and, in the latter case, the numbers of the lines adopted as premises, too). (Repl) is the Replacement of (provable) Equivalents, i.e.,

$$\frac{\alpha, \beta \leftrightarrow \gamma}{\alpha[\beta/\gamma]} \quad \text{(Repl)}$$

where $\alpha[\beta/\gamma]$ is the result of uniformly replacing all occurrences of β in α with γ. (Df$\leftrightarrow$) applies the usual definition of the biconditional, i.e., $\alpha \leftrightarrow \beta =_{df} (\alpha \to \beta) \wedge (\beta \to \alpha)$. (Ass) is the rule of Assumption, which allows us to introduce any formula at any line, depending on itself. ($\wedge$E) is Conjunction Elimination or Simplification:

$$\frac{\alpha \wedge \beta}{\alpha} \quad \frac{\alpha \wedge \beta}{\beta} \quad (\wedge E)$$

($\to$E) is Conditional Elimination, or Detachment, or *modus ponens*:

$$\frac{\alpha \to \beta, \alpha}{\beta} \quad (\to E)$$

($\wedge$I) is Conjunction Introduction or Adjunction:

$$\frac{\alpha, \beta}{\alpha \wedge \beta} \quad (\wedge I)$$

($\vee$E) is Disjunction Elimination:

$$\frac{\alpha \vee \beta, \overset{[\alpha]}{\underset{\gamma}{\vdots}}, \overset{[\beta]}{\underset{\gamma}{\vdots}}}{\gamma} \quad (\vee E)$$

2.2 Two Lines of Attack

Since rejecting condition no. 1 for semantic closure is out of question for ordinary English, we might try to attack no. 2 or no. 3. Notice that both of them have a decisive role in the formal proof given above, which can be easily retranslated into informal English. At line three, we have derived the paradoxical thesis – *not* an explicit contradiction, though – that the liar is true iff it is false. This has been obtained from (FP$_\lambda$) by replacement of equivalents, via an instance of the T-schema introduced at line two. You may remember that the theory is supposed to be able to provide a materially adequate definition of truth, so any instance of the T-schema should be a theorem of the theory. The corresponding situation for natural language, as Tarski puts it, is that "all sentences which determine the adequate usage of [the truth predicate] can be asserted in the language",[19] which means precisely that all instances of the T-schema can be asserted. Besides using basic logical rules, such as Simplification and *modus ponens*, the proof is an application of Disjunction Elimination, and the disjunction assumed (at line five) is an instance of the Principle of Bivalence.

It seems that we are forced either to dismiss the idea that a definition of truth for a given language L which conforms to Tarski's convention (i.e., from which all instances of the T-schema are deducible) can be phrased *within* L, or to abandon bivalence, i.e., the idea that all sentences of L are either true or false. These are, as a matter of fact, the two main lines followed by those who have tried to save the (LNC) from the liar paradox. I shall discuss them in the following §§.

2.3 Parameterisation, Part One

The semantic paradoxes led Tarski to conclude that the notion of truth for a given language should not be expressible within that language. Therefore, Tarski attacked condition no. 2 for semantic closure. In point of fact, he did *not* propose his cure as solving the problem for natural languages. On the contrary, he did take ordinary language to be semantically closed, and manifested scepticism on the possibility of providing a scientific se-

---

Assumptions of the disjuncts α and β are within square brackets, to point out that they are discharged.

[19] Tarski (1944): 124.

mantics for it. Actually, sometimes he considered the question somehow moot, due to the "inexactly specified structure" of ordinary language. But he also concluded that we may "risk the guess" that "a language whose structure has been exactly specified and which resembles our everyday language as closely as possible would be inconsistent".[20] Others, though, have believed the Tarskian solution to work for natural languages. After having exposed the solution, I shall therefore turn to considerations on its applicability to ordinary English, and finally report some well-known intrinsic difficulties of the theory.

The basic idea is that the truth predicate cannot be *univocal*. A single expression of the surface grammar, "is true", has an ambiguous function for different languages, each of which is semantically open at the level of some deep logical grammar. Instead of a unique language, we would have a hierarchy, more or less with the following structure. For any ordinal n, we have a language L_n, and n is the *order* of L_n. Let us begin with L_0, taken as the language for which we are seeking a (good) definition of truth. Such a definition shall not be expressed within L_0, but in a language, L_1, which is its *metalanguage*: L_1 will contain predicates that refer to the semantic concepts of L_0, and by means of which we can provide a definition of truth for L_0. L_1 is itself semantically open: it cannot express *its* own semantic concepts. So a definition of truth for L_1 shall be expressed in a language, L_2, which is the metalanguage of L_1; and so on.[21]

Each instance of the T-schema is now essentially metalinguistic. To repeat the Tarskian example:

"Snow is white" is true if and only if snow is white

is a metalinguistic sentence about a sentence of the object language, where (the above token of) "Snow is white", appearing on the left hand side of the biconditional, is a metalinguistic name (quotation marks included) of the corresponding sentence of the object language. And (the above token of) "snow is white" (quotation marks excluded) constituting the right-hand side of the biconditional is the metalinguistic translation of that object language sentence.[22]

---

[20] Tarski (1944): 125.
[21] See Sainsbury (1995): 118-9.
[22] "Calling it a translation may seem funny, since it looks like exactly the same sentence, but it is technically a translation" (Kirkham (1992): 279).

This is a clear example of the *parameterisation* strategy I have mentioned in § 1.1.2 – one of those "distinctions of respects" which are often employed in the presence of a challenge to the (LNC). The Tarskian solution parameterises the semantic predicates along the hierarchy of the metalanguages: the metalinguistic "true" and "false" are now abbreviations for "true in the object language", "false in the object language". In particular, the standard liar turns out to be "This sentence is false in the object language". It belongs to the metalanguage, and it is just *false* there, not paradoxical: since metalinguistic sentences do not belong in the object language at all, the liar does not have the property it claims to have.[23]

We will find hierarchies and distinctions of respects also in some attempts to solve the set-theoretic paradoxes. Meanwhile, let us wonder: does the Tarskian strategy provide a good representation of how our ordinary language works?

2.3.1 Its Failure

A decisive difference between such a hierarchy and English is that there does not seem to be *any* metalanguage for English – which becomes obvious, if we accept the principle according to which ordinary language is, so to speak, "transcendental": anything that is linguistically expressible can be expressed within ordinary language – there is no *limit* to it.[24] Furthermore, there is no evidence that the predicate "is true" performs some ambiguous function along some hidden hierarchy of languages, metalanguages, metametalanguages, etc. This makes the proposal look more like a form of *revisionism*: a suggestion to the effect that ordinary English be somehow regimented. If the idea came to Tarski's mind, he certainly found it unsatisfactory:

> Whoever wishes, in spite of all difficulties, to pursue the semantics of colloquial language with the help of exact methods will be driven first to undertake the thankless task of a reform of this language. He will find it necessary to define its structure, to overcome the ambiguity of the terms which occur in

---

[23] See Kirkham (1992): 280.

[24] In Tarski's own words: "A characteristic feature of colloquial language (in contrast to various scientific languages) is its universality. It would not be in harmony with the spirit of this language if in some other language a word occurred which could not be translated into it; it could be claimed that 'if we can speak meaningfully about anything at all, we can also speak about it in colloquial language'" (Tarski (1956): 164).

it, and finally to split the language into a series of languages of greater and greater extent, each of which stands in the same relation to the next in which a formalized language stands to its metalanguage. It may, however, be doubted whether the language of everyday life, after being 'rationalized' in this way, would still preserve its naturalness and whether it would not rather take on the characteristic features of the formalized languages.[25]

Tarski's reluctance is easily understood. First, the level or order of a truth- or falsity- ascription in English depends on unpredictable contingent and contextual factors. This makes extremely difficult to believe that one could, in principle, index the truth predicates to the appropriate level, even implicitly. For instance, the level of:

(9) Everything the Pope said in the *Angelus* yesterday is true,

depends on what the Pope said yesterday and, particularly, on the Pope's asserting something on the truth of some other sentence. Therefore, one should check all the relevant claims. If, among other things, the Pope has declared: "Brothers, let it be perfectly clear that everything Dan Brown has claimed in *The da Vinci Code* is false!", one would have to check all the sentences included in the book, etc.[26] A Tarskian hierarchy also rules out quite a lot of meaningful and unproblematic sentences of ordinary English, so that any language satisfying the conditions of the hierarchy would not be English (anymore), but something quite weaker from the expressive point of view. Consider, for instance:

(10) All the sentences printed in *How to Sell a Contradiction* are true.

This looks like a perfectly meaningful English sentence. To establish its falsity, it would suffice to discover one false sentence printed in this book (not an impossible task, apart from any issue concerning the famous "paradox of the preface": I am rather absent-minded and, statistically, I commit some mistake in all of my books). But (10) cannot belong to a Tarskian hierarchy, since it ascribes truth to itself.

Things are even worse, as Kripke has pointed out. Suppose that the only truth/falsity ascription uttered by Silvio Berlusconi on 05/01/06 was

---

[25] *Ibid.*: 267.
[26] See Kripke (1975): 59.

the following, making a prediction on some official speech to be given by Romano Prodi:

(11) Everything Mr. Romano Prodi will say tomorrow is false;

and suppose that the only truth/falsity ascription uttered by Romano Prodi on 05/02/06 was a comment on Berlusconi's previous declarations:

(12) Everything Mr. Silvio Berlusconi said yesterday is false.

Now (11) and (12) would be meta- to each other.[27] However, if we allow object language and metalanguage to be meta- to each other, a looped liar resurfaces again:

(13a) (13b) is false in the object language.

(13b) (13a) is true in the metalanguage.

According to Kirkham, this "destroys the possibility of the object language/metalanguage distinction as a solution to the paradox".[28]

2.3.2 Liar II – the Revenge

Apart from its inadequacy with respect to ordinary language, the Tarskian solution faces some intrinsic problems. First, it faces a revenge liar, albeit different from the ones mentioned above (and which shall be addressed later). Consider the following sentence:[29]

(14) (14) is false at its level.

Formally, it would be a sentence λ_3 such that:

(FP$_{\lambda_3}$) $\lambda_3 \leftrightarrow F_{lev(\lambda_3)}(\ulcorner \lambda_3 \urcorner)$,

---

[27] See *Ibid.*: 59-60. Kripke's original example, of course, did not involve Italian politicians, but, as usual for him, Richard Nixon.

[28] Kirkham (1992): 281.

[29] The following argument is taken from Priest (1987): 19-20.

If λ_3 belonged to a Tarskian hierarchy, it would place itself at a certain level, say, i. Then, via the T-schema for level i (and replacement), (FP$_{\lambda 3}$) would give us:

$$T_i(\ulcorner\lambda_3\urcorner) \leftrightarrow F_{lev(\lambda 3)}(\ulcorner\lambda_3\urcorner).$$

Since $lev(\lambda_3) = i$, we have:

$$T_i(\ulcorner\lambda_3\urcorner) \leftrightarrow F_i(\ulcorner\lambda_3\urcorner),$$

and given Bivalence (which is not contentious here) an explicit contradiction follows, via a formal proof structurally identical to the one given above for the standard liar. Of course, the hierarchist can deny that the revenge liar (14), or λ_3, *is* expressible in any language of the hierarchy. First, this shows that we are not dealing with (a semantic theory for) English, since we can meaningfully express the notion of order or level in English, refer to, and quantify into, different orders or levels. But things are even worse, because the hierarchic theory itself requires such a quantification *in order to be expressed*. While in the business of explaining the hierarchy, we are forced to say such things as:

(15) For each level n, we have a different truth predicate, T_n.

(15) expresses precisely a central thesis of the hierarchic theory, and does it by quantifying into levels. However, this is precisely what one could not do, according to the theory. The hierarchic view entails that we are always talking at some level, and at the same time talks in general about levels or orders. Can one reply that such assertions should be interpreted as themselves *ambiguous*, or schematic in n? Priest argued not – as soon as one recognizes that "what such a typically ambiguous assertion means, what we are supposed to understand by it, is just what is expressed by a single sentence which quantifies universally over [n]".[30]

We are, then, in front of a case of the situation envisaged in § 1.4.2. An attempt to solve the logical paradoxes and save the (LNC) has to face a new paradox, formulated by means of the characteristic concepts introduced by the theory itself (specifically, the concept of *order*, or linguistic

---

[30] Priest (1987): 20.

level). This new paradox can be addressed by excluding that the concepts at issue are expressible. In the case of the hierarchic theory, such exclusion is poisonous for the theory itself. In addition, it seems that such a trouble can affect any attempt to solve the paradoxes that plays heavily on parameterisation (as we shall see when we investigate some approaches to the set-theoretic paradoxes). Once different respects, or parameters, have been introduced, we can quantify on them and rebuild a paradox. Excluding the meaningfulness of such a quantification rules out the meaningfulness of the theory itself, that ends up self-refuting. For instance, Priest has claimed that the same fate awaits the Burge-Barwise-Etchemendy approach,[31] based on the idea that English semantic predicates have an extension that depends on context (so that, more than as ambiguous, they may be considered as indexical). Barwise and Etchemendy's approach is developed within so-called situation semantics; this time, the parameterisation is with respect to contexts, or situations, and the revenge liar now claims to be false in all contexts (tokenings, situations, etc).[32] Therefore, these strategies, too, "face the dilemma of the choice between inconsistency and a self-refuting inexpressibility".[33]

2.4 Gaps

If attacking condition no. 2 for semantic closure is problematic, we may try with condition no. 3. Their dismissing the Principle of Bivalence unifies various approaches to the liar, also quite different from one another: truth value *gaps*, i.e., sentences which are neither true nor false, are admitted, and the liars are put among them.[34] Some of these approaches have been grouped under the label of *non-hierarchical*, since they try to stick to the idea that, contrary to the hierarchical view, there has to be only one truth predicate. The attempt to avoid hierarchies, though, is in the end doomed to fail, as we will see.

Typically, a key move in a non-bivalent semantics lies in an adjustment of the semantic clause for negation. As we know, standard negation

---

[31] See Burge (1979); Barwise and Etchemendy (1987).
[32] See Priest (1993b) and (1995): 151-5.
[33] Priest (1987): 263.
[34] This line is followed by Bar-Hillel (1957), Martin (1967), van Fraassen (1968), Kripke (1975), Parsons (1984), and Sainsbury (1995).

conforms to (Neg2), that is, its characteristic clause sounds (now in a different and "meta-" notation):

(Neg2) $T(\ulcorner \neg \alpha \urcorner) \Leftrightarrow \text{Not } T(\ulcorner \alpha \urcorner)$.

(Neg2) says that the truth of the negation of α, i.e., the falsity of α, amounts to the untruth of α. The equivalence has to go in a mainstream non-bivalent approach; in particular, against the right-left direction of (Neg2), untruth does not entail falsity. Therefore, it may happen that α is not true, but not false either. Since truth and falsity are made partly independent, now we have to give both truth and falsity conditions. We may adopt, then, the following two typical semantic clauses for a so-called *choice* (or *internal*) *negation*:

(S¬1) $T(\ulcorner \neg \alpha \urcorner) \Leftrightarrow F(\ulcorner \alpha \urcorner)$

(S¬2) $F(\ulcorner \neg \alpha \urcorner) \Leftrightarrow T(\ulcorner \alpha \urcorner)$.

The negation of α is true iff α is false, and *vice versa*. The intuitive idea concerning the paradox is the following: even though the liar is a sentence such that, *if* it were true, it would be false, and *vice versa*, no explicit contradiction (of the (2a)-kind) according to which it is both true and false follows. We can avoid the contradiction by rejecting the idea that truth and falsity are the only two options for a sentence, and maintain that the liar is neither.[35]

The first problem for non-bivalent approaches to the semantic paradoxes is to provide independent reasons to consider the liars as gaps. It is fair to say that the general case for gaps is independently attested. Philosophers of language have renounced to bivalence in order to take care of sentences containing non-denoting terms, or vague predicates, or affected by presupposition failures, etc. (I have already mentioned van Fraassen's, Fine's, and others' supervaluational approach). However, why should *the liars* be allowed to join the club? As we observed in the last Chapter, it is not too difficult to pick one premise of the paradoxical argument at random and discard it. A good solution avoiding *ad-hocness*

---

[35] See e.g. Sainsbury (1995): 112.

is supposed to do something more. Now, the motivations proposed by the gappers do not appear to be very convincing.

According to R.L. Martin's *category solution*, the liars are category mistakes and, as such, truth-valueless. This is also taken as a kind of *meaninglessness*. The initial idea (dating back at least to Gilbert Ryle)[36] is that, besides being syntactically well-formed, sentences of ordinary language should also respect rules of "semantic appropriateness", so to speak. In particular, each predicate has a proper field of application, that is, a range of objects to which it *can* be (truly or falsely) applied. A sentence $P(m)$ can be truth-valueless and meaningless, despite being syntactically well-formed, if m is not in the range of P – if it is not, as it were, the right *kind* of thing which can be (truly or falsely) claimed to be P.[37] Consider the famous Chomskian example:

(16) Colourless green ideas sleep furiously.

Unlike:

(17) Green sleep and is but and,

(16) contains no grammatical mistake. However, we feel there is something wrong with (16), due to semantic reasons. Meaning postulates strongly rooted in our lexical competence tell us that, if something is an idea, i.e., an immaterial being, then it has no colour, and it cannot sleep; that if something is green, then it is not colourless; and so on. Fair enough. But why should the liars be category mistakes, since they do not look as something similar to (16) at all? Exactly in what sense are the liars meaningless? In the standard liar, "This sentence is false", the denotation of the subject, i.e., the standard liar, seems to be just the *right* kind of thing to which semantic predicates can apply: a sentence. As Kirkham has observed, "if there is something objectionable about the [liar] sentence, it is far from obvious what it is, for the sentence is grammatically correct, it is neither vague nor ambiguous, and it commits no category mistake: sentences are among the things that can bear truth values".[38] Why should we consider it as a category mistake? Martin tried to provide a decision procedure for category

---

[36] See Ryle (1950).
[37] See Martin (1967): 288ff.
[38] Kirkham (1992): 271-2.

correctness; but, first, it is directly applicable just to atomic sentences – therefore, adjustments are required for non-atomic sentences, such as the strengthened liars; second, paradoxical sentences would pass the test, if a special clause for them were not added to the procedure – which move is completely *ad hoc*.

Kripke's (1975) theory of truth does not seem to fare any better. The Kripkean approach is based upon the notion of *groundedness*, which may be illustrated very informally as follows. Take a sentence, ε, which claims that (some of) the sentences of a certain set A have a semantic property – e.g., they are true. Then, a truth value can be assigned to ε only if the values of the sentences in A are determined. Suppose now that one of the sentences in A, $ε_1$, claims that (some of) the sentences of a set B have that semantic property. Now a truth value can be assigned to $ε_1$ only if the values of the sentences in B are determined. Suppose now that one of them, $ε_2$, ..., and so on. If the chain ends, eventually, with sentences that do *not* speak of semantic properties, then ε is *grounded*, in the sense that we have, as it were, some good reason to assign it some truth value or other. Otherwise, ε is *ungrounded*: it does not reach base, and it remains floating in the rarefied air of sentences talking about the semantics of sentences.[39] This happens, particularly, with the liar. There is no non-semantic fact to rely on in order to evaluate "This sentence is false", whereas the underlying idea of the notion of groundedness is precisely that the truth value of a sentence should be grounded in something outside the semantic world.[40]

You may notice that, given this intuitive characterization, the property of being ungrounded is not rigidly determined, as it were, *a priori* by means of syntactic or semantic conditions, but normally depends on the empirical facts[41] (recall the looped liar (2d), i.e., "(2e) is true". What makes it paradoxical and, in the Kripkean framework, ungrounded, is the fact that (2e) in its turn attributes a semantic property to (2d). Were (2e) "Snow is white", then (2d) would be plain true). Besides, the property does not overlap with that of being self-referential. For instance:

---

[39] On fn. 8 of his essay, Kripke claims to have borrowed the terminology from Herzberger (1970). The ungrounded Kripkean sentences, though, do not coincide with the Herzbergerian ones.

[40] See Sainsbury (1995): 114-5.

[41] See Kripke (1975): 57.

(18) This sentence is composed of seven words

is self-referential, but grounded (and true). Finally, ungrounded sentences do not even coincide with *paradoxical* sentences. The famous Kripkean example is the "truth-teller":

(19) (19) is true.

The truth-teller appears to be a close relative of the standard liar: it is self-referential; it attributes to itself the semantic property *par excellence*. But (19) gives origin to no paradox, since if it is true, then it is true; and if it is false, then it is false. Our being seemingly unable to find a reason to favour one option over the other gives us an intuitive idea of why (19) sounds ungrounded, and truth-valueless. However, here also emerges all the *ad-hocness* of the decision to assign the liar the same status. Take our initial standard liar again:

(2) (2) is false.

In spite of the affinities, (2) and (19) are deviant in very different ways. In the case of (19), our *bona fide* intuitions appear to be insufficient to assign a truth value, which is what would make of it an intuitive candidate to the status of a truth value gap: its truth-status looks *under*determined. But those very intuitions, in the case of (2), appear to be *more* than sufficient, in the sense that they give rise to a *glut* of truth values:[42] (2) is intuitively *over*determined, true *and* false.[43] It seems that the one reason for not considering (2) as a glut, but as a gap, is precisely the *ad hoc* one of protecting the (LNC). Therefore, the Kripkean framework seems unable to fulfil the requisite no. 2 for good solutions of the paradoxes, listed at § 1.4.1. As Kirkham has acknowledged, despite being "an implication of a theoretical construct", which does not simply *state* that sentences have no truth value when their having one would entail a contradiction, "Kripke's solution is no more and no less ad hoc than is [...] Tarski's. He has no independent reasons, other than to solve the paradox,

---

[42] The gaps and gluts terminology is due to Fine (1974).
[43] See Priest (1987): 14-5.

for placing the restrictions he does on what can and cannot have a truth value".[44]

2.4.1 More Revenge Liars

The most difficult problems for all non-bivalent approaches to the paradoxes, though, come – as Kripke himself has acknowledged – from the revenge liars. Here the strengthened liars I introduced at § 2.1 begin to play:

(6) (6) is not true,

(7) (7) is false or neither true nor false.

The key idea of the gapper's strategy is to admit truth-valueless sentences (and declare that the liars should join the club). Can the gapper express *this* idea within the language for which she is providing her theory? It seems not. Tyler Burge has made the point.[45] To begin with, if a sentence α is not true and not false, *a fortiori*, it is not true, so "α is not true" should be true.[46] Therefore, we have:

(20) If α is not true, then "α is not true" is true.

Now let us take the strengthened liar (6) and reason by cases. (6) is either true, or false, or neither true nor false. If it is true, given what it says, it is not true. If it is not true, i.e., either false or not true and not false, then it is true. Can the gapper claim that the inference from the lack of truth value of (6) to its truth is illicit? He is committed to (20), and an instance of (20) is:

(21) If (6) is not true, then "(6) is not true" is true.

Now, since "(6) is not true" is nothing but (6), if (6) is not true, then (6) is true. The same kind of thing happens with (7). The gapper can exclude that the notion of truth-valueless or gappy sentence, that his, the

---

[44] Kirkham (1992): 291.
[45] See Burge (1979).
[46] See Sainsbury (1995): 117.

main notion she employs, is expressible within the language for which she is providing her theory. At worst, "the gappy solution is one of silence: Whereof one cannot speak, one should be silent".[47] At best, the language for which the theory is proposed, once again, is not the language in which the theory is expressed – and, as Kripke admits, "we still cannot avoid the need for a metalanguage".[48] However, we *can* express the whole situation in English (I have just done it, and the gappers do it in their theories). Therefore, the language for which the solution of the paradoxes is proposed is not English. More generally, all the approaches to paradoxes, which resort to semantic gaps, have to impose limits on what can be expressed within the language, say, L, for which the semantics is given. In particular, the lack of truth value of truth-valueless sentences of L cannot be truly expressed within L. This limitation is not dependent upon Kripke's particular choices: the same problem resurfaces within Martins' framework as well. The need for a metalanguage is not avoided in the gappers' approach either.

A pragmatic way out for the gapper has been proposed by Parsons (1984), and recently retrieved by Restall (2004). You may recall the controversial equivalence between rejection/denial and acceptance/assertion of negation I talked about at § 1.1.1:

(Acc) $\dashv_x \alpha \leftrightarrow \vdash_x \neg \alpha$.

(Acc) possesses the field among philosophers, being sometimes presented as something hardly worth arguing for:

> To deny a statement is to affirm another statement, known as the negation or the contradictory of the first.[49]

> After all, disbelief is just belief in the negation of a proposition.[50]

Nevertheless, the gappers have suggested that the equivalence holds, at most, as a default principle, and (Acc) can be overcome in peculiar circumstances, one of them being the presence of a truth value gap. As we

---

[47] Beall and van Fraassen (2003): 123.
[48] Kripke (1975): 81.
[49] Quine (1951): 1.
[50] Sorensen (2003): 153.

know, the strengthened liar (6) is a λ_1, which is equivalent to $\neg T(\ulcorner\lambda_1\urcorner)$. However, the gapper cannot just *assert* that it is not true, on pain of falling foul of the extended paradox. In order to avoid the "ineffable" situation, Terence Parsons has suggested that, in the presence of λ_1, a denial is *just* a denial, and the gapper can *deny/reject* it, $\dashv_x\lambda_1$, without thereby asserting/accepting anything – in particular, without asserting that λ_1 is not true, $\vdash_x \neg T(\ulcorner\lambda_1\urcorner)$.

It will not work. The point is that the gapper is strictly committed to the thesis that *there are truth value gaps*. Since the liar sentence is acknowledged as belonging to the club, the gapper is also committed to the claim that the liar is a truth value gap, i.e., neither true nor false, hence, *a fortiori*, not true. As Priest has pointed out, "the theory itself, then, entails that $\lambda_{[1]}$ is not true; so we *should* assert this".[51] Besides, gap theorists cannot just utter plain rejections of the liars. They aim at reasoning on them, so the claim that λ_1 is *not* true should be embedded within larger sentences (e.g., conditionals) in the gapper's own discourse. This shows that the "not" cannot be interpreted as a mere denial: force operators simply make no sense embedded in larger sentences.

This does not entail that (Acc) should be restored; as we will see in § 14, its rejection has a pivotal role also within the strategy of some of the foes of the (LNC).

2.5 The Essence of the Liar

Different strategies of solution of the semantic paradoxes, then, appear to face a single structural problem – which, as may be quite clear, is the one anticipated at § 1.4.2. The solution introduces such concepts as those of *order*, or *gap*, which can be exploited to produce new paradoxes. The coherence of the theory can be saved only by denying that those concepts are expressible within the language for which the theory has been provided. This amounts to admitting that the language in question is not ordinary English, in which, on the contrary, those concepts can be expressed, formulated, and defined. According to Priest, this shows not only that the problem of showing that the semantics of English is free from contradiction has not been solved so far, but also that it *cannot* be solved

---

[51] Priest (1987): 264.

by any approach that attempts to retain the (LNC). This is quite a dramatic move – it takes us from a *de facto* statement to *de iure* one: when medieval logicians called the paradoxes of self-reference *insolubilia*, they should have been taken literally. Therefore, Priest's argument for such a move is definitely worth listening.

According to Priest, the strengthened liars show that a single *essence* of the semantic paradox underlies its different formulations. The totality of sentences is divided into two subsets: the true ones, and their "*bona fide* complement" – call it the Rest. Now the essence of the liar is "a particular twisted construction which forces a sentence, if it is in the *bona fide* truths, to be in the Rest (too); conversely, if it is in the Rest, it is in the *bona fide* truths".[52] The standard liar, "This sentence is false", is but a particular instance of the schema, producing a contradiction within the bivalent framework, in which the Rest is identified with the set of the false sentences. Now, we can try to resolve the problem by admitting sentences that are neither true nor false, so that the false ones become a proper subset of the Rest. However, the strengthened liars show that we can use the notions introduced to solve the previous paradox to *re-describe* the Rest. In a framework in which the set of sentences undergoes a trichotomy (true, false and neither true nor false), "This sentence is false or neither true nor false" embraces with its disjunction precisely the whole Rest, i.e. the (new) complement of the set of the true sentences. Adding more values is, of course, useless. If there is a fourth thing that a sentence can be, besides true, false, and neither true nor false, we can always take the notion *fourth thing* and produce another strengthened liar:

(22) (22) is false, or neither true nor false, or the fourth thing,[53]

and so on. Therefore, Priest, concludes, "the extended paradoxes are not really novel paradoxes, but merely manifestations of one and the same problem, suitable to different contexts".[54] They force us, as has been observed, to admit that the proposed theory was formulated in a language different from, and more powerful than, the one for which it has been built. And we are back into the object language/metalanguage distinction, whose inadequacy has been underlined when we dealt with the hierarchic

---

[52] Priest (1987): 23.
[53] See Kirkham (1992): 293-4.
[54] Priest (1987): 24.

approaches. As Kripke has admitted at the end of *Outline of a Theory of Truth*, "the ghost of the Tarski hierarchy is still with us".[55]

---

[55] Kripke (1975): 80.

3 Unbounded Abstraction

> In order to draw a limit to thinking we should have to be able to think both sides of this limit (we should therefore have to be able to think what cannot be thought).
>
> Ludwig Wittgenstein, *Tractatus logico-philosophicus*

3.1 Existence and Objectivity, or the Abstraction Principle

Analytic philosophers of language have focused mainly on semantic paradoxes, but set-theoretic ones are no less interesting, also from a metaphysical point of view. So-called naïve set theory is based upon two principles capturing our intuitive conception of set. The *Extensionality Principle* spells the sufficient conditions for identity between sets:

(EP) $\forall x(x \in y \leftrightarrow x \in z) \rightarrow y = z$,

informally: "If y and z have precisely the same elements, they are the same set". A set, then, is wholly determined by its elements. We will be particularly interested in the *Abstraction* or *Comprehension Principle*, which we have previously glimpsed at, and which can now be formulated (without set abstracts) as:

(Abs) $\exists y \forall x(x \in y \leftrightarrow \alpha[x])$

(y being not free in α).[1] (Abs) captures the notion of set given in Cantor's celebrated definition:

> By an "aggregate", we are to understand any collection into a whole M of definite and separate objects m of our intuition or of our thought.[2]

---

[1] Frege's original formalization was a higher order one, whereas I will always take (Abs) as a schema. This difference is of minor importance, though.
[2] Cantor (1895): 85

Therefore, the basic insight is that *any* "collection into a whole" is a set, which means that each condition $\alpha[x]^3$ is taken as defining one. This is intuitive if anything is: when we think of something *as* a thing of a certain sort (taking "sort" in a broad sense, not in the strict sense of sortal concepts), at the same time we appear to think of it as being one of a group, which is, itself, a thing of a certain sort. More specifically, given any multiplicity with some characterizing condition, (Abs) seems to guarantee:

(1) that *there is* a set *y* of all and only those objects (I will therefore call this fact Existence); furthermore,

(2) *y* is itself an *object*. "Object" should be taken as meaning, more or less: something we can *refer to* as a *unity*, which is the subject of attributions and predications, and has *properties* (I will therefore call this fact Objectivity).

The one protest ever made against (Abs), as far as I know, is that it produces contradictions. In fact, set-theoretic paradoxes are typically produced by the consideration of particular conditions $\alpha[x]$ which, to put it in the way we were talking about at § 1.4.2, constitute some limit-cases of conceptual abstraction. Because of this, all the standard set-theoretic frameworks deny either Existence or Objectivity for such limit cases. You may notice that these two strategies correspond roughly to the two Kantian solutions of (some of) his antinomies: (1) deny the existence of the limit-object, i.e., the relevant Unconditioned; (2) admit that it exists, but consider it as a noumenal thing (which amounts to a denial that it can be altogether treated as an *object*, i.e., that some properties can be meaningfully ascribed to it). This Chapter will be concerned with the difficulties of both ways out. Let us begin by producing, by means of (Abs), some celebrated contradictions.[4]

3.1.1 Notable Paradoxes

The most celebrated set-theoretic paradox is unquestionably the one produced by the famous Russellian set:

---

[3] It is common to use "condition" in this sense within set theory: an open formula with a free variable *x* is called a *condition on x*.

[4] In the following, I will not enter into the details of the proofs. They can be found in any good textbook, such as, e.g., Fraenkel, Bar-Hillel and Levy (1973).

$R = \{x \mid x \notin x\}$.

R is the set of all and only those things that are not members of themselves. Since α[x] in (Abs) stands for any condition, by taking $x \notin x$ we have:

$\exists y \forall x (x \in y \leftrightarrow x \notin x)$.

So (1) by Existence, *there is* a set – or better, given (EP), *the* set, corresponding to such a condition, i.e., y is precisely R:

$\forall x (x \in R \leftrightarrow x \notin x)$;

but, furthermore, (2) by Objectivity, R is itself an *object*, that is, something about which we can wonder, given any property or condition, whether it has that property, or satisfies that condition, or not. This is the case also for the property of not being a member of itself. Since what holds for any x holds for R, we have:

$R \in R \leftrightarrow R \notin R$;

hence, assuming the Law of Excluded Middle, we get an explicit contradiction:

$R \in R \land R \notin R$.

This is "*the* contradiction". The argument uses the (LEM). Unlike philosophers of language, set theorists have not thoroughly investigated the possibility of avoiding the paradoxes by means of a dismissal of bivalence and/or of the (LEM) (a fact that may be explained by the Platonic attitude dominating set theory). Some attempts, though, have been made. Bochvar used a three-valued logic (the values being True, False and Indeterminate) to build formal theories that avoid such paradoxes as Russell's or Grelling's by making the paradoxical sentences take the value Indeterminate.[5] As Church noticed quite soon, though, such a resistance to paradoxes was due to the use of a *negatio deminuta*: first, one introduces

---

[5] See Bochvar (1939). See also Haack (1978): 207-8, Beall and van Fraassen (2003): 126ff, for short presentations of the logic.

a weak or internal negation (say, ~) with a non-bivalent semantics; and, by means of it and an assertion operator (say, A) which forces bivalence, one defines an external negation, which is the (weak) negation of assertion ($\neg\alpha =_{df} {\sim}A\alpha$).[6] Now, if R is characterized by means of the weak negation, as:

$$R = \{x|\sim(x \in x)\},$$

then the paradoxical biconditional:

$$R \in R \leftrightarrow\, \sim(R \in R)$$

does not produce an explicit contradiction, precisely because $R \in R$ is allowed to take the value Indeterminate. Nevertheless, if we have:

$$R_1 = \{x|\neg(x \in x)\},$$

the contradiction is quickly regained.[7] Now, a reformulation of the point with an external negation amounts to a production of a set-theoretic "revenge paradox". Therefore, the situation is not too different from the one of non-bivalent approaches to semantic paradoxes we met in the previous Chapter. Finally, various set-theoretic paradoxes can be produced,[8] which have direct proofs and do not use the (LEM).

Russell's paradox is a simplified variant of a paradox known to Cantor since 1899, even though it was published only in 1932. This begins with the consideration of the universal set, which is indicated (since Peano) as V. V is usually characterized by means of a condition anything is supposed to satisfy, such as self-identity:

$$V = \{x \mid x = x\}.[9]$$

---

[6] Since the only designated value of the logic is True, the system with supplied only the "weak" connectives had no tautologies. The addition of the assertion operator was required in order to repair this feature.

[7] See Church (1939): 98-9.

[8] Such as Burali-Forti's (which we shall meet soon) and Mirmanoff's.

[9] See e.g. Fraenkel, Bar-Hillel and Levy (1973): 124.

But V can be taken as the set of all *sets* if we take a *pure* theory of sets, that is to say, if we assume that the domain does not include *Urelemente*, objects which are not sets. In this case, one usually assumes as the only "basic" object the empty set, ∅, and, roughly, the desired set are obtained via the iteration of certain operations generating sets from set. This is a bit of a peculiarity (at least from the philosophical, if not from the mathematical point of view), due mainly to Fraenkel. Following the spirit of the Cantorian characterization of *Menge*, Zermelo thought that entities of any kind may be assumed as the basic objects of the theory. On the contrary, ∅ is, as it were, both an "improper object" (in the sense that it is a set) and an "improper set" (in the sense that it has no elements). The idea of a pure theory of sets, in which the processes captured by the axioms of the theory guarantee the existence of the desired sets having as their only base the empty set, comes down to this: one can describe the whole realm of the abstract without any basic stuff, except the idea that there is no basic stuff (such an "idea" being captured by ∅).

Now, a little bit of terminology. Two sets x and y are said to have the same cardinality ($x \cong y$), or to be equinumerous, or equipollent, iff there is a one-to-one correspondence between them: a function mapping each member of x to a member of y, and such that each member of y is mapped to by a single member of x. Then, x is bigger than or equal to y ($x \geq y$) iff there is a subset of x which has the same cardinality of y; and x is bigger than y ($x > y$) if $x \geq y$ but it is not the case that $x \cong y$.

The core of the Cantorian theory of the transfinite is *Cantor's Theorem*: there are no functions from x onto its power set, P(x) (the set of all subsets of x), that is to say, P(x) has a higher cardinality than x: P(x) > x. It is intuitive that P(x) $\geq x$, so one has to show that it is not the case that P(x) $\cong x$. Cantor's proof begins by assuming that there is a one-to-one function ϕ from x to P(x), so that they have the same cardinality, for the sake of a *reductio*. Let us take into account the set z of all the elements of x that are not members of the set assigned to them by ϕ – so $z = \{y \in x \mid y \notin \phi(y)\}$. z is a member of P(x), since it is a subset of x. So there must (by supposition) be an element w of x, such that $z = \phi(w)$. Now the question is: is w a member of z, i.e., $\phi(w)$, or not? We have:

$$w \in \phi(w) \leftrightarrow w \in \{y \in x \mid y \notin \phi(y)\} \leftrightarrow w \notin \phi(w).$$

Given the (LEM), either w is a member of $\phi(w)$ or not, hence:

$w \in \phi(w) \land w \notin \phi(w)$.

The contradiction so deduced leads us to conclude that there cannot be such a one-to-one ϕ. This is called the *diagonal argument*.[10] Cantor initially used it to show that the set of natural numbers is not the largest infinite set, for it is surpassed by the set of the real numbers. This is just the beginning. Given any infinite set, Cantor's theorem assures that one can always jump into a larger infinite by iterating the power set operation. Now, consider the universal set V, i.e., the set of all (pure) sets, and take its power set, P(V). Since all members of P(V) are sets, P(V) is a subset of V. But V is itself a subset of P(V). Therefore, P(V) = V. So there is a one-to-one correspondence between V and P(V), namely, identity, and P(V) $\cong$ V. But Cantor's theorem rules this out for any set, so we have:

P(V) $\cong$ V $\land \neg$(P(V) $\cong$ V).

Even more rapidly: given Cantor's theorem, P(V) is bigger than V. This is absurd, since V is, by definition, the most inclusive of all sets.[11] Notice that Cantor's paradox is nothing but Russell's, once one chooses as ϕ the identity function (therefore, sometimes scholars speak of a unique Cantor-Russell paradox). In this case, $\phi(w)$ just *is* w, i.e., $\{y \in x \mid y \notin y\}$, and the thing goes like:

$w \in w \leftrightarrow w \in \{y \in x | y \notin y\} \leftrightarrow w \notin w$.

Despite being initially controversial, the diagonal argument is now ordinary mathematics. Cantor's theorem and paradox have had fundamental upshots, not only in set theory and mathematics, but also within philosophy (for instance, they are at the heart of Patrick Grim's arguments mentioned in§ 1.4.2 for the non-existence of a totality of truths, etc.).

Another paradox, which I shall introduce very quickly, is Burali-Forti's. This is important both historically, since it was the first one to be

---

[10] "The essence of Cantor's proof is as follows. Given a list of objects of a certain kind (in this case, subsets of x), we have a construction which defines a new object of this kind (in this case z), by *systematically destroying the possibility of its identity with each object* on the list. The new object may be said to 'diagonalise out' of the list" (Priest (1995): 119, my italics).

[11] See Fraenkel, Bar-Hillel and Levy (1973): 7.

discovered; and theoretically, since its proof is a direct one (no (LEM) required). The paradox concerns ordinal numbers. Cantor's initial idea was that ordinals should index *well-ordered* sets. A well-ordered set is a set such that each of its non-empty subsets has a least element (following von Neumann's later idea, an ordinal can correspond to the set of the preceding ordinals: so if 0 is $\emptyset$, 1 is $\{0\}$; 2 is $\{0, 1\}$; 3 is $\{0, 1, 2\}$, and so on). Now consider the set Ω of *all* ordinals. One can give independent arguments for both $\Omega \in \Omega$ and $\Omega \notin \Omega$. By construction, Ω is itself well-ordered, so since any well-ordered set has an ordinal number, Ω must have an ordinal, too. However, this ordinal must be greater than any member of the set, therefore it cannot be in the set.[12]

3.2 Logical Types

3.2.1 The Vicious Circle Principle

I mentioned the fact that the distinction between semantic and set-theoretic paradoxes is posterior to the *Principia mathematica*. Russell believed that all the logical paradoxes had their root in some form of circularity, or self-referentiality, he named "reflexiveness":

> In all the above contradictions [...] there is a common characteristic, which we may describe as self-reference or reflexiveness. The remark of Epimenides must include itself in its own scope. If *all* classes, provided they are not members of themselves, are members of [R], this must also apply to [R]; and similarly for the analogous relational contradiction. [...] In the case of Burali-Forti's paradox, the series whose ordinal number causes the difficulty is the series of all ordinal numbers. In each contradiction something is said about *all* cases of some kind, and from what is said a new case seems to be generated, which both is and is not of the same kind as the cases of which *all* were concerned in what was said.[13]

Here Russell grasps the common structure underlying an entire group of paradoxes. In the third part of the book, we will see how such a general structure can be represented within a paraconsistent set-theoretic framework. For the moment, let us look at the solution he proposed. Russell rooted the paradoxes in a particular kind of definitions, named *impredica-*

---

[12] See *Ibid*: 8.
[13] Russell and Whitehead (1910-25): 61-2.

tive. Everyone knows that a good definition has to avoid circularity, i.e., the *definiendum* must not be used as a part of its own *definiens*. An impredicative definition, though, hosts a subtler circularity: it is the definition of an object, which includes a reference to a totality to which the object to be defined itself belongs. Russell regarded such circularity as *vicious*, and formulated the Vicious Circle Principle, whose most useful formulation for us is the negative one:

(VCP) If, provided a certain collection had a total, it would have members only definable in terms of that total, then the said collection has no total.[14]

Russell's strategy consists in a denial of the first fact we met at the beginning of this Chapter, namely, Existence: the sets corresponding to the "illegitimate totalities" cannot *exist*. The tool employed to obtain such a limitation was the *theory of logical types*.[15] In order to get a general idea of how it works, we have to consider that it is not phrased in terms of sets, but of *propositional functions*. At a certain point of his philosophical development, Russell came to believe that assertions concerning sets had to be reduced to assertions concerning propositional functions. Even though the paradoxes concern sentences, ordinals, sets, etc., the "most fundamental case"[16] of the (VCP) had to be expressed in terms of propositional functions. Actually, Russell had a gift for the use/mention confusion, which affects the Russellian notion of propositional function so thoroughly, that according to some it is not even clear whether what the theory deals with are linguistic expressions, or the things signified by them.[17] Anyway, the theory builds a complex hierarchy of functions, structured by a double distinction of *orders* and *types*.[18] We do not need,

---

[14] *Ibid*: 37.

[15] Which was drafted in the *Principles of Mathematics* (Russell (1903)), and, after having been developed in Russell (1908), was incorporated in the *Principia*.

[16] Russell (1910): 6.

[17] On the problems and obscurities following from the confusion between propositional functions as linguistic entities, and as attributes or properties, see Quine's introduction to Russell (1908): 151-2.

[18] After the publication of the *Principia*, Ramsey and L. Chwistek proposed a simplification of the Russellian theory, including only plain levels and therefore called *simplified* theory of types. The version including both orders and types came to be called the *ramified* theory. We need not go into this, though.

though, to enter into the details. To keep things straightforward, let us simplify the original theory, consider only one-place functions, and reduce their ordering to a simple hierarchy.[19] The universe is divided into levels. Individuals which are not functions belong to order 0; functions having individuals as arguments belong to the first order; functions having as arguments first-order functions belong to the second order; and so on. Functions are therefore divided into separate levels, and the variables they contain range only over the appropriate orders – which can be expressed by indexing: the variables for order n are x^n, y^n, etc. Which is the appropriate order? The order of a function must always be one more than the order of its argument. And the (VCP) is incorporated in the hierarchy on the basis of the idea that "a function is not a well-defined function unless all its values are already well-defined. It follows from this that no function can have among its values anything which presupposes the function".[20]

The whole construction aims at ruling out the "illegitimate totalities". Translating from propositional functions to the more familiar sets, the point is that the theory allows any set to contain only things of *one* order – it allows only set composed, so to speak, of objects that are *homogeneous* with respect to the hierarchy. Therefore, there is no set of all sets, or of all ordinals, or of all propositional functions, etc. The Abstraction Principle can now be formulated as:

(Abs$_T$) $\exists y^{n+1} \forall x^n (x^n \in y^{n+1} \leftrightarrow \alpha[x^n])$,[21]

and, for instance, Russell's paradox disappears because a set can neither be, nor not be, a member of *itself*. The relation *member of*, $\in$, can occur only between something that belongs to order n and something that belongs to order $n + 1$. The very phrase "the set of all things which are not members of themselves", which described R, becomes meaningless. We can meaningfully talk of all things that fulfil a given condition only if they all belong to the same level. For the same reason, Cantor's paradox disappears because there can be no set V of all sets: we cannot even *say* within the theory that a set contains *all* sets.[22]

---

[19] I have taken the following exposition from Priest (1995): 136-7.
[20] Russell (1910): 5-6.
[21] See Fraenkel, Bar-Hillel and Levy (1973): 158.
[22] See *Ibid*: 158-9.

The principle according to which no set can be a member of itself dates back to Aristotle's *Topics* (144a28ss). In the context of the axiomatic set theories we shall deal with soon, the idea is built in by means of a principle called *Axiom of Foundation*. The axiom forbids not only direct self-membership, but also "circular" chains of membership, such as $x \in y \wedge y \in x$. The sets that conform to the axiom of foundation are called *well-founded*.

3.2.2 Parameterisation, Part Two

At this point, someone may have the feeling of a *déjà-vu*. The strategy underlying the theory of types, in fact, is fundamentally a pervasive parameterisation. The theory parameterises not only the truth predicate, but also all the expressions that may give rise to the "illegitimate totalities":[23]

> That the words "true" and "false" have many different meanings, according to the kind of proposition to which they are applied, is not difficult to see. [...] [There is] a systematic ambiguity in the meanings of "not" and "or", by which they adapt themselves to propositions of any order.[24]

> In all [the paradoxes], the appearance of contradiction is produced by the presence of some word which has systematic ambiguity of type, such as *truth, falsehood, function, property, class, relation, cardinal, ordinal, name, definition*. Any such word, if its typical ambiguity is overlooked, will apparently generate a totality containing members defined in terms of itself, and will thus give rise to vicious-circle fallacies.[25]

To be sure, these ambiguous words encompass most of the concepts logic and mathematics have to deal with. We can therefore expect the Russellian hierarchy to face problems structurally similar to the ones of the Tarskian hierarchy of metalanguages. This, indeed, is the case. First, the *ad-hocness* of parameterisation is, again, clear:

---

[23] Of course, the temporal order is the converse of the one of my exposition: Tarski's work came after Russell's.
[24] Russell and Whitehead (1910-25): 42-3.
[25] *Ibid*: 64.

Russell's solution appears to consist in an ad hoc declaration that anything that violates the VCP is meaningless. He pushes back the ad hoc element to a deeper level by postulating a theory of types [...] which *entails* that such propositions are meaningless. But it must be remembered that the theory itself is ad hoc. Russell has no reason to create this theory save that by doing so he can resolve the paradoxes. He says as much himself: "The division of objects into types is necessitated by the reflexive fallacies which otherwise arise" (1908b, 61).[26]

Of course, the charge of *ad-hocness* can be mitigated by observing that the theory aims at solving *all* the logical (semantic and set-theoretic) paradoxes at once – and providing a uniform solution is no doubt an epistemic virtue (we will come back to this point in § 11). However, Quine already emphasized the many "unnatural and inconvenient consequences" of the theory of types several years ago.[27] The theory requires an infinite multiplication of all the basic set-theoretic notions. Now each level has its own (pseudo-)universal set and its own empty set. The whole Boolean algebra is multiplied and, because of this, the natural complement of each set is lost and replaced by a quasi-complement having little to do with our intuitive idea of complementation. The "complement" of x now does not include all the non-members of x, but only those that belong to the same level as the members of x.[28]

Furthermore, by ruling out all impredicative definitions, the theory also rules out some mainstream mathematical theorems within which various impredicative totalities occur not problematically – for instance, the theorem that each set of real numbers has a least upper bound.[29] In order not to lose such results, Russell added to the theory a famous principle, the Axiom of Reducibility, according to which, given any function ϕ with arguments of order n, there is an extensionally equivalent function (named *predicative*), whose order is $n + 1$:

---

[26] Kirkham (1992): 276.

[27] See Quine (1937).

[28] And "many mathematicians will find this reduplication repugnant, not only for intuitive reasons but also, perhaps mainly, because it severely restricts their accustomed freedom of expression and notation and often requires complicated technical operations in order to overcome the unwanted notational restrictions" (Fraenkel, Bar-Hillel and Levy (1973): 160).

[29] See Gödel (1944): 25.

We assume, then, that every function is equivalent, for all its values, to some predicative function of the same argument. This assumption seems to be the essence of the usual assumption of classes; at any rate, it retains as much of classes as we have any use for, and little enough to avoid the contradictions which a less grudging admission of classes is apt to entail. We will call this assumption the *axiom of classes*, or the *axiom of reducibility*.[30]

The axiom was supposed to recover some obligatory impredicative notions. Russell himself was never fully satisfied with it; first, because it was introduced with a clearly *ad hoc* move; second, because even its status of a purely *logical* principle was dubious – whereas the initial aim of Russell and Frege's enterprise was to fully reduce mathematics to logic by means of set theory.[31]

Philosophically, the edifice of orders and types also betrays the Platonic vision underlying the Cantorian conception of set. Sets were supposed not to depend on our mental constructions and, in this sense, were taken as objectively existent. A set-theoretic structure should not come into being by means of our definition procedures. What makes an impredicative definition unwelcome, on the other hand, is precisely a constructivist attitude. From a constructivist point of view an impredicative definition, by including reference to something that presupposes the *definiendum* – therefore, by presupposing the existence of what should be built by the very definition, is *viciously* circular. However, if the object in question pre-exists, then the duty of the definition is a descriptive one. This is what the Platonist Gödel observed in *Russell's Mathematical Logic*, so rehabilitating impredicativity:

> Even if "all" means an infinite conjunction, it seems that the vicious circle principle in its first form applies only if the entities involved are constructed

---

[30] Russell (1908): 168. Russell uses the typical term "class" here; in the following, I shall give it a more specific meaning, on the basis of the subsequent distinction between sets and classes.

[31] "That the axiom of reducibility is self-evident is a proposition which can hardly be maintained. But in fact self-evidence is never more than a part of the reason for accepting an axiom, and is never indispensable. The reason for accepting an axiom, as for accepting any other proposition, is always largely inductive..." (Russell and Whitehead (1910-25): 59). In the *Tractatus*, Wittgenstein claimed that "propositions like Russell's 'axiom of reducibility' are not logical propositions [...]. We can imagine a world in which the axiom of reducibility is not valid. But it is clear that logic has nothing to do with the question whether our world is really of this kind or not" (6.132-3). See also Fraenkel, Bar-Hillel and Levy (1973): 173-4.

by ourselves. In this case there must clearly exist a definition (namely the description of the construction) which does not refer to a totality to which the object defined belongs, because the construction of a thing can certainly not be based on a totality of things to which the thing to be constructed itself belongs. If, however, it is a question of objects that exist independently of our constructions, there is nothing in the least absurd in the existence of totalities containing members, which can be described (i.e., uniquely characterized) only by reference to this totality.[32]

Now the Axiom of Reducibility looks like a late attempt to reconcile with a betrayed lover: mathematical Platonism. But things get worse. The axiom is supposed to hold for all functions. Therefore, when Russell expresses it by talking about "every function", or about "any given function ϕx", he is violating the hierarchy of orders, which would allow meaningful statements to quantify only on things of a single order. Again, you may experience a *déjà-vu*: just like the Tarskian hierarchy, the theory prescribes us, as it were, to always talk from a single level in the hierarchy. However, in order to expose it, we have to talk about orders, or levels, in general. When we try to expose the theory, we are forced to claim, among other things, that each propositional function belongs to a single order. But since no variable can range over all propositional functions, and each is confined in its own order, "each propositional function" is ruled out as meaningless by the theory. Attempts to formulate the theory of types violate the theory (a self-refutation pointed out by Fitch a long time ago).[33] In addition, the (VCP) itself cannot be expressed without contravening the (VCP). In Priest's words:

> Even decent statements of the VCP cannot be made without violating the VCP since they must say that *for any function* f, any propositional function which 'involves' f cannot be an argument for f. Such statements are impossible by Russell's own admission. [...] By his own theory, Russell's theory cannot be expressed [...]; but he does express it [...]. Russell cannot explain his account without talking about all propositions.[34]

One may try to solve the problem (as Russell did) by exploiting the very idea of systematic ambiguity in the meanings of the key notions, and

---

[32] Gödel (1944): 25.
[33] See Fitch (1946). See also Gödel (1944): 31.
[34] Priest (1995): 138-40.

treat the relevant sentences of the theory, e.g., the Axiom of Reducibility, as schematic, producing an infinity of substitution instances (one for each order). The point made by Priest, and already observed in the previous Chapter with regard to the Tarski-type hierarchies, is that "however we express it, what we are supposed to understand by a systematically ambiguous formula [...] is exactly what would be obtained by prefixing the formula with a universal quantifier".[35] This illustrates once again the general point I made in the preceding Chapter with respect to the Aristotelian "distinctions of respects": once different respects, or parameters, have been introduced, we can (implicitly, or explicitly) quantify on them. This allows rebuilding a paradox, unless such quantification is doomed as meaningless. But this renders the parameterising theory meaningless, or self-refuting, since it cannot but employ that quantification itself.

3.3 Aristotle, ZF, and the Limitation of Size

3.3.1 The ZF strategy

The most established axiomatic set theory nowadays is probably the one proposed by Zermelo, developed and modified by Fraenkel, known as **ZF**, and expressed in a first-order language with variables ranging on pure sets (let us assume here the version of the theory without *Urelemente*). The theory is based upon the intuitive principle that has come to be called of the *limitation of size*, and which, roughly, prohibits some very comprehensive sets. Actually, set theorists usually talk about limitations of size with respect to the von Neumann-Bernays theories we shall deal with soon. However, the idea, which, as we will see, is captured within these approaches by an *internal* axiom, dominates "from the outside" also **ZF**.[36]

ZF denies Existence. This was also Russell's strategy, as we have seen. In the theory of types, though, the restriction concerns the *members* of the sets: sets have to be composed only of things that are homogeneous with respect to the hierarchy of levels. Zermelo, on the other hand, believes that restrictions should be specified directly with respect to the nature of the sets: existence is denied to such sets as V, the universal set, or

---

[35] *Ibid*: 139.
[36] See Fraenkel, Bar-Hillel and Levy (1973): 32 and 135. A beautiful book on the subject is Hallett (1984).

Ω, the set of all ordinals, because of their being "too big", and only sets authorized by some pre-determined processes expressed by the axioms of the theory are admitted.

It is usually claimed that the first explicit formulation of the limitation of size should be ascribed to Russell (1905): "there is not such a thing as the class of all entities".[37] But the idea, to begin with, is almost as ancient as philosophy, and dates back to Aristotle. He observed, *contra* his predecessors (Parmenides, and Plato) that some very comprehensive notions cannot produce a genus, or a set, on pain of contradiction. In a famous passage of the *Metaphysics*, he shows that being is not a genus – and, therefore, that "being" is a pollacîj legÒmenon, it is spoken of in many ways:

> It is not possible that that either unity or being should be a genus of things; for the differentiae of any genus must each of them both have being and be one, but it is not possible for the genus to be predicated of the differentiae taken apart from the species (any more than for the species of the genus to be predicated of the proper differentiae of the genus); so that if unity or being is a genus, no differentia will either be one or have being.[38]

To deny that being is a genus is to deny that something like V exists, if (leaving aside the restriction to pure sets for a moment) we take V to be not only the set of all sets, but also the set of all things.[39] According to the Scholastics, then, the argument had to work for all those totalities that were labelled as *transcendental*, since they transcended, i.e., surpassed, the Aristotelian distinction between the categories of being. Such were (the extensions of) concepts like *ens*, *unum*, *verum*, etc. I shall henceforth take the liberty of lyrically calling "transcendentals" our illegitimate totalities.

3.3.2 Aussonderung

According to Zermelo, (Abs) had to be replaced by a group of principles capable of ruling out sets corresponding to the totality of sets, or of ordinals, etc.[40] Beginning with the empty set (the only "basic" element in the

---

[37] See Russell (1908): 150.
[38] *Met.* 998b22-7.
[39] See also Gödel (1944): 24.
[40] See Zermelo (1908).

purified version of the theory), such principles should guarantee that we have sets of a sufficient high cardinality to reconstruct (the safe part of) Cantor's vision, without ever reaching the dangerous *transcendentals*. The first version of Zermelo's system, though, turned out to be too weak for the purpose: it could not guarantee certain sets essential to the theory of the transfinite. Hence came the important developments provided by Fraenkel, such as the addition of the Axiom of Replacement, to guarantee the desired increase in power. We shall not deal with these, though (nor shall we talk about the specificities of **ZFC**, the extension of **ZF** obtained by adding to it the famous Axiom of Choice).[41] The deepest philosophical question posed by **ZF** concerns the main heir of (Abs) within this system, namely, the Axiom of Separation (*Axiom der Aussonderung*, as Zermelo called it):

(Sep) $\forall x \exists y \forall z (z \in y \leftrightarrow z \in x \wedge \alpha[z])$

(as usual, y is not free in α). As Fraenkel, Bar-Hillel and Levy point out, the *Aussonderung* principle "has the awkward property of being impredicative", but "naturally, a Platonistic attitude would judge this situation quite differently than a constructive attitude".[42] Paradoxes are not tackled here by ruling out impredicativity *tout-court*, but via a different stratagem. (Sep) assures that there is a set y corresponding to a condition $\alpha[z]$ only insofar as the zs are taken from a *presupposed* set x, or, as Zermelo tersely puts it, "sets [...] must always be *separated* as subsets from sets already given".[43] This fact rules out the illegitimate totalities. In particular, it excludes the possibility that the totality of the objects of set theory is, itself, a set: the *domain* of the totality of objects is not an object of the theory. Such a totality would be V, the universal set. Therefore, there is no set of all sets.

Of course, the first issue is, again, which independent reasons are there for assuming that such sets as V, Ω, etc., do not exist. Denying a premise of the paradoxes is easy but, according to the conditions considered at § 1.4.1, the move avoids *ad-hocness* only if an independent explanation of why precisely that premise has to be abandoned is provided. Of course, if the set of all sets existed, then all the paradoxical sets would

---

[41] See Fraenkel, Bar-Hillel and Levy (1973): 53ff, for an extensive discussion.
[42] Fraenkel, Bar-Hillel and Levy (1973): 38.
[43] Zermelo (1908): 202.

follow by means of (Sep). Why there is no V, then? Let us listen to Zermelo:

> THEOREM. Every set M possesses at least one subset M_0 that is not an element of M.
> *Proof.* [...] If now M_0 is the subset of M that, in accordance with Axiom [of Separation], contains all those elements of M for which it is not the case that $x \in x$, then M_0 cannot be an element of M. For either $M_0 \in M_0$ or not. In the first case, M_0 would contain an element $x = M_0$ for which $x \in x$, and this would contradict the definition of M_0. Thus M_0 is surely not an element of M_0, and in consequence M_0, if it were an element of M, would also have to be an element of M_0, which was just excluded.
> It follows from the theorem that not all objects x of the domain D can be elements of one and the same set; that is, *the domain D is not itself a set,* and this disposes of the Russell antinomy so far as we are concerned.[44]

3.3.3 Cantor's Domain Principle

Now, the fundamental philosophical trouble for **ZF** is that it presupposes an unrestricted notion of set, which is incoherent given **ZF** itself. Various authors have delineated such a problem.[45] This comes from a very straightforward and intuitive principle due to a Cantorian insight. It has been called the Domain Principle, and it can be phrased in the jargon of contemporary model-theoretic semantics (which, by the way, is also a set-theoretic jargon) by simply saying:

(DP) Each variable presupposes the existence of its domain of variation.[46]

By "variable" we just mean the one of mainstream logical languages: something which can be assigned different values. Therefore, the point of (DP) is, as Priest claims, that "for a sentence containing a variable to have a determinate meaning, the range of the quantifiers governing the variable (which may be implicit if the variable is free) must be a determinate totality".[47] In particular, a sentence about a *variable quantity* has a determi-

---

[44] Zermelo (1908): 203.
[45] See Parsons (1974); Lear (1977); Mayberry (1977); Moore (1990).
[46] This is close to the formulation provided by Priest (1995): 158.
[47] Priest (1995): 125.

nate meaning only if the domain of its variability is a determinate totality. This takes us closer to the original, and more suggestive, formulation given by Cantor in terms of potential and actual infinity:

(DP) For every potential infinity, there is a corresponding actual infinity.[48]

Let us listen to Cantor's argument for (DP):

> There is no doubt that we cannot do without *variable* quantities in the sense of the potential infinite; and from this can be demonstrated the necessity of the actual-infinite. In order for there to be a variable quantity in some mathematical study, the 'domain' of its variability must strictly speaking be known beforehand through definition. However, this domain cannot itself be something variable, since otherwise each fixed support for the study would collapse. Thus, this 'domain' is a definite, actually infinite series of values.
> Thus, each potential infinite, if it is rigorously applicable mathematically, presupposes an actual infinite.[49]

We are learning to appreciate the analogies between logical paradoxes and the Kantian antinomies. Now, it seems that we are dealing exactly the Kantian idea that, given an indefinitely extensible series of conditions, the thought of the Unconditioned as the totality of the series is *inevitable*. In Kant's own words:

> If the conditioned is given, the whole series of conditions, subordinated to one another – a series which is therefore itself unconditioned – is likewise given, that is, is contained in the object and its connection. [...] The transcendental concept of reason is, therefore, none other than the concept of the *totality* of the *conditions* for any given conditioned. [...] The pure concepts of reason – of totality in the synthesis of conditions – are thus at least necessary as setting us the task of extending the unity of understanding, where possible, up to the unconditioned, and are grounded in the nature of human reason.[50]

Undoubtedly, from a Kantian point of view, treating such a thought as a regulative ideal, with no objective existence, should be sufficient to

---

[48] See Hallett (1984): 7, principle (*a*).
[49] *Ibid*: 25.
[50] Kant (1781): 306 and 315-16.

avoid the paradoxes. The principle according to which "the series of conditions (whether in the synthesis of appearances, or even in the thinking of things in general) extends to the unconditioned" should be taken as just "a logical precept, to advance towards completeness by an ascent to ever higher conditions".[51] Moreover, self-declared heirs of Kant within the philosophy of mathematics, such as intuitionists[52] or, more generally, those who adopt a constructivist stance, may think they can reject (DP) tout-court. A constructivist mathematician typically discards higher cardinalities, accepting only denumerable infinities, and, in a strictly finitary case, only the indefinite iteration of construction procedures. However, to begin with, it is doubtful that a constructivist approach *can* escape (DP). Hallett and Priest think not.[53] According to Priest, one may conceive the domain in question as an intuitionistic species, instead of a classical set; but in any case, "it is conceptualized as a unity, for just the same reason as in the classical case".[54] More specifically, "intuitionist quantification presupposes a set (or species) of all sets. (And Zermelo's argument that there is no universal set is intuitionistically valid, too.)".[55]

Apart from this, no way out of this kind could be available to classic *set theory*, which is Platonist and realist in its spirit (that spirit Russell's ramified theory betrayed, save from a late reconciliation by means of the Axiom of Reducibility). *A fortiori*, it is not available to **ZF**. Now, (DP) requires that a sentence containing a variable x has a determinate meaning only if the domain of x is a determined totality. However, in **ZF** variables range over *all* (pure) *sets*. Therefore, **ZF** presupposed *this* domain of variation: the extension of the term "set", over which extension the set quantifiers range. And this is nothing but V. Therefore, the intelligibility of **ZF** presupposes a set whose existence has to be denied by the theory, on pain of incoherence of the theory itself. The very intelligibility of the concept *set* – something whose extension should be the totality of sets – according to Manuel Bremer, is threatened:

---

[51] *Ibid*: 307.

[52] Brouwer famously called his own position *neo*-intuitionism, explicitly linking it to the notion of *intuition* of Kant's transcendental idealism, and to the Kantian conception of arithmetics as based upon the pure intuition of time, "the immediate condition of inner appearances" (Kant (1781): 77).

[53] Priest (1995): 159-60; Hallett (1984): 26ss.

[54] Priest (1995): 126.

[55] *Ibid*: 159.

Now, what is a set? It cannot be the extension of "( ) is a set", since this extension would be a universal set [...] but there is none. So in standard set theory there is no set/extension corresponding to our usage of "( ) is a set". [...] Standard set theory is using a fundamental notion that cannot be explained *by this theory*! Or uses a fundamental notion that is *incoherent* given that very theory![56]

3.4 Von Neumann's Proper Classes

Since the rejection of Existence does not appear to lead to spectacular successes, we may try, instead, to deny Objectivity. This way has been taken by set theories based upon the distinction between sets and *classes*. The mainstream approaches here are due to von Neumann, Bernays, and Gödel, but the initial intuition is usually ascribed to a famous letter written by Cantor to Dedekind. Cantor proposed to divide multiplicities into two groups, the *consistent* and *inconsistent* ones:

> For a multiplicity can be such that the assumption that *all* of its elements "are together" leads to a contradiction, so that it is impossible to conceive of the multiplicity as a unity, as "one finished thing". Such multiplicities I call *absolutely infinite* or *inconsistent multiplicities*.
> As we can readily see, the "totality of everything thinkable", for example, is such a multiplicity; later still other examples will turn up.
> If on the other hand the totality of the elements of a multiplicity can be thought of without contradiction as "being together", so that they can be gathered together into "*one* thing", I call it a *consistent multiplicity* or a "set". [...]
> Two equivalent multiplicities either are both "sets" or are both inconsistent.[57]

Cantor had the clear intuition that the status of "finished things", i.e., real, full-fledged objects, had to be denied *precisely only* to the dangerous *transcendentals*, the illegitimate totalities (such as the totality of the thinkable, or the totality of all sets, etc). The Cantorian insight has been independently developed and made precise by John von Neumann.

---

[56] Bremer (2005): 141.
[57] Cantor (1899): 114.

3.4.1 Axiom (IV.2)

The basic idea is that the real danger does not come from assuming the existence of an extension for any condition. It comes from treating any such extension as a (mathematical) *object*. Von Neumann's initial system was formulated in terms of functions.[58] Since sets are inter-definable with their characteristic functions, this is a minor issue, and I will rephrase the core of the theory in terms of sets. The totality of things is divided into I-objects and II-objects. I-objects are those that can be a member or element of something, and II-objects are those of which something can be an element.[59] The two groups partly overlap, i.e., there are I-II-objects that, besides having elements, can be elements. II-objects are called *classes* nowadays, and the point of the construction is that not all II-objects or classes are I-II-objects, that is: some classes cannot be members or elements of anything. Gödel These have been later named *proper classes* by Gödel. Classes that can be members of classes are nowadays called *sets*, and they correspond to the sets admitted within **ZF**. Put it another way: all sets are classes, but not all classes are sets.

Which classes are not sets or, equivalently, which classes are proper classes? The celebrated von Neumann Axiom (IV.2) provides the answer. Stated in terms of functions, it says:

(IV.2) A II-object f is not a I-II-object iff there exists a function g such that for every I-object x there exists a y for which both $y \in f$ and $g(y) = x$ hold.[60]

Translating into sets, (IV.2) claims that proper classes, i.e., II-objects that are not I-objects, are all and only *those that can be mapped one-to-one with the universal class* **V**.

(The formalization of) (IV.2) has impressive deductive power: it makes at once superfluous the axioms of Separation, Choice and Replacement and, besides them, the Well-Ordering Principle (i.e., the principle according to which each set can be well-ordered): they all can be

---

[58] Von Neumann (1925).

[59] The Neumannian approach involved *Urelemente*, but the theory can be easily "purified".

[60] See von Neumann (1925): 400 (I have modified the notation into the standard functional one).

obtained by means of (IV.2).[61] The point that concerns us, though, is the intuition underlying (IV.2) that the necessary and sufficient condition for being a (proper) class is to be the same size as V. As Fraenkel, Bar-Hillel and Levy explain, the main idea behind von Neumann's approach is "the discovery that the antinomies do not arise from the mere existence of very comprehensive sets, but from their elementhood, i.e., from their being able of being members of other sets".[62] According to this strategy, therefore, the transcendentals, such as V, Ω, etc., do exist. The mistake lies in giving them the status of real substances, so to speak – "finished things", to repeat Cantor – which can be members or elements of further multiplicities. For instance, the Russell set R does not seem to produce contradictions *simpliciter*: intuitively, the set of all elephants is not an elephant, so it is an element of R. The set of all things which are not elephants, is itself not an elephant, so it is not an element of R. Troubles seem to begin only when we wonder whether R itself is or is not an element of R. What is denied now, therefore, is not Existence, but Objectivity:

> If we compare [our theory] with Fraenkel's axioms and definitions, we see that most of our axioms have analogues among his. There are, however, some quite essential differences, That we speak of "functions" rather than of "sets" is no doubt a superficial difference; it is essential, however, that the present set theory deals even with sets (or "functions") that are "too big", namely, those II-objects that are not I-II-objects. Rather than being completely prohibited, they are only declared incapable of being arguments (they are not I-objects!). This suffices to avoid the antinomies.[63]

3.4.2 Bernays and Classes Taken (Not) Seriously

Paul Bernays has developed Von Neumann's approach in a series of papers constituting probably the most articulated analysis of the issues of classical set theory[64] - so that it is now common to talk of a **VNB** (von Neumann-Bernays) theory. Bernays' main contribute from the philosophical point of view consists in having clearly distinguished the properly mathematical aspect of set theory from the one we may label as logical-

---

[61] See *Ibid*: 400-2.
[62] Fraenkel, Bar Hillel and Levy (1973): 135-6.
[63] Von Neumann (1925): 401.
[64] See Bernays and Fraenkel (1958).

theoretical. Von Neumann took sets as a sub-species of classes (those that are not proper classes), whereas in Bernays' approach sets and classes are completely disjoint groups: if something is a class, then it is not a set, and *vice versa*. The philosophical insight is that classes, after all, are logical abstractions. Only sets are the real objects of mathematics, having properties and standing in relations studied by the discipline. A relation of *representation* now captures the connections between the two groups. Each set represents a class, but some classes are not represented by sets. These correspond to von Neumann's proper classes. Also in this version, therefore, V exists as a class, but no set can correspond to it: the totality of all mathematical objects is not a mathematical *object*. Formally, Bernays' systems typically express the difference between sets and classes by adopting two sorts of variables and by duplicating the membership relation. The Abstraction Principle can be taken in its unrestricted form for classes, so any condition determines a class. Paradoxes are still avoided in von Neumann's fashion: sets can belong to sets and classes, whereas classes belong to nothing. Mature versions of the Bernays systems also forbid the quantification of variables for classes. The underlying attempt, of course, is to avoid any "closure" in the realm of classes, which is allowed for sets by allowing quantification of the variables for sets.

Can one claim that the class of all sets should do justice to our intuition that all sets can be collected together? One may doubt it. Von Neumann's Axiom (IV.2) *incorporates* in the axiomatic theory the principle of the Limitation of Size. In the **ZF** approach, this is an informal rule regulating, as it were, "from the outside" the formulation of such axioms of the theory as (Sep) – a rule which works as a Kantian super-ego: "Thou Shalt Avoid the Unconditioned". (IV.2) explicitly states the rule by claiming that something is a real set, an object, not a mere abstraction, if there is no one-to-one correspondence between it and the universal aggregate, V. Given all this, **VNB** set-theoretic approaches are not as alternative to **ZF** as von Neumann hoped. On the contrary, "set theory with classes and set theory [scil. **ZF-ZFC**] with sets only are not two separate theories; they are, essentially, different formulations of the same underlying theory"[65] – not only because the von Neumann-Bernays approach does not prove anything more than **ZF** concerning *sets*; so that "if one is interested only in sets and regards classes as a mere technical device, one

---

[65] Fraenkel, Bar-Hillel and Levy (1973): 119.

should regard ZF and VNB as essentially the same theory".[66] But, above all, because, as Priest has observed, introducing classes (or proper classes) violates again, just as in **ZF**, the Domain Principle:

> The variables of von Neumann's theory range over the domain of all collections (I-objects and II-objects). The totality of all collections is not, therefore, V (the collection of all sets), but V', the collections of all sets and classes. And it is this, and similar collections, whose existence cannot be consistently admitted in von Neumann's theory. Hence, the theory violates the Domain Principle just as much as Zermelo's. (Sometimes, von Neumann's theory is represented as a two-sorted theory with different kinds of variables ranging over I-objects and II-objects. This makes no essential difference to the point being made: the Domain Principle is violated by the variables ranging over II-objects).[67]

The *ad-hocness* of introducing classes as separate entities is also clear, since the only motivation for such a move is to avoid the antinomies. A further philosophical point with the rejection of Objectivity is that when we explain the theory we treat classes, or proper classes, as full-fledged *things* ("objects", von Neumann said).[68] By referring to them, we treat them as a unity, and as something having properties and satisfying conditions – for instance, the property of being "too big"; which is specified formally by claiming that they can be put into one-to-one correspondence with the universal class. *Having the same size as V* definitely looks like a property (proper) classes have in common, so it would be quite natural to think of the class of all classes having the same size as V... More generally, it is difficult to understand why classes cannot be members of any other collection (e.g., their singletons). However, if we assume that classes can be elements of further aggregates – say, hyper-classes standing to classes in the same relation as classes stand to sets[69] – we have just pushed the trouble up one level. In Manuel Bremer's words:

---

[66] *Ibid*: 131.

[67] Priest (1995): 165.

[68] "We shall always speak of 'I-objects', 'II-objects', and 'I-II-objects' instead of 'arguments', 'functions', and 'argument-functions', respectively" (von Neumann (1925): 399).

[69] See Fraenkel, Bar-Hillel and Levy (1973): 142.

There seems to be no reason why classes could not be elements of (some) other classes. The general inhibition to elementhood seems to be an exaggeration.
If one then allows for classes to be elements of (some) classes we arrive at a second hierarchy (a hierarchy of classes), and now our old troubles return: There cannot be a universal class.[70]

This is close to what happens, according to Hegel, with Kant telling us lots of things concerning the unknowable noumena.

3.5 The Cumulative Hierarchy

Considering the three requirements for good solutions of the paradoxes listed at § 1.4.1, the mainstream set-theoretic strategies hardly reach level no. 2, i.e., they barely manage to independently motivate the rejection of one of the premises of the paradoxical arguments. As Crispin Wright has claimed:

> We have learned of a variety of strategies which seem to keep us out of trouble; but none of them has the simple intuitive appeal originally possessed by the 'naïve' assumptions concerning class existence and predication, and what constitutes an admissible range of quantification, which featured in, for example, Frege's foundational theory. It is difficult to defend a notion of 'error' in this context for which the criterion is not precisely the potential to generate paradox; and this criterion, naturally, fails to discriminate in point of preferability between the alternative, seemingly successful strategies for avoiding paradox.[71]

And Benardete:

> What we want to know is precisely what it is about sets as such, sets *qua* sets, that precludes there being a set of all sets that are not members of themselves.

---

[70] Bremer (2005):141-2. See also Mayberry (1977): 10-1: "once having admitted proper classes as *objects*, clearly individuated by means of unambiguous identity conditions, what can we put forward as an argument against accepting *collections* of such objects? After all, since we can distinguish proper classes, we can presumably count them, which strongly suggests that we should be able to form them into aggregates or collections. Surely the idea of a definite *thing* which cannot be put into collections with other things is completely unintelligible. But what is now to stop us from forming collections of classes, families of collections of classes..."

[71] Wright (1980): 297.

Insight being above all what we seek, Zermelo would seem to be the least promising source to consult; for he was positively ostentatious in insisting that this was no more than a home remedy, suited to the needs of the working mathematician on the most pedestrian level.[72]

However, are **ZF** or **VNB** actually *suited* to the needs of the working mathematician? One may doubt it. We can have a general overview of the problem by looking at the set-theoretic hierarchy to which most of the ordinary set theories are reducible, the so-called (transfinite) cumulative hierarchy. Intuitively, this is obtained by taking the power set of a given set, the power set of the power set, and so on, and collecting up as we go. Less roughly, the hierarchy can be defined by recursion on the ordinals. We get a transfinite sequence by beginning with a set of individuals or, in the "purified" version, with the empty set. We take the power set of a given set K_n at each successor ordinal $n + 1$ (and collect up, if we have *Urelemente* at the base level); and we collect up everything at each limit ordinal l:

$K_0 = \emptyset$ or $\{x \mid x \text{ is an individual}\}$

$K_{n+1} = K_n \cup P(K_n)$

$K_l = \bigcup_{n<l} K_n$[73]

The hierarchy is called *cumulative* because a set appearing at any level is a proper subset of any upper level set. Therefore, unlike the theory of types, sets can have members of various levels; but the level of each set is greater than that of any of its members. All the sets in the hierarchy are well-founded. Most of the current theories have a natural model in an initial segment of the hierarchy.[74] For instance, the typical **VNB** holds in the cumulative hierarchy just as **ZF** does, but goes one level further – the level at which we find the (proper) classes.[75] The idea that only those sets

---

[72] Benardete (1989): 124.

[73] See e.g. Hallett (1984): 188-9; Woods (2003): 158; Potter (2004): 186ff.

[74] I say "most of", because there are exceptions. A notable one is Quine's system **NF** (and its class extension, **ML**). We shall deal with **NF** in the third part of the book, when I will speak of certain paraconsistent set theories inspired by it.

[75] See Martin (1984): 238.

exist, that belong to the cumulative hierarchy, is nowadays seen as *the* solution of all traditional set-theoretic paradoxes. This is, of course, the most general set-theoretic shape of the Kantian strategy, trying to put up boundaries to conceptual abstraction in some way or other.

3.5.1 Logical and Mathematical Problems

It is easy to observe that the hierarchy does not recapture the Cantorian insight according to which a set as the extension of an arbitrary condition. One independent issue has to do with non-well-founded sets: the developments of non-well-founded set theory[76] cast doubts on the cumulative hierarchy quite independently of the issue of set-theoretic paradoxes. Apart from this, the main philosophical problem concerning us now is that the transfinite sequence *itself* cannot be a set, which would be, again, the set of all (pure, well-founded) sets, on pain of contradiction.[77] Therefore, the problem of the violation of the Domain Principle resurfaces in its most general shape.

According to Mayberry, Bremer, and others, the inadequacy of the cumulative hierarchy surfaces at the level of ordinary elementary logic, whose semantics is typically framed in set-theoretic terms. The violation of the (DP) here goes as follows. Consider the characterization of logical consequence one can find in any textbook of logic. Typically, it goes like this (where Γ is a given set of formulas):

(Cons) $\Gamma \vDash \alpha$ iff, in every interpretation in which, for all $\beta \in \Gamma$, β is true, α is true.

Now, an *interpretation* is a set-theoretic structure. In the simplest case, it is a pair, $\langle D, i \rangle$, where D is some (non-empty) set, and i is a function assigning denotations in D (members of D, subsets of D, sets of ordered *n*-ples, etc.) to the non-logical expressions of the formal language. It is not very important that, for instance, we may need a different function for variables (i.e., that we may have to consider different assignments to the variables within the same interpretation). The fact that we can impose more and more ontological complexity on the models, e.g., by dealing with intensional interpretations (assigning intensions to the

---

[76] See Aczel (1988).
[77] Precisely, Mirmanoff's paradox would follow. I shall not deal with it, though.

sub-sentential expressions, as functions from possible worlds to extensions, or as sets of possible worlds, etc.), makes little difference, too. Since the achievements of model-theoretic semantics have conquered formal logic, the standard procedure consists in the specification of a model determining the relevant domain(s); of a function assigning meanings to the descriptive vocabulary (and, if this is the case, of assignments of values to the variables). Then comes the recursive characterization of truth conditions. Then come the definitions of the notions of logical truth and logical consequence.

Talking about "every interpretation", (Cons) quantifies over the universe of sets. However, if we stick to the cumulative hierarchy, this is *not* a set. How are we to give the semantics of *that* expression, then? To put it another way: consider the language in which the definition of logical consequence is given. This is usually an informal, or semi-formal, language adopting set-theoretic notions. Of course, this is supposed to be perfectly meaningful. But if only those sets exist, that belong in the cumulative hierarchy, as Priest remarks, "every interpretation" points to a totality which does not exist (and you may notice that it is essential that we all dealing with *all* interpretations. Otherwise, (Cons) would not define *logical* consequence, but some semantic consequence restricted to particular domains). So "assuming the cumulative hierarchy to be the correct account of set, the semantics of set theory thus becomes impossible to specify in a coherent fashion".[78] Bremer makes the point with reference to **ZFC**:

> [When defining logical consequence] one now talks about *any* interpretation. And the domain of interpretation is arbitrary. It may be a set of arbitrary high rank. So the supposed definition talks about *all* sets of an arbitrary high rank (i.e. of the completed hierarchy), but in **ZFC** we can never get *all* sets! So it seems that our understanding of consequence cannot be modelled by **ZFC**.[79]

Conversely, if (Cons) and similar central claims of the semantics for first-order logic do have a determinate meaning, then it seems that we have implicitly resorted to the *naïve, unrestricted* conception of set, which was officially dismissed.[80]

---

[78] Priest (1987): 37.
[79] Bremer (2005): 141.
[80] On this point, see particularly Mayberry (1977): 5ff.

The *extrema ratio* now is to claim that set-theoretic paradoxes are not a real *mathematical* problem, having to do only with the circumlocutions of philosophers. I think this was also Gödel's idea, at least at a certain stage of his career. In *What is Cantor's Continuum Problem?*, he claimed that the paradoxes of set theory "are a very serious problem, not for mathematics, however, but for logic and epistemology". No professional mathematician has ever dealt with such things as Russell's paradox. Mathematics has to do with sets of integers, or reals, or of their functions, etc. Therefore, nobody ever reaches such sets as the ones obtained by considering the totality of things.[81]

Things have changed nowadays. Even those authors who stick to the idea that **ZF** and **VNB** "provide a framework in which practically all of current mathematics can be systematically worked out" have to admit that there is a notable exception, namely, "the informal general theory of mathematical structures, particularly the theory of categories".[82] The point is usually expressed with respect to **ZF-ZFC**, but things do not get better with a von Neumann-Bernays approach. One cannot just treat large categories (like the categories of all groups, all sets, etc.) as classes or proper classes: within category theory, one wants not only to admit these things, but also to *operate on* them. Therefore, given two large categories, A and B, for instance, one may want to have the functor category A^B of all the functors from B to A, which is precisely what is ruled out by the hierarchy. As J. Bell has claimed, "the operations on large categories which appear so natural to category-theorists are not justified by current set-theoretic foundations"[83] (of course, one might resort to hyper-classes but, as we know, this would only push the problem up one level).[84] Given all this, it seems that *any* Kantian attempt to limit the Cantorian idea of sets as extensions of arbitrary conditions is doomed to fail in some way or other:

---

[81] See Gödel (1947): 262.

[82] Feferman (1984): 238; see also Feferman (1977); Bell (1981).

[83] Bell (1981): 356.

[84] "Categories are classes, which may also be proper classes, and category theory also deals with functions defined on classes of categories and with other kinds of objects which are unavailable in [standard set theories]. [...] This process of adding bibber classes and hyper-classes has to stop somewhere; and we have to decide where to do so. [...] [A] choice is as arbitrary as any other. This arbitrariness is, to some extent, due to the antinomies and hence unavoidable" (Fraenkel, Bar-Hillel and Levy (1973): 143-5).

> The failure of set theory to justify the unlimited application of category-theoretic operations is a consequence of its success in eschewing the over-comprehensive collections which were originally deemed responsible for the paradoxes [...] In fact, set theory's failure to embrace the notion of arbitrary category (or structure) is really just another way of expressing its failure to capture completely the notion of arbitrary property.[85]

---

[85] Bell (1981): 356.

4 The Gödel Paradox

> Just think: in mathematics, this paragon of reliability and truth, the very notions and inferences, as everyone learns, teaches, and uses them, lead to absurdities.
>
> David Hilbert, *On the Infinite*

4.1 Peano Arithmetic, PA

In an influential series of essays, Priest and Routley have proposed a bold interpretation of Gödel's First Incompleteness Theorem, which should provide a further argument against the (LNC).[1] Priest has described it is a "non-constructive argument", that is, one which does not produce an explicit true contradiction, but shows the existence of some.[2] In order to understand how it is supposed to work, we need to become familiar with the First Incompleteness Theorem. This will also give us a chance to clarify the ideas introduced in § 2 to explain the non-contextual formulations of the semantic paradoxes. Furthermore, the things we are about to learn will turn out useful when we meet the peculiarities of paraconsistent arithmetics, in the third part of the book. In that context, I shall also say something about Gödel's Second Incompleteness Theorem.

Let us begin by introducing the exemplary case of a theory to which Gödel's Theorems apply: Peano arithmetic, **PA**. The theory can be formulated as a second-order one, with the five famous axioms, the so-called *Peano* (or Dedekind-Peano) *Axioms*. The more common first-order formulation of **PA** can be obtained by simply adding to the ordinary axioms of first-order logic with identity the following principles:

(PA1) $\forall x(Succ(x) \neq 0)$
(PA2) $\forall xy(Succ(x) = Succ(y) \rightarrow x = y)$
(PA3) $\forall x(x + 0 = x)$
(PA4) $\forall xy(x + Succ(y) = Succ(x + y))$

---

[1] Priest (1979), (1984a), (1987): Ch. 3; Routley (1979a), (1979b); Priest and Routley (1989b) and (1989d).
[2] See Priest (1987): 39.

(PA5) $\forall x(x \times 0 = 0)$
(PA6) $\forall xy(x \times Succ(y) = (x \times y) + x)$
(PA7) $\alpha[x/0] \to (\forall x(\alpha[x] \to \alpha[x/Succ(x)])) \to \forall x\alpha[x])$.[3]

The theory holds in the so-called *standard model* of arithmetic (be it $\mathbb{N}$), i.e., in the model constituted by natural numbers and the operations on them we have learned when we were children. Variables are therefore supposed to range on natural numbers, and "0" is the name of number zero. One of the strangest consequences of Gödel's First Theorem is that (first-order) **PA** is not, as model theorists say, *categorical*. This means, more or less, that from Gödel's results follows the existence of certain models, called *non-standard*, which satisfy **PA** despite being structurally different from the standard model $\mathbb{N}$. In particular, there is no way to constrain the variables of **PA** so that they range exclusively on ordinary natural numbers. I shall talk about this in § 12. Meanwhile, let us stick to the intended interpretation, that is, to what the teacher has made us take for granted about arithmetic when we were children. The intended reading of the one-place functor $Succ(x)$ is "the (immediate) successor of x". Therefore, $Succ(0)$ is 1, that is, the (immediate) successor of zero in the series of natural numbers; $Succ(Succ(0))$ is 2, that is, the successor of 1; etc. + and ×, of course, are read as addition and multiplication. Therefore, (PA1) claims that zero is the successor of no number; (PA2) claims that if x and y have the same successor, they are the same number. (PA3)-(PA6) characterize addition and multiplication. (PA7) is the schematic formulation of the so-called (mathematical) Induction Principle, which claims that if some $\alpha[x]$ holds for zero and for the successor of a given number x for which it holds, then $\alpha[x]$ holds for all numbers.

4.2 Gödel, First Half

As we have seen in § 2, a gödelization allows us to univocally associate a natural number to each symbol, formula and sequence of formulas of **PA**, so that we can always move back and forth between an expression of the language of **PA** and the number to which it has been paired (its Gödel number). You may recall that a sufficiently powerful theory can "talk" about some of its syntactic properties, and that, in particular, the property

---

[3] See e.g. Potter (2004): 98-100.

of *being a theorem of the theory* can be expressed within the theory itself. More generally, a *k*-ary relation (whose extension consists in a set of ordered *k*-ples) R is said to be *expressible* in **PA**, iff there is a formula α[x_1, ... , x_k], such that, for any ordered *k*-ple of numbers <n_1, ... , n_k>, we have that:

(a) If <n_1, ... , n_k> ∈ R, then ⊢$_{PA}$ α[x_1/**n**$_1$, ... , x_k/**n**$_k$];
(b) If <n_1, ... , n_k> ∉ R, then ⊢$_{PA}$ ¬α[x_1/**n**$_1$, ... , x_k/**n**$_k$],

where **n** is the numeral of number *n*.[4] We have also maintained that **PA** is the typical case of a sufficiently powerful theory. Logicians usually call "sufficiently powerful" a theory capable of representing the (primitive) recursive functions described in the seminal Gödelian paper.[5] Recursive functions-relations have precisely the role of codifying the syntax of the theory. Metalinguistic claims on **PA** are mirrored within the, as it were, "officially" arithmetic language of **PA** (hence this situation is called, for obvious reasons, *arithmetization of syntax*).[6] Now, the arithmetic predicate no. 45 in Gödel's paper is:

Prf(*x*, *y*),

whose informal reading is: "*x* is (the Gödel number of) a proof of the formula (whose Gödel number is) *y*".[7] *Prf* holds between those pairs of numbers which are, respectively, the Gödel number of a sequence of formulas of **PA**, and the Gödel number of a formula of **PA**, such that the former is a proof of the latter. Predicate no. 46 is defined by means of no. 45, thus:

Th(*y*) =$_{df}$ ∃*xPrf*(*x*, *y*);

therefore, it holds of those numbers which are the Gödel numbers of formulae of **PA** for which there is a proof in **PA** (properly, being a theorem of **PA** is not an expressible, but a *semi-expressible* or *weakly expressible*

---

[4] See Gödel (1931): 607.
[5] The qualification "primitive" is due to Kleene: see Kleene (1976).
[6] See e.g. Boolos, Burgess and Jeffrey (2002): Ch. 15.
[7] See Gödel (1931): 606.

notion, but this is a minor point). The fundamental condition in order to prove Gödel's First Theorem concerns provability within **PA**:

(D) $\vdash_{PA} \alpha \Rightarrow \vdash_{PA} Th(\ulcorner\alpha\urcorner)$.

(D) claims that, if α is a theorem of **PA**, then the formula expressing this fact *within* **PA** is itself a theorem **PA**. Now, before the Fixed Point Lemma was employed to obtain fully formalized liars, Gödel used it to build a sentence (be it γ) attributing to itself not falsity, but non-theoremhood:

(FP$_\gamma$) $\gamma \leftrightarrow \neg Th(\ulcorner\gamma\urcorner)$.

γ is an arithmetic sentence, whose informal reading is: "I am not a theorem". Given the definition above, it is equivalent to: $\neg\exists x Prf(x, \ulcorner\gamma\urcorner)$, that is, "I am not provable".

We have to assume, then, that **PA** is both consistent and ω-*consistent*. Consistency has already been glimpsed at here and there, and I will have to refer to it quite often in the following. Briefly, a given formal system **S** supplied with negation is said to be (syntactically) *consistent*, or non-contradictory, iff for no formula α of its language L it cannot be the case that $\vdash_S \alpha$ and $\vdash_S \neg\alpha$, that is, iff **S** cannot prove both a formula and its negation. ω-consistency is more typically arithmetical. A system is called ω-*consistent* iff for no formula $\alpha[x]$ of its L it is possible to prove both $\alpha[x/\mathbf{n}]$ for each natural n, and $\exists x\neg\alpha[x]$. Now, Gödel demonstrated that:

(1) If **PA** is consistent, then $\nvdash_{PA} \gamma$;
(2) If **PA** is ω-consistent, then $\nvdash_{PA} \neg\gamma$.

As for (1): if γ were a theorem of **PA** then, given (D), also $Th(\ulcorner\gamma\urcorner)$ would be. Hence, given (FP$_\gamma$), the provability of $\neg\gamma$ would follow. We would have, then, $\vdash_{PA} \gamma$ and $\vdash_{PA} \neg\gamma$, against the assumption that **PA** is consistent. As for (2): since the notion of proof is (primitive) recursive, we have that for each n either $\vdash_{PA} Prf(\mathbf{n}, \ulcorner\gamma\urcorner)$, or $\vdash_{PA} \neg Prf(\mathbf{n}, \ulcorner\gamma\urcorner)$. The former case is ruled out by the fact that, as (1) claims, γ is not provable – therefore, for each n it is not the case that n is the code of a proof of γ in

PA. Hence, the latter case holds. It follows, given the assumption that **PA** is ω-consistent, that $\exists x Prf(x, \ulcorner \gamma \urcorner)$ is not a theorem. But $\exists x Prf(x, \ulcorner \gamma \urcorner)$ is nothing but $\neg\gamma$.[8] The conjunction of (1) and (2) gives us Gödel's First Incompleteness Theorem. This tells us that Peano arithmetic includes a sentence, γ (its own so-called *Gödel sentence*) which is constitutively *undecidable* within **PA**, that is, not provable and not refutable (i.e., such that its negation is not provable either).

The exposition of the First Incompleteness Theorem usually goes together with the following story. Since γ "claims" (via arithmetization) to be not provable, and we have just proved that it is not provable, then γ just is what it claims to be; hence, it is *true*. It is usually maintained, then, that incompleteness results establish a fundamental gap between provability and truth (and precisely because of this, they have been taken as a cornerstone of mathematical realism, based on an interpretation encouraged by Gödel himself).[9] If a formal system has to be correct, i.e., it must capture *only* mathematical truths, then it cannot capture them *all*. Furthermore, the truth predicate for a given language L, were it expressible within L, under the usual conditions would originate the liar paradox (therefore, if Tarski is right, it should *not* be expressible within L), whereas the provability predicate is (weakly) expressible within **PA** with no danger for the (LNC). The liar, "I am false", under the usual *bona fide* assumptions, generates the antinomy. However, with the Gödel sentence, "I am not provable", no contradiction is expected. Or, at least, this is the standard reading of the situation.[10]

Enough for the established facts. Let us move on to the bold interpretation.

---

[8] See e.g. Smullyan (1992): Ch. V; Boolos, Burgess and Jeffrey (2002): 225-7.

[9] Gödel's realist stance emerged, as is well known, only several years after his 1931 paper (see Gödel (1947)); but he declared that his mathematical Platonism had been the heuristic key for the discovery of the incompleteness results. On this point, see Feferman (1983).

[10] See e.g. Kleene (1976).

4.3 The Argument

4.3.1 The Naïve Mathematical Theory

How exactly do we prove that γ is true? It is usually claimed, "with a wave of hands",[11] that it is proved by means of an informal, "intuitively correct" argument external to the system **PA**. Now, according to Priest we obtain the relevant case against the (LNC) by applying the First Theorem itself to the theory that formally captures our *intuitive, naïve notion of proof*. By "naïve notion of proof" Priest apparently means the one underlying ordinary mathematical activity: "proof, as understood by mathematicians (not logicians), is that process of deductive argumentation by which we establish certain mathematical claims to be true".[12] When we want to settle the question whether some mathematical sentence is true, we try to deduce it, or its negation, from other mathematical sentences which are already known to be true. The process cannot go backwards *in infinitum*. So we should reach, eventually, mathematical sentences which are known to be true without having to be proved – e.g., because they are "self-evident". But this is not really important (nor is important to establish *which* are the primal truths; concerning arithmetic, they may be, for instance, principles such as those of Peano, that is, claims according to which every number has a successor, etc.).

Given this rough characterization, it is clear that the naïve-intuitive theory Priest links to the naïve-intuitive notion of proof is rather informal. Nevertheless, "it is accepted by mathematicians that informal mathematics could be formalized if there were ever a point to doing so, and the belief seems quite legitimate".[13] One could regiment the fragment of English in which the naïve theory is expressed, and turn it into a formal language. Then, the primal truths would be written down in the (now) formalized language and taken, say, as axioms; and proofs would be expressed as formalized arguments. Priest also claims that, after having been so translated, the naïve theory would certainly be sufficiently pow-

---

[11] Priest (1979): 222.

[12] Priest (1987): 40. "The argument concerns our naïve proof procedures. These are those informal methods of proof which are used by working mathematicians [...] to settle the truth of some matter" (Priest (1984a): 165).

[13] Priest (1987): 41.

erful in the sense explained above, i.e., capable of representing all (primitive) recursive functions.

Is the naïve notion of proof really recursive? This is less straightforward. Dummett (1978) has famously stressed that it belongs in the very notion of *proof* that one can recognise a proof when one sees it, at least in principle – which is not the case with *truth*. In addition, given Church's Thesis (that is, the thesis according to which the effectively computable functions coincide with the recursive ones) the recursiveness of the naïve notion of proof is entailed by its being effective. As for this, let us listen directly to Church:

> Consider the situation which arises if the notion of proof is non-effective. There is then no certain means by which, when a sequence of formulas has been put forward as a proof, the auditor may determine whether it is in fact a proof. Therefore he may fairly demand a proof, in any given case, that the sequence of formulas put forward is a proof; and until the supplementary proof is provided, he may refuse to be convinced that the alleged theorem is proved. This supplementary proof ought to be regarded, it seems, as part of the whole proof of the theorem.[14]

A major objection still to be faced is that the established standards for accepting a deductive argument as a proof within mathematics may vary across time, so the very naïve notion of proof may *change*. Gödel himself is considered as having suggested that proof may not be recursive because now and then we add new principles or rules to the old ones, after having ascertained that they entail various things accepted as true, and nothing known to be false. The way in which we perform such additions would not be itself rule-governed. To this, Routley and Priest have two main replies. The first one simply proposes to eliminate the alleged diachronic variability by restricting our attention to the naïve notion of proof that conforms to the *current* standards.[15] The second, and more radical, one is that Gödel's hypothesis may be refuted by the fact that the naïve notion of proof is *socially* learned and taught. Since the set of ordered pairs constituting the extension of the proof relation is potentially infinite, one cannot just teach and learn the notion by means of a list. If the notion were not recursive, the process whereby mathematics is learnt, and the general agreement of working mathematicians on what counts as a

---

[14] Church (1956): 53.
[15] See Priest (1987): 43.

mathematical proof, would turn out to be a mystery. Of course, this is a particular case of a famous and more general argument to the effect that *language* can only be learnt recursively.[16] On the contrary, as Routley claims, if the notion is recursive we can understand "how one generation of mathematicians learns what counts as true from the previous generation, namely they learn certain basic mathematical truths and how to prove others by making deductions".[17]

4.3.2 Its Inconsistency

Let **T** be, then, the formalisation of our naïve, intuitive mathematical theory. Since **T**, just like **PA**, is sufficiently powerful and its notion of proof is recursive, *if* **T** is consistent, then it satisfies the conditions of Gödel's First Theorem: "if *T* is consistent there is a sentence φ which is not provable in *T*, but which *we can establish as true by a naïve proof,* and hence *is* provable in *T*".[18] Of course, anything that is naïvely-intuitively provable is provable within the naïve-intuitive theory. "Assuming its consistency, it would, therefore, seem to be both complete and incomplete in the relevant sense".[19] Now we have no way to avoid a contradiction. Either we accept *this* one, i.e., $\vdash_T \varphi$ and $\nvdash_T \varphi$, which spells troubles for the (LNC) anyway; or we have to admit that our naïve mathematical theory, with its naïve notion of proof, is not consistent, that is, it allows the derivation of contradictions. However, our naïve proof procedures are, by definition, the deductive arguments by means of which we establish certain sentences to be true. Therefore, by this "non-constructive" line of argumentation, there are true contradictions.

Routley and Priest claim that the Gödel sentence φ for the (formalisation of) the naïve theory can be proved within **T** itself, together with its negation – so **T** is inconsistent in the sense that $\vdash_T \varphi$ and $\vdash_T \neg\varphi$. The point is that "This sentence is not provable" now has its "provable" understood as meaning "demonstrated to be true", and Gödel's proof becomes a real *paradox*. According to Priest, such a conclusion is deeply illuminating on the nature of our naïve mathematical theory:

---

[16] On which see, famously, Davidson (1984): Ch. 1.
[17] Routley (1979a): 327.
[18] Priest (1987): 44, my italics.
[19] Priest (1984a): 165.

In fact, in this context the Gödel sentence becomes a recognisably paradoxical sentence. In informal terms, the paradox is this. Consider the sentence 'This sentence is not provably true'. Suppose the sentence is false. Then it is provably true, and hence true. By *reductio* it is true. Moreover, we have just proved this. Hence it is provably true. And since it is true, it is not provably true. Contradiction. This paradox is not the only one forthcoming in the theory. For, as the theory can prove its own soundness, it must be capable of giving its own semantics. In particular, [every instance of] the T-scheme for the language of the theory is provable in the theory. Hence [...] the semantic paradoxes will all be provable in the theory. Gödel's "paradox" is just a special case of this.[20]

To get the point, we may recall that, as the sloppy story on the First Theorem goes, the truth of γ should be settled by means of some intuitive reasoning external to **PA**. More correctly, one may say that it is provable within a theory that can deal with the semantics via the notion of *truth*, which is not definable, given the Tarskian result, within **PA**. "Ask: how is [γ]'s truth established? The answer is: by a metamathematical *proof* of [γ]",[21] that is, by means of a detour through the metatheory. However, **T**, formalizing as it does our naïve notion, according to Priest and Routley should absorb the metatheory within the theory. **T** is semantically closed in the Tarskian sense, and inconsistent: the reasoning behind the proof of the truth of the Gödel sentence is now carried out within the formal system itself. After all, mathematicians use ordinary English, and ordinary English, given the considerations of our § 2, may well be semantically closed. As Routley stressed, "everyday arithmetic as presented within a natural language like English appears, unlike say first-order Peano arithmetic, appropriately closed". And "is provable in arithmetic" and "is arithmetically true" are "English, and in a good sense arithmetical, predicates".[22]

The Priest-Routley argumentation agrees with the aspect of the officially Gödelian position, according to which consistency and completeness become incompatible once a theory reaches a certain expressive power. However, according to Priest and Routley, only by giving up the former we can get to grips with actual mathematical practice. A theory that captures mathematical rationality should be inconsistent and capable

---

[20] Priest (1987): 46-7; see also Priest (1984a): 172.
[21] Routley (1979a): 325.
[22] *Ibid*: 326.

of expressing its own semantics: "we might say that our naïve proof procedures are not just contingently inconsistent, but *essentially* so". And this fact "can be seen as vindicating the Kant/Hegel thesis that Reason is inherently, by its very nature, inconsistent".[23]

It is easy to see how audacious the argument is: it places true contradictions at the very core of mathematical practice. Therefore, its having been variously questioned is no surprise. Charles Chihara and Neil Tennant have claimed that it is much more reasonable to interpret the discourse on naïve provability as showing that there cannot be a complete formalization of our naïve proof procedures.[24] This seems to beg the question against Priest and Routley, though, since they independently (i.e., apart from the Gödelian issue) question with various arguments that consistency has to be privileged over any other (broadly) epistemic virtue of a theory.[25]

Seemingly more dangerous criticisms have been put forward by Stewart Shapiro. Shapiro has proposed to specify the idea of "provability in the naïve theory", which Chihara and others found less than precise,[26] by building a semantically closed formal arithmetic **PA***, including its truth predicate. The theory proves its Gödel sentence (say, γ^*) and the truth predicate is used in the proof. But γ^* is itself purely arithmetic and, given the soundness of **PA***, embarrassing consequences follow: primitive recursive relations turn out to be inconsistent, too. Therefore, ordinary **PA** and even Robinson's arithmetic **Q** are inconsistent; and some number n both is and is not the code of a proof of γ^*,[27] which is hard to swallow. This critique, however, also seems to be an *inconveniens*, whose importance depends on how far opponents of the (LNC) are prepared to go. As we will see in the following Chapters, some of them are willing to bite many bullets.[28] Furthermore, the feeling of disease one may have in front of Priest and Routley's shocking claims to the effect that real arithmetic

---

[23] Priest (1987): 47.
[24] See Chihara (1984): 121-2; Tennant (2004): 383.
[25] See e.g. Priest and Routley (1989d); Priest (2006): part III.
[26] See Chihara (1984): 119.
[27] See Shapiro (2002): 817.
[28] For instance, Priest has replied to Shapiro, among other things, that "if one endorses an inconsistent arithmetic, but tries to hang on to either a consistent computational theory or a consistent metamathematics of proof, one is in for trouble. The solution to Shapiro's problems is therefore not to be half-hearted, and to accept that these other things are inconsistent too" (Priest (1987): 243).

is inconsistent may be mitigated nowadays. Recent developments in the application of paraconsistent logics have shown that formal inconsistent arithmetics are technically feasible. They may even have *palatable* consequences. I shall talk about them in the third part of the book.

Part II.
The Theories

5 "On Detonating"[1]

> When the unmovable object meets the unstoppable force, there's nothing you can do about it.
>
> Pulp, *Seductive Barry*
>
> But a man who denies the Law of Non-Contradiction, or asks for a proof of it, is not committed to the assumption that *everything* is contradictory.
>
> Jan Łukasiewicz, *Aristotle on the Law of Contradiction*

5.1 The Scotist Conception of the Absurd

In the preceding part of the book, we have examined some cases on behalf of contradiction. In this Chapter, I shall take from the relevant literature some methodological constraints any logic or theory that admits contradictions is supposed to meet. These constraints will provide criteria for evaluating several such logics and theories in the subsequent Chapters. Let us begin with a little bit of terminology, precisely, with syntactic definitions of the notions of *(in)consistency* and *triviality*.

5.1.1 Inconsistent and Trivial

We have already met consistency and inconsistency. A given formal system **S** is said to be (syntactically) consistent iff, for no formula α of the language L on which it is built, it is the case that $\vdash_S \alpha$ and $\vdash_S \neg\alpha$: iff **S**, in other words, does not allow to prove both a formula and its negation – therefore, a contradiction of the $(C1_{dist})$-kind. If, on the contrary, this is the case, **S** is said to be inconsistent. If both α and $\neg\alpha$ are provable, and the system includes some axiom or rule of inference expressing Adjunction, that is, the intuitive idea that if formulas $\alpha_1, \ldots, \alpha_n$ hold, then their

---

[1] This comes from the title of Schotch and Jennings (1989).

conjunction $\alpha_1 \wedge \ldots \wedge \alpha_n$ holds, too, we will have $\vdash_S \alpha \wedge \neg\alpha$, a collective (C1)-contradiction. A formal system **S** on L is said to be *trivial* iff it proves all the formulas of its L; conversely, a **S** is not trivial iff there is at least one formula α of L, such that it is not the case that $\vdash_S \alpha$.[2]

The terminology is not completely standardized. What I have defined as inconsistency is sometimes called *negation-inconsistency*, and what I have defined as triviality, that is, the deducibility of anything, is sometimes called *absolute inconsistency*.[3] Sometimes (in)consistency is referred to as *(non-)contradictoriness*.[4] Furthermore, the notions are normally applied also to *theories*, taken in the wide sense as sets of sentences closed under logical consequence; and also to (sets or systems of) *beliefs* (whose closure under logical consequence, as is well known, is at the very least dubious). It is claimed, then, that a theory is inconsistent if it *includes* both α and $\neg\alpha$... and so on.

5.1.2 Ex falso quodlibet, *Paraconsistency*

Inconsistency and triviality are so strictly connected that they are usually taken as simply equivalent. The definition of triviality does not refer to negation but, obviously, any trivial logic or theory supplied with negation is inconsistent. The problem is the converse entailment. This is based upon a famous logical law, classically and intuitionistically valid, named *Scotus'* (or *Pseudo-Scotus'*) *Law* – henceforth: (PS). The law bears such a name because it was studied in the *In universam logicam quaestiones*, attributed at one time to Duns Scotus, now to a logician belonging to the Scotist school (some say, John of Cornwall). It states the so-called Scotist conception of the absurd. Medieval authors expressed it by saying: *ex falso* (paradigmatically: *ex contradictione*) *sequitur quodlibet*; anything follows from the *absurdum*, and contradiction is taken as the height of absurdity.

As a matter of fact, just like "contradiction" and "Law of Non-Contradiction", also Scotus' Law is spoken of in many ways: it can be phrased syntactically or semantically, as an axiom, an axiom schema, a rule of inference, and so on. A recurrent formulation is the following:

---

[2] See e.g. da Costa (1974): 497.
[3] See e.g. Woods (2003): 85.
[4] See e.g. Carnielli and Marcos (2002): 4.

(PS1) $\neg\alpha \to (\alpha \to \beta)$,

but the "imported" form is sometimes favoured, since it makes the fact that a (C1)-contradiction entails everything immediately evident:

(PS$_i$1) $\alpha \wedge \neg\alpha \to \beta$.

Of course, β can be itself a contradiction. Therefore, we have:

(EQ) $\alpha \wedge \neg\alpha \leftrightarrow \beta \wedge \neg\beta$,

a principle stating the equivalence of all contradictions. The pregnancy of a given implicative-negative law such as (PS1) depends, of course, on what we take $\neg$ and $\to$ to mean. With reference to the standard material conditional, various versions of (PS) are usually labelled as a *paradox of material conditional*, for reasons I shall deal with mainly in § 9 on relevance logics. However, it is worth noting since the beginning that (PS) holds also for C.I. Lewis' strict implication (be it $\prec$, the "fishhook"), as one of the so-called paradoxes of strict implication. The reason is intuitively clear. Lewis introduced the fishhook as a primitive symbol but, given very weak assumptions, $\alpha \prec \beta$ can be taken as equivalent to $\Box(\alpha \to \beta)$. This holds iff it is not possible that α and not β. Therefore, the mere factual falsity of the antecedent is not sufficient to make a strict implication true, as it happens with the material conditional. The *impossibility* of the antecedent, nevertheless, is sufficient, i.e., an impossible sentence strictly entails anything ("If there is a greatest prime, then Rome is on Mars"). Now, a contradiction is usually taken as the peak of impossibility: $\neg\Diamond(\alpha \wedge \neg\alpha)$. Therefore, $\alpha \wedge \neg\alpha \prec \beta$.[5]

I shall talk about conditionals at length. But a conception of absurdity is also, and mainly, linked to a conception of *negation*. For instance, it is quite common within sub-classical logics to define negation as the entailment of absurdity, in the style of Johansson:

(Df$\neg$) $\neg\alpha =_{df} \alpha \to \bot$,

---

[5] See Haack (1978): 197-8.

which leaves it open to fix the exact meaning of the absurdity (*falsum*) symbol. In the classical and intuitionistic case, we have a Scotist negation for a Scotist absurd: a negated formula is a formula which entails absurdity, which itself consists in entailing anything:

$$\bot \to \beta.$$

Therefore, (PS) is sometimes taken within formal system as a primitive principle, directly regulating the conduct of negation. We can find it as an inference rule, sometimes bearing the (a bit confusing) name of *Negation Elimination*:

$$\frac{\alpha,\ \neg\alpha}{\beta}\quad (\neg E)$$

Of course, there is no danger in deriving a contradiction from certain *assumptions* or hypotheses within a logical calculus: on the contrary, such a derivation is essential in proofs conducted by means of *reductio*. Nevertheless, if a contradiction is obtained as a *theorem* within a theory whose underlying logic includes something like (PS), or (PS_i), or ($\neg E$), calamities follow: the system allows us to prove anything, so it is deductively useless. A single contradiction produces an *explosion*. We can therefore suggestively call *explosive* any logic within which (PS) holds in some form or other. (PS) is itself called Explosion, mainly by those who object to it and to the logics including it.

So far, I have been giving a mainly proof-theoretic description of the situation. However, intuitively enough, a logic can be characterized as explosive also semantically, via its logical consequence relation. Therefore, an explosive logical consequence is one in which, for any α and β:

$$\{\alpha, \neg\alpha\} \vDash \beta.[6]$$

Given the T-schema, via replacement of equivalents (PS_i1) becomes:

(PS_i2) $T(\ulcorner\alpha\urcorner) \wedge T(\ulcorner\neg\alpha\urcorner) \to T(\ulcorner\beta\urcorner),$

---

[6] See e.g. Priest and Routley (1989c): 151; Priest (2001): 67, (2002c): 288; Beall (2004a): 3.

i.e., from a true sentence with a true negation follows the truth of anything. The antecedent of (PS$_i$2) is a semantic (C2b)-contradiction, therefore we can also say, via (Neg1):

(PS$_i$2) $\quad T(\ulcorner\alpha\urcorner) \wedge F(\ulcorner\alpha\urcorner) \to T(\ulcorner\beta\urcorner)$,

if something is both true and false, then everything is true. *Trivialism* is often referred to semantically, as the position according to which everything is true. And this seems to be unacceptable if anything is. Any inconsistent theory is trivialized by an underlying explosive logic. The main problem of any logic that aims at serving as the basis for inconsistent theories, therefore, is to avoid explosion. Let us say that a logic satisfying this (still non-specific) requisite meets the *Weak Condition of Non-Triviality* (we shall soon understand in which sense this condition is "weak"). This is what *paraconsistent logics* aim at.[7] We can therefore give a general characterization of paraconsistency by saying: a logic is paraconsistent iff it is not explosive.[8]

5.1.3 "...Understood By Every Thinking Man"

I said that the Pseudo-Scotus can be taken as an axiom or as a primitive rule. However, (PS) can also be *derived* in various ways, and the point is that the logical principles from which it follows appear to be all quite intuitive. Karl Popper used the proof of (a version of) (PS) in his essay *What is Dialectics?*, in order to argue against Hegel and Marx's dialectics that any theory which accepts contradictions is doomed to triviality. Popper presented the argument as a fact that deserves to be "known and understood by every thinking man".[9] An inessential variant of Popper's proof is often referred to as "Lewis' proof", or "Lewis' independent argument", having been produced by C.I. Lewis with respect to his strict

---

[7] The term "paraconsistent", taken as meaning "beyond the consistent", is due to the philosopher F. Miró Quesada (see Priest, Routley and Norman (1989): xx). For some history, see Arruda (1989).

[8] Priest's definition is: "Paraconsistent logics are those which permit inference from inconsistent information in a non-trivial fashion" (Priest (2002c): 287). I will come to the issue of *information* in due course.

[9] See Popper (1969): Ch. 15.

implication.[10] A natural deduction proof of (PS$_i$1) I shall often refer to is the following:

$$
\begin{array}{llll}
(1) & 1 & \alpha \wedge \neg\alpha & \text{Ass} \\
(2) & 1 & \alpha & 1, \wedge E \\
(3) & 1 & \neg\alpha & 1, \wedge E \\
(4) & 1 & \alpha \vee \beta & 2, \vee I \\
(5) & 1 & \beta & 3, 4, \text{DS} \\
(6) & & (\alpha \wedge \neg\alpha) \to \beta & 1, 5, \to I
\end{array}
$$

($\vee$I) is Disjunction Introduction, or Addition:

$$\frac{\alpha}{\alpha \vee \beta} \qquad \frac{\beta}{\alpha \vee \beta} \quad (\vee I)$$

(DS) is Disjunctive Syllogism or *modus tollendo ponens*:

$$\frac{\alpha \vee \beta,\ \neg\alpha}{\beta} \quad (\text{DS})$$

The last step of the proof is an application of Conditional Introduction, or Conditional Proof, i.e.:

$$[\alpha]$$
.
.
$$\frac{\beta}{\alpha \to \beta} \quad (\to I)$$

([α] indicating, as usual, that the assumption is discharged). We met the remaining rules, namely, Assumption and Simplification, in § 2.1.3.

What could one question in the derivation? It is intuitionistically valid: Strong Double Negation, $\neg\neg\alpha \to \alpha$, and the (LEM), $\alpha \vee \neg\alpha$, have not been employed. Conditional Proof mimics within natural deduction (a half of) the Deduction (Meta)Theorem, i.e.:

---

[10] In Lewis and Langford (1932).

(DT) $\alpha_1, \ldots, \alpha_n \vdash \beta \Leftrightarrow \alpha_1, \ldots, \alpha_{n-1} \vdash \alpha_n \to \beta$.

Nevertheless, even if we adopted a logic without (DT), we would still reach step 5, having (PS) as a derivable sequence, or a derived rule, etc. Therefore, we are left with few options: reject Simplification, Addition, or the Disjunctive Syllogism. *Prima facie*, the first two alternatives seem to be out of question. ($\wedge$E) and ($\vee$I) capture a part of the truth-functional meaning of conjunction and disjunction: if a conjunction is true, its conjuncts taken separately definitely are, too; and if a sentence is true, it will remain true for sure when embedded in a disjunction, which fact suffices to make the disjunction true.[11] If we straightforwardly reject both (PS) and (DS), though, we obtain Johansson's minimal logic:[12] we stick to the idea that $\neg \alpha$ is $\alpha \to \bot$, but we drop the Scotist view of $\bot$. Now, minimal logic has its own reputable models: we have completeness proofs of the minimal calculus with respect to semantic structures which, besides invalidating the (LEM), provide counterexamples to (PS) and (DS). These are possible worlds structures including so-called *non-standard* or *non-normal* worlds. I shall talk of non-normal worlds at length in the following. A non-normal minimal world satisfies all negated formulas, that is to say, for any α it makes both $\neg \alpha$ and $\neg\neg \alpha$ true. Therefore, all so-called *strong* contradictions of the form $\neg \alpha \wedge \neg\neg \alpha$ hold. Since the negation of any formula is minimally derivable from a contradiction, i.e., since the Minimal Pseudo-Scotus holds:

(PS1$_M$) $\neg \alpha \to (\alpha \to \neg \beta)$,

as Routley and Priest claim, even if minimal logic is "technically paraconsistent", it is not "*interestingly paraconsistent*".[13] Even if it is not the case that anything follows from a contradiction, an entire collection of

---

[11] A logic invalidating Addition is Bochvar's three-valued logic (we met it in § 3.1.1). The system is based on the so-called weak Kleene three-valued tables, and the value Indeterminate has the characteristic of being "infectious": a single jot of rat's dung spoils the soup, that is, any formula with a sub-formula taking the value Indeterminate is itself indeterminate (see Haack (1978): 207). Therefore, α can be true and $\alpha \vee \beta$ indeterminate, if β is indeterminate. Since Indeterminate is not a designated value, ($\vee$I) is invalid (it does not preserve the designated values). On the other hand, Bochvar's logic is explosive, not satisfying the Weak Condition of Non-Triviality (see Bremer (2005): 35).

[12] See Johansson (1936).

[13] Priest and Routley (1989c): 156.

merely syntactically individuated formulas (namely, negations) does. Sometimes this is expressed by claiming that minimal logic is *partially explosive*, or that it delivers partial triviality in the presence of contradictions.[14] Consequently, some authors, such as Diego Marconi, simply avoid including minimal logic into the paraconsistent club:

> Presumably, the idea is that a paraconsistent system should not justify any derivation which appeals – in one way or another – to the alleged 'untenability' of contradiction: it ought not to be possible to derive a formula from a contradiction *just because* it is a contradiction.[15]

5.2 Disjunctive Syllogism and the Conditional

We have to try a different strategy. In most (though, admittedly, not all) paraconsistent approaches the Lewis-Popper proof is blocked by rejecting the Disjunctive Syllogism. We will deal with this topic mainly when we come to talk of relevant logics: relevant logicians have most thoroughly investigated the point, and most strenuously argued against (DS). We need a small preview, though, which should allow us to understand why the rejection of (DS) has major consequences for the semantics of the conditional.

Paraconsistency does not have to do with the conditional as such, but it is no accident that many paraconsistent logics have non-standard approaches to the conditional. A good reason to reject (DS) is *precisely* the admission of contradictions. Suppose that both α and $\neg\alpha$ are true. Given ($\vee$I), this makes $\alpha \vee \beta$ true too, even if β is false. Then (DS) is a bad rule of inference, quite simply because *it does not preserve truth*: it can lead us from the true premises $\alpha \vee \beta$ and $\neg\alpha$ to a false conclusion, β. I will discuss this line of argument in the following. In the meantime, let us see what follows for the conditional. Is it well known that within classical logic one can define the material conditional as follows:

---

[14] See Carnielli and Marcos (2002): 29-30.

[15] Marconi (1981): 409. In a funny Chapter of *On the Principle of Contradiction*, on the other hand, Łukasiewicz envisages a community of individuals who follow some sort of minimal logic and take any negated sentence as true. They believe, therefore, that the sun does not shine, that man does not die, that two and two is not four, etc., even when the sun shines, people die, and numbers still behave (see Łukasiewicz (1910a): Ch. XVI). Łukasiewicz tries to show how such a community could still reason deductively and inductively, and even build a non-Aristotelian scientific world view.

(Df→) $\alpha \to \beta =_{df} \neg\alpha \vee \beta$.[16]

A material conditional $\alpha \to \beta$ rules out only that its antecedent holds when its consequent does not, which means precisely that either it is not the case that α, or it is the case that β. Now, given Double Negation (and replacement), (DS) is equivalent to:

$$\frac{\neg\alpha \vee \beta, \ \alpha}{\beta}$$

which, given (Df→), is nothing but *modus ponens* (or Detachment, or Conditional Elimination):

$$\frac{\alpha \to \beta, \ \alpha}{\beta} \quad (\to E)$$

Now, it has sometimes been proposed that certain logical paradoxes be faced by means of some extreme weakening of the logic, leading even to a rejection of *modus ponens*.[17] And, as we shall see, paraconsistent logics sometimes adopt decidedly non-standard interpretations of logical operators. Nevertheless, (→E) seems to express *the* basic inferential feature of the conditional. We can translate the point in terms of truth-preservation: a minimal condition for something to be a conditional is that it preserves truth forwards, that is, if $\alpha \to \beta$ is true, then if α is true, the truth has to be transmitted to β (whether a conditional should preserve falsity backwards is another issue, to which I will come in due course). If a connective does not play the game of *modus ponens*, it does not seem to be a *conditional* in an intuitively acceptable sense. Because of this, any paraconsistent logic with *modus ponens* seems to be forced to reject (Df→) or, equivalently, it has to deny that the connective that is captured by that definition is detachable. A distinctive account of the (real) conditional, consequently, should be specified, independently of that of dis-

---

[16] This may not be the case with some non-classical logics, within which logical operators are not inter-definable, such as intuitionistic logic.

[17] See e.g. Fitch (1953).

junction. Following Manuel Bremer, let us call this methodological constraint the *Modus Ponens Condition*.[18]

5.3 "Change of Logic, Change of Subject"

The above concern on the non-negotiable semantical and/or inferential features of the conditional leads us directly to a very common topic within the debate on paraconsistent logics, having to do with the famous Quinean slogan: "Change of logic, change of subject". Non-classical logics are notoriously under the suspicion of trading on concealed alterations of the meanings of logical vocabulary. This holds particularly for paraconsistent logics, since those who question in some way or other such a fundamental principle as the (LNC) typically endorse them. When someone says: "For some sentence α, both α and not-α are true", one wonders what is meant by "true",[19] and, of course, by "not":

> The fact that a logical system tolerates *A* and *~A* is only significant if there is reason to think that the tilde means 'not'. Don't we say 'In Australia, the winter is in the summer', 'In Australia, people who stand upright have their heads pointing downwards', 'In Australia, mammals lay eggs', 'In Australia, swans are black'? If 'In Australia' can thus behave like 'not' […], perhaps the tilde means 'In Australia'?[20]

It is worth noting that, when Quine made his point in *Philosophical Logic*, he was targeting exactly paraconsistent logics:

> To turn to a popular extravaganza, what if someone were to reject the law of non-contradiction and so accept an occasional sentence and its negation as both true? An answer one hears is that this would vitiate science, Any conjunction of the form 'p . ~p' logically implies every sentence whatever; therefore acceptance of one sentence and its negation as true would commit us to accepting every sentence as true, and thus as forfeiting all distinction between true and false. […]
> My view of the dialogue is that neither party knows what he is talking about. They think that they are talking about negation, '~', 'not'; but surely the notion ceased to be recognisable as negation when they took to regarding some

---

[18] See Bremer (2005): 36-9.
[19] See e.g. Slater (1995).
[20] Smiley (1993): 17.

conjunctions of the form 'p . ~p' as true, and stopped regarding such sentences as implying all others. Here, evidently, is the deviant logician's predicament: when he tries to deny the doctrine he only changes the subject.[21]

Philosophers often disagree on the content of basic logical and metaphysical concepts (such as *identity, existence, necessity*, etc.), or on the validity of some very basic principles of inference (such as Contraposition, *reductio ad absurdum*, or Disjunctive Syllogism). It is well known that this kind of discussion often faces an impasse, or seems to turn into a hard conflict of intuitions. This may be due, among other things, to the fact that we cannot examine such concepts as *predication, negation*, etc., without using them. It is very difficult to establish when some party or other begins to beg the question, and it is not an easy issue whether a non-standard explanation of a basic notion involves a real disagreement with a classical account of *that* notion, or its principles simply describe a different thing using the same name or symbol. Is intuitionistic negation the real negation, or does it simply mean, though typographically identical, something else than the classical one? Do non-truth-functional theories of conjunction and disjunction, such as supervaluationism and non-adjunctive logics[22] describe *conjunction* and *disjunction*? Authors like Michael Resnik consider at least some of these puzzles simply unsolvable.[23]

5.3.1 Three Battlefields for Intuitions

One may find three major (variously overlapping) battlefields for conflicting intuitions, when we are dealing with the meaning of the logical vocabulary.

(1) First, we have a very general issue concerning standard formal semantics. As is well known, the recursive clauses expressing truth conditions for sentences may be taken as characterizing (assuming that Tarski was right) the notion of truth for the object-language (the characterization

---

[21] Quine (1970): 81.

[22] We will meet non-adjunctive logics in the following Chapter.

[23] "I take a dim view of the idea that revising our logic entails using so-called logical words with new meanings. Suppose that until now my mathematical proofs used non-constructive principles, but now I announce that I will restrict myself to constructively acceptable proofs. Have I revised my logic, while continuing to mean the same by 'not' and 'or' or have I decided to use those words with a different meaning? I don't perceive a fact of the matter here." (Resnik (2004): 180).

is materially adequate, Tarski taught us, if we can infer from it any instance of the T-schema). Alternatively, one may take the formal recursion as giving the meanings of the logical constants. However, it seems that we cannot have it both ways. By paying in an *independent* understanding of the logical vocabulary, we may buy a characterization of truth. Conversely, by paying in an independent understanding of the concept of truth, we may buy a characterization of the logical vocabulary. Such a cross-dependence[24] is one of the first sources of equivocation and question-begging cross-charges in discussions on deviant logics that seem to provide a non-standard account of logical vocabulary. Did they begin with an alteration of the meaning of connectives and/or quantifiers, or did they move from a change in the notion of truth (probably supported by different metaphysical intuitions)?

Some authors reject the idea that we can go from truth conditions to the meanings of the logical constants. For instance, Michael Tye observed that homophonic semantic clauses presuppose that we grasp the meaning of the connective used in the metalanguage, in order to understand truth conditions.[25] One may simply reply that recursive clauses are not necessarily homophonic: a typical choice negation can be defined, as in § 2.4, via two clauses:

(S¬1) $T(\ulcorner \neg \alpha \urcorner) \Leftrightarrow F(\ulcorner \alpha \urcorner)$

(S¬2) $F(\ulcorner \neg \alpha \urcorner) \Leftrightarrow T(\ulcorner \alpha \urcorner)$,

and, as we can see, no negations appears in the metalanguage here. On the contrary, one may conjecture that to understand how "¬" is characterized by (S¬1) and (S¬2) for the object-language we have to know something about what the truth and falsity predicates mean in the metalanguage.

---

[24] On which see Copeland (1986) – an essay I will deal with in later Chapters.

[25] Tye draws the consequences of this with reference to disjunction: "It is […] a mistake to suppose that the truth-conditions for disjunctive sentences analyse the meaning of the term 'or'. Rather it is because 'or' means what it does, that the truth-conditions obtain. [fn. 24:] One who lacks the concept of disjunction, for example, will not come to understand it by being shown the truth-conditions for disjunctive sentences. Rather, the purpose of a formal statement of truth-conditions is to explain rigorously how the truth-value predicates are to be applied" (Tye (1990): 547).

(2) A second source of disagreement comes from the fact that we currently have (at least) two competing characterizations of the logical vocabulary itself: the one in terms of truth values (via Tarskian clauses or truth tables), and the Gentzen-style presentation in terms of introduction/elimination rules (just like when we claimed that a connective is a conditional only if the rule of Detachment, or *modus ponens*, holds for it). An author may therefore consider the fact that some commonly accepted pattern of inference involving a given connective fails within a system as a decisive sign that something has gone wrong with it. Her *opponens* may reply that the fact that a truth-table presentation yields the "intuitively expected" truth values provides sufficient evidence that we have hit the target, or even question the inferential approach in general (*das Tonkproblem* easily comes to mind).[26] It is sometimes said that the inferential account provides the constructive meaning of logical vocabulary, whereas the truth-table account gives us the classical meaning. However, as is well known, we may have perfectly acceptable natural deduction presentations of classical logic and, *vice versa*, truth-table presentations of non-classical (e.g., many-valued) logics.

(3) A third source of disagreement comes from the issue whether sentential connectives (or which among them) should be truth-functional. Supervaluational treatment of disjunction is paradigmatic. Suppose both α and β come out neither true nor false in some evaluation, because different ways of filling the gaps due to vague predicates, or non-denoting singular terms, yield different truth values. Then, it may be the case that $\alpha \vee \beta$ is truth-valueless, but it may also be the case that it comes out true – particularly, if β is $\neg \alpha$. According to some, such a failure of truth-functionality is a sign that supervaluationism misses the point of the meaning of disjunction.

It is clear that the three kinds of clash are variously intermingled.[27] For instance, the features of supervaluationism can produce a battle of intuitions, not on the concept *disjunction*, but, again, on the concept *truth*. Since supervaluational semantics validates the Law of Excluded Middle while dismissing the Principle of Bivalence, it has to reject the inference from the former to the latter. As we have seen in § 1.2, this amounts (mi-

---

[26] See Prior (1960); Belnap (1962). On these issues, also see Haack (1978): Chs. 3 and 11.

[27] See Haack (1978): Chs. 11 and 12.

nor provisos apart) to a rejection of a half of the T-schema, which has been taken as a minimal condition on a truth predicate.

5.3.2 The Avoid Overload Condition

The following Chapters will show us how the three clashes of intuitions recur in the discussion on most paraconsistent logics. In order to keep the debate under control, Manuel Bremer has recently proposed, as a criterion for evaluating paraconsistent logics, the two following methodological criteria:

> The other [i.e., different from the conditional] usual connectives (like negation, conjunction and disjunction) may not be redefined in a way which is at least as implausible as the paradoxes of material implication. This we call the "Extensionality condition". [...]
> Rules of deduction we commonly associate with the conditional (or the extensional connectives [...]) are not to be given up lightly (for example the transitivity of that connective). This we call the "Minimal Damage Condition". It allows of degrees in meeting it.[28]

Without doubt, these are not absolutely clear-cut requirements, not only because they can be met in different degrees. They mirror, though, our (equally non-clear-cut) intuitions on what can be more or less implausible, or what can be "given up lightly" within logic. I think the two conditions can be taken together, since they conjunctively amount to the request that we make as small changes as possible with respect to the mainstream (truth-functional and/or inferential – via "rules of deduction") treatment of the logical vocabulary. Therefore, I will label them as a unique *Avoid Overload Condition*. This can be made more precise by reference to the remaining conditions: the abandonment of reliable inferential patterns, reasonable semantic clauses, etc., must be minimal *compatibly* with the fulfilment of the other methodological constraints (think about Asimov's Laws of Robotics, the latter law being mandatory for the robot insofar as it does not conflict with the former).

As we will see, some paraconsistent logics unquestionably appear to violate the Avoid Overload Condition. On the other hand, it is equally clear that *something* in a paraconsistent logic has to behave in a non-standard fashion – otherwise we would just have classical logic, with its explosive nature. Evolution requires some departure from tradition.

---

[28] Bremer (2005): 39.

5.4 The Curry Paradox

An interesting paraconsistent logic should allow us to develop inconsistent *theories* based on it. We want a given (naïve-like) set theory, or a given semantic theory for a semantically closed language, not to be trivialized by the underlying logic because of their being inconsistent. This entails that the logic in question should respect a stronger requisite than the one expressed by the Weak Condition of Non-Triviality. In order to understand why, we have to take into account the sentence producing the Curry paradox I mentioned in § 2.1.1, which goes thus:

(1) If (1) is true, then β.

Formally, it is a sentence λ_2, such that:

$(FP_{\lambda 2})$ $\lambda_2 \leftrightarrow (T(\ulcorner \lambda_2 \urcorner) \rightarrow \beta)$.

Now, suppose that $(FP_{\lambda 2})$ is derivable within a semantically closed theory. Then, by means of the following argument:

(1) $\lambda_2 \leftrightarrow (T(\ulcorner \lambda_2 \urcorner) \rightarrow \beta)$ $(FP_{\lambda 2})$
(2) $T(\ulcorner \lambda_2 \urcorner) \leftrightarrow \lambda_2$ (T) (with $\alpha = \lambda_2$)
(3) $T(\ulcorner \lambda_2 \urcorner) \leftrightarrow (T(\ulcorner \lambda_2 \urcorner) \rightarrow \beta)$ 1, 2, Repl
(4) $T(\ulcorner \lambda_2 \urcorner) \rightarrow (T(\ulcorner \lambda_2 \urcorner) \rightarrow \beta)$ 3, Df$\leftrightarrow$
(5) $T(\ulcorner \lambda_2 \urcorner) \rightarrow \beta$ 4, Contr
(6) $(T(\ulcorner \lambda_2 \urcorner) \rightarrow \beta) \rightarrow T(\ulcorner \lambda_2 \urcorner)$ 3, Df$\leftrightarrow$
(7) $T(\ulcorner \lambda_2 \urcorner)$ 5, 6, $\rightarrow$E
(8) β 5, 7, $\rightarrow$E

the theory would be trivialized (and it is easy to build a version of the argument that uses set-theoretic notions, instead of semantic ones). The peculiarity of the Curry paradox is that neither negation, nor the falsity predicate, appears in it; it delivers triviality without using any version of the horrible *ex falso quodlibet*. And the remaining logical assumptions of the proof are quite modest. Besides using an instance of the T-schema (which we certainly want to keep in a semantically closed theory), the usual definition of the biconditional, and replacement of equivalents, the proof employs only two rules concerning the conditional: Detachment or *modus ponens*, and, at line no. 5, the following one, usually called Contraction, or Absorption:

$$\frac{\alpha \to (\alpha \to \beta)}{\alpha \to \beta} \quad \text{(Contr)}$$

Therefore, the logic underlying an inconsistent set theory or semantic theory cannot include both *modus ponens* and Contraction (as a rule, or as the corresponding principle: $(\alpha \to (\alpha \to \beta)) \to (\alpha \to \beta)$). *Modus ponens*, though, was claimed to be a minimal inferential constraint on a real conditional. Therefore, we are forced to abandon (Contr).[29] This is not such a drastic move as it may appear *prima facie*, since various quite intuitive semantics for the conditional provide counterexamples to (Contr) independently of any topic concerning paraconsistency: for instance, semantics for linear logic, some continuum-valued semantics for fuzzy logics, etc. Let us say, then, that a logic respecting such a proviso conforms to the S*trong Condition of Non-Triviality*. We should pay attention to what this means: a logic can avoid Explosion, and therefore join the paraconsistent fellowship, even if it does not fulfil this requirement. In this case, though, it is defective from the point of view of its *applications*, that is to say, of what we are allowed to build on top of it. We will see in the following that there are logics which satisfy the Weak, but not the Strong Condition of Non-Triviality.

5.5 The Classical Recapture

Let us sum up. We have listed the following methodological constraints in the light of which paraconsistent logical theories should be evaluated.

(1) Weak Condition of Non-Triviality: inconsistency, or the admission of contradictions, must not trivialize the system by means of some version or other of (PS) or *ex falso*, that is to say: our logic must not be explosive.

---

[29] In its strongest form, the Curry paradox forces us to dismiss a principle usually called Assertion, i.e., $\alpha \wedge (\alpha \to \beta) \to \beta$. The version involving Assertion has been proposed by Meyer, Routley and Dunn (1979).

(2) Modus Ponens Condition: our logic has to include a detachable conditional (i.e., a conditional for which *modus ponens* holds), different from the standard material conditional.

(3) Avoid Overload Condition: logical operators (and the rules of inference governing them) should deviate as little as possible from their common treatment, compatibly with the need to respect the remaining conditions.

(4) Strong Condition of Non-Triviality: inconsistent theories must not be trivialized by our underlying logic because of some version or other of the Curry paradox.

A quite general requirement can be added to these four constraints. Paraconsistent theorists have labelled it as the issue of the *classical recapture*.[30] The constraint is a broadly epistemological one (the Avoid Overload Condition being, as it were, its strictly logical counterpart). Contemporary epistemology has widely studied the problem of the preservation of *problem solving ability*, and we can put the point by means of typical Lakatosian notions.[31]

The discovery of set-theoretic paradoxes and the difficulties of their Kantian solutions adopting standard logic; the incompleteness of formal arithmetic granted by Gödel's results; the paradoxes of material conditional: paraconsistent logicians regard all these phenomena as decisive signs that the research program of classical logic – taken as the program inaugurated by Frege and Russell – is in a degenerating stage. Typically, a degenerating research program is just defending itself, so to speak: no interesting new results and discoveries appear; only minor, specialized, and almost esoteric, developments of what has been on the market for too long are produced; the anomalies emerged within the program remain unsolved; and so on. Paraconsistency, given its logical radicalism and the wide range of its (actual and predicted) applications, presents itself as a research program alternative to the classical one, and capable of beginning a stage – to use Kuhn's categories – of "revolutionary science". In Priest's own words:

> We have seen that by both Kuhn's and Lakatos' standards, logical theory is in a state of crisis, i.e. a time when the conditions are objectively ripe for revolution. Now if during a crisis a new theory emerges which a) solves some

---

[30] See for instance Routley (1979b): 892ff; Priest (1987): Ch. 8, and (2002c): 347ff.
[31] See Lakatos (1970).

of the anomalies of the old paradigm, b) preserves a substantial amount of the problem-solving ability of the old paradigm, and c) poses new open problems with promise of solution (i.e. promises to be the basis of a new and fruitful research program), a revolution is liable to occur. But such a theory is already emerging: paraconsistent logic.[32]

This should be obtained, according to Priest, by rejecting the general, unconditioned validity of the (LNC), or the mutual exclusiveness of truth and falsity. What we are interested in now is point (b): *preserve* the good achievements of the research program to be replaced. A philosophically interesting paraconsistent strategy should illustrate in detail how its logical theory can recapture the virtues (results, proof procedures, mathematical theorems, etc.) of the classical paradigm. It should explain how we can regain fundamental theorems of classical mathematics whose proof uses rules of inference that are (typically) paraconsistently invalid, such as the Disjunctive Syllogism: (DS) is a case of resolution usually adopted, for instance, within standard theorem-provers. We will see that this kind of explanation is spelt out in different (and not always satisfactory) ways in the various paraconsistent approaches. Most of the authors, anyway, felt they had something to say on the classical recapture. And it is a general opinion among paraconsistent logicians, that reasoning in accordance with classical logic is legitimate in consistent situations.

5.6 Weak Paraconsistency, Dialetheism, and Something in the Middle

So far I have been talking of paraconsistency in a broad fashion. I shall end the Chapter, though, by introducing an internal distinction of great importance. It is common in the literature to discern between (at the very least) two degrees of paraconsistency, namely, *weak* and *strong* paraconsistency.[33] The weak variant is sometimes introduced (albeit a bit misleadingly) as a *proof-theoretic* one,[34] being based on the remark that there are inconsistent but interesting and non-trivial theories also without having to admit the *reality* of contradictions, or the possibility of *true* contradictions. We met some such theories: naïve set theory and "intuitive se-

---

[32] *Ibid*: 140-1.
[33] See for instance Beall (2004a); Bremer (2005): *Introduction*.
[34] See Priest and Routley (1989c): 151ff.

mantics", for instance, are inconsistent. The unrestricted Abstraction Principle of naïve set theory entails, among other things, that Russell's set both belongs and does not belong to itself, $R \in R \wedge R \notin R$. Even supporters of consistency frequently – though grudgingly – admit that the naïve theory does capture our natural intuitions on what a set is (that is to say, the extension of an arbitrary condition). It follows that our intuitions are inconsistent. However, the theory is intuitively not trivial: our intuitions also lead us to *reject* numerous claims concerning sets, e.g., that the singleton of the empty set is a member of that set, $\{\varnothing\} \in \varnothing$. Likewise, intuitive semantics includes the semantic paradoxes, but it leads us to reject, e.g., that the conjunction of α and β is true iff either α is true or β is true. Other famous theories I shall not deal with in this book have proved useful, despite being born inconsistent, or maybe still being such. The Newton-Leibniz infinitesimal calculus was originally inconsistent. Bohr's atomic model contradicts Maxwell equations, but these are essential to Bohr's theory. Now the latter claims that an electron can orbit around the nucleus of an atom without radiating energy, whereas the former require an electron to radiate energy.[35] Further examples will easily come to mind to anyone who has read Feyerabend. It is therefore natural to suspect that the logic underlying these theories should not be explosive.

An essential motivation for weak paraconsistency is that almost anyone would agree on the fact that our *system of beliefs* is inconsistent – so this position is also called *doxastic* paraconsistency.[36] As John Woods said, "inconsistency is no *rara avis*, not the defection from logical rectitude of an occasional abstract theory. Inconsistency routinely dogs us in the management of our beliefs, the manipulation of our memories, and the organization of our desires". And "the [orthodox logician's] canons are not, or not always, the ones that people resort to as strategies for inconsistency-*management*. The last thing that a competent reasoner would do is to submit himself to *ex falso* as such a strategy".[37] Of course, there is no *algorithm* for consistency; but we cannot just stop and wait until the inconsistency is removed before we draw any further conclusion. This entails that any explosive logic is structurally unsuitable if we want to represent most ordinary inferential practices. A weak paraconsistentist can therefore motivate her rejection of (PS) by observing that inferences are

---

[35] See Brown (1993b).
[36] See Mares (2004): 265.
[37] Woods (2003): 95-6.

commonly made from bodies of information containing inconsistencies, but people are clearly not at liberty to conclude anything they like.[38] Typically, though, she assumes paraconsistent models as useful mathematical tools, but refuses the idea that they represent genuine possibilities. The development of a paraconsistent theory is independent of the idea that contradictions can be *true*, or (given a minimally realist view on truth) that there can be inconsistent objects, or states of affairs. In this sense, weak paraconsistency is the more conservative position – the "right wing", as Mares calls it.[39]

Strong paraconsistency constitutes the left wing: it is characterized by the admission of true contradictions, against some version or other of (LNC2), the semantic Law of Non-Contradiction; it is also called *dialetheism* (some prefer to say *dialethism*).[40] The term is due to Routley and Priest, who expressed the idea of a true contradiction by coining the expression *di-aletheia*, "double truth": like "a Janus-faced creature which faces both truth and falsity".[41] In my opinion, dialetheism is philosophically the far more interesting position. Its motivations have been investigated in the previous Chapters, being based upon the idea that some contradictions are *provable*, and *inevitable*: they are entailed by evident facts and analytic principles concerning ordinary language (e.g., its semantic closure) and our thought processes (e.g., ascent to the "transcendental" limits and the Domain Principle). And attempts to avoid them implode

---

[38] "Once we start considering inconsistent sets of beliefs or inconsistent theories, [classical logic] turns out to be completely misguided even as an extensional logic. [...] Indeed, even if the world is consistent and even if inconsistent theories have to be transformed into consistent ones [...], this does not help us out in the meantime. As long as we are not able to rework an inconsistent theory into a consistent one, or to replace it by a consistent one, rejecting the inconsistent theory leaves us with no theory at all in the domain" (Batens (1980): 197). "Anyone working as a knowledge engineer, assembling and managing knowledge databases, will be perfectly aware that gathering inconsistent information is the rule rather than the exception. And again, either if you assume, by some sort of methodological requirement, inconsistent theories to be problematic [...] or not [...], this does not prevent you from also assuming them to be, in general, quite *informative*, and wanting to *reason* about them in a sensible way" (Carnielli and Marcos (2002): 4)

[39] See Mares (2004): 265.

[40] Beall (2004a): 6, also distinguishes between a strong paraconsistentist and a dialetheist: the former admits "real possibilities" in which contradictions can be true; the latter makes the final step, and accepts true contradictions, that is to say, contradictions that are true in *this* world.

[41] Priest, Routley and Norman (1989): xx. This has been inspired by a famous passage of Wittgenstein's *Bemerkungen* on the foundations of mathematics.

under the weight of extended contradictions, revenge liars, and the other difficulties we have examined at length.

Now, dialetheism is not automatically committed to a specific conception of truth. Nevertheless, if we accept even a mild form of *realism*, the truth of some contradictions, i.e., the violation of (LNC2), entails the existence of inconsistent *objects* and/or *states of affairs* (those that make the contradictions true), in violation of (LNC3), the metaphysical principle.[42] I shall discuss issues concerning realism at length in the following Chapters. For the purposes of this book, though, we do not need a technical notion of realism. In the broadly accepted use, to be a realist (as opposed to an idealist, or an anti-realist, or a constructivist) *about* some kind of entities is to maintain that such entities objectively exist apart from, and antecedently to, anyone's thought of them; and, therefore, that our thoughts, beliefs and theories concerning such entities are made objectively true or objectively false by them, apart from what we think of them.

In § 1.2, we followed Aristotle and Łukasiewicz in declaring the equivalence of the ontological and the semantic (LNC). We also noticed, though, that the T-schema, which is the essential tool in order to obtain such an equivalence, is a minimal constraint on the truth predicate that could be assumed independently of a correspondence theory. So there may be room for a further intermediate paraconsistent position, that is to say, an *anti-realist dialetheism* which accepts true contradictions without inconsistent objects or states of affaires as their truth-makers: contradictions are produced by the relationship between language and the world, but they do not inhabit the world as such. This position has been surprisingly little explored in the literature, but I would consider Kroon (2004) and Mares (2004) as initial and very interesting efforts in this direction.[43]

A dialetheist can envisage the possibility of *observable* contradictions. For instance, Priest (1999a) and (2006): Ch. 3, explores the idea that some of them may lurk in certain paradoxes of perception, such as those pro-

---

[42] See for instance Beall (2000b) for a dialetheic version of the correspondence theory of truth, providing a clear example of an explicitly realist dialetheism.

[43] One may also provide an anti-realist or quasi-instrumentalist account of "true" in terms of "told true" or "believed true", and claim that *truth* is nothing more than this. See J.C. Beall's *constructive methodological deflationism*, in Beall (2004b): "after all, if truth is a mere (human) construction, introduced to play a given expressive role, then it is not surprising – indeed, it is likely – that the construction should turn out to be inconsistent" (208).

voked by Escher-like drawings. Dialetheists seem to disagree on this point, though: some of them believe it is safer to limit the possibility of true contradictions within the realm of semantics and set theory, therefore having at best (at worst?) *abstract* inconsistent objects, such as sets.[44] Another way to let the possibility of inconsistency spread into the empirical world is to give a dialetheic interpretation of the phenomenon of vagueness. I will come to all these issues in due course.

Should we favour weak paraconsistency or dialetheism? According to Bremer:

> If dialetheism claims that some contradictions can be proven and weak paraconsistency denies the (ultimate) truth of contradictions that can only mean that weak paraconsistency has to reject the systems and logics that allow these proofs.
> That might mean weak paraconsistency has the shortcomings of both worlds: having neither the strength of [standard first-order logic] nor semantic universality (or other benefits of dialetheism).[45]

According to Woods, on the contrary:

> Although weakly paraconsistent logics will accommodate a dialetheic truth value, they do not require it. If dialetheism is true, these are candidate logics for it; but dialetheism need not be true for such logics to have a competent rationale. Therefore [...] I think that it must be said that weakly paraconsistent logics are in general better models of strategies for the management of cognitive dissonance than are their dialetheic cousins.[46]

Woods is referring mainly to the so-called *preservationist* logical approaches. I shall say something about them at the end of the following Chapter. Many non-explosive logics can work both in a weakly and in a strongly paraconsistent environment. Furthermore, we will see how, in some cases (e.g., relevance logics), paraconsistency has been a by-product within research programs whose initial aim was to tackle different problems. For the time being, let us just bear in mind that paraconsistency and dialetheism are not the same thing. The latter requires the for-

---

[44] Issues of this kind are widely discussed in J.C. Beall's works: see Beall (2000), (2001a); Beall and Colyvan (2001).
[45] Bremer (2005): 16.
[46] Woods (2003): 112.

mer, i.e., it requires non-explosive logics, on pain of lapsing into triviality for mere logical reasons; but the converse may not be inevitable. All of this, in any case, does not rule out the danger that dialetheism lapses into some form of triviality for *non*logical reasons. I will come to this, too, in due course.

6 Non-Adjunction and Impossible Worlds

> Impossible is possible tonight.
>
> Smashing Pumpkins, *Tonight, Tonight*

6.1 Discussive Logic

Let us label as "non-adjunctive approaches" a group of paraconsistent logical theories, having in common the thought that one can keep detonation via contradictions under control by rejecting Adjunction. The idea behind Adjunction, as we now know, is that if sentences $\alpha_1, \ldots, \alpha_n$ separately taken hold, then their conjunction $\alpha_1 \wedge \ldots \wedge \alpha_n$ holds, too. Within non-adjunctive theories, the distinction introduced in § 1.1.1 between contradictions taken in the *distributive* and in the *collective* sense comes into play. "Adjunction", too, is actually spoken of in many ways, so that its rejection also takes on different shapes in this kind of approach.

Let us begin with an old and pioneering essay by Stanislaw Jaskowski, named *Propositional Calculus for Contradictory Deductive Systems*. Many authors in the paraconsistent tradition have been inspired by it, or at least by its declaration of intents:

> The task is to find a system […] which: 1) when applied to the contradictory systems would not always entail their overcompleteness, 2) would be rich enough for practical inference, 3) would have an intuitive justification.[1]

The opportunity of admitting contradictions into a system is motivated by the unsatisfactoriness of the alleged hierarchical solutions to semantic and set-theoretic paradoxes (Russell and Chwistek are explicitly mentioned). The target is the "law of overcompleteness", as Jaskowski calls (PS), but, as we shall see, the proposed system avoids only one of the versions of *ex falso*. It is usually called *discussive* (or *discursive*) *system*, and labelled as D_2, but it is also known in the literature as **J**. The intuitive justification for D_2, i.e., requirement no. 3 of Jaskowski's program, is provided by referring the rise of contradictions to *dialogical* situations.

---

[1] Jaskowski (1948) (quoted in Woods (2003): 80); "overcompleteness" corresponds to our triviality.

Suppose we want to represent within a formal system a dispute between different and opposed viewpoints. Examples of the situation would be contradictory testimonies presented in a trial; but also, a database that is sometimes fed with inconsistent information provided by different sources. Such conflicting theses will be introduced as axioms within the system, so it will include reciprocally inconsistent principles. It should nevertheless allow us to draw consequences in a non-trivial fashion.

Now, $\mathbf{D_2}$ is a disguised propositional modal logic. First, a thesis can be included in the system only if it is classically *possible*. The rule of the dialogic game, that is, maintains that if some source of information (witness, or participant in a debate, etc.) x claims (that) α, α is admitted within $\mathbf{D_2}$ iff (classically) $\Diamond\alpha$. It is required that each participant puts forward claims which are at least possible from the point of view of what Jaskowski calls a "impartial referee"; a single source of information, therefore, is assumed self-consistent. The implicit modal flavour surfaces in the definitions of the *discussive connectives* of $\mathbf{D_2}$. Jaskowski defines *discussive implication* (be it $\supset$, whereas $\to$ is the standard material conditional) thus:

(DI) $\alpha \supset \beta =_{df} \Diamond\alpha \to \beta$.

Further connectives are discussive equivalence (be it $\equiv$) and discussive conjunction ($\bullet$):

(DE) $\alpha \equiv \beta =_{df} (\Diamond\alpha \to \beta) \wedge (\Diamond\beta \to \Diamond\alpha)$

(DC) $\alpha \bullet \beta =_{df} (\alpha \wedge \Diamond\beta).$[2]

The fact that discussive connectives are defined by means of the standard ones and the diamond points to an obvious proximity between $\mathbf{D_2}$ and propositional modal logic. A precise relationship with the traditional $\mathbf{S_5}$, in which the discussive system can be easily translated, has been outlined. It is easy to prove by recursion on the complexity of $\square$ that:

(TR) $\vdash_{\mathbf{D2}} \alpha \Leftrightarrow \vdash_{\mathbf{S5}} \Diamond\alpha$,

---

[2] See *Ibidem*. See also the axiomatic presentations proposed by da Costa and Dubikajtis (1968), Kotas and da Costa (1979).

i.e., a given formula is a theorem of $\mathbf{D_2}$ iff its possibility is a theorem of $\mathbf{S_5}$.[3] Thereby, the system gets such nice properties as completeness and decidability. Algebraic models for $\mathbf{D_2}$ directly derived from those of $\mathbf{S_5}$ were proposed.[4] Furthermore, Jaskowski claimed that $\mathbf{D_2}$ also satisfied requirement no. 2 of his program, having quite respectable inferential power. What about no. 1? Within $\mathbf{S_5}$, the following is not a logical law:

$$\lozenge(\lozenge\neg\alpha \rightarrow (\lozenge\alpha \rightarrow \beta)).$$

To have an intuitive counterexample, as Jaskowski points out, it is sufficient to take something possible, but not necessary, for α, and something impossible for β.[5] Correspondingly, given (TR) and the definition (DI) of discussive implication:

$$(PS_d) \neg\alpha \supset (\alpha \supset \beta)$$

(the "implicative law of overcompleteness") does not *non* hold in $\mathbf{D_2}$. Therefore, it is claimed, it is not the case that anything follows within $\mathbf{D_2}$ even though two "inconsistent" theses are taken on board. This should allow contradictory theses to coexist without "overcompleteness". I will discuss to what extent the proposal works in a few pages. Before doing it, though, we have to deal with another non-adjunctive theory.

6.2 Impossible Worlds

6.2.1 Rescher and Brandom's Logic of Inconsistency

Probably the most developed essay in the non-adjunctive tradition is still *The Logic of Inconsistency*, by Nicholas Rescher and Robert Brandom. The sub-title of their book, "A Study in Non-Standard Possible-Worlds Semantics and Ontology" gives us an idea of their approach, which is not a proof-theoretic one: it is entirely dedicated to semantic and ontological issues. Rescher and Brandom do not propose a real non-explosive logic, in the sense of some set of paraconsistent axioms, theorems and/or infer-

---

[3] See Marconi (1979): 281; da Costa and Marconi (1989): 12.
[4] See Kotas (1971) and (1975).
[5] *Ibidem*.

ence rules. They explain that their aim is to question the (LNC) directly in its semantic and metaphysical versions – therefore placing themselves on the side of strong paraconsistency: "our concern is with hard, *existential* inconsistency, not with soft, *epistemological* inconsistency".[6] This makes the book quite interesting, despite its looking a little bit outdated in some of its parts. One of the most fascinating fields of study within paraconsistency, in fact, has to do with the so-called *impossible worlds*. Let us begin, then, with a general overview of the subject.

6.2.2 Impossible Worlds in General

It is not too difficult to envisage worlds in which some fundamental biological or physical law does not hold: for instance, a world where John Kennedy dies both on November 22, 1962, and on August 18, 1967, or where on November 22, 1962 he is both in Dallas and in Las Vegas. Such worlds may be taken as biologically or physically impossible. *Logically* impossible worlds, on the other hand, should be worlds where some logical law does not hold. These deserve the title of impossible worlds *simpliciter*, since logical laws are supposed to be the most general and topic-neutral ones. It is sometimes claimed that to consider worlds where a logical law does not hold is preposterous, precisely because logical laws hold in *all* worlds, by definition. Now the question is, laws of *which* logic?

I have already mentioned non-normal minimal worlds in which the (LEM) and *ex falso* fail. The former also fails in models for intuitionistic logic (we shall have a quick look at them in the following). Now, it seems that we refer to these worlds when we evaluate such conditionals as: "If intuitionist logic were right, then the (LEM) would fail" – which is true; and "If intuitionist logic were right, then *ex falso* would fail" – which is false. A classical logician would consider a world where the (LEM) fails as a strictly impossible world, since she takes classical logic as the correct logic. Nevertheless, anyone who understands intuitionism, or minimal logic, or quantum logic, etc., knows how things would be if one of these logics were correct (assuming that they are not). One does not even have to take impossible worlds as worlds where the logically impossible *happens*. It may be the case that some logical law does not hold in a world, even though no counterexample to it actually inhabits that world (just as

---

[6] Rescher and Brandom (1980): 2.

within a physically impossible world the light speed limit may not hold, even though no object in that worlds actually moves that fast).[7]

What we are interested in now are worlds where *the* logical law, i.e., the (LNC), which holds also minimally and intuitionistically, fails. Now, even those who are not willing to question the general and unconditioned validity of the (LNC) at all may admit that, just as there are various ways the world could be, there are various ways the world could *not* be. According to authors like Greg Restall,[8] if there were no ways the world could not be, it would follow that *anything* is possible.[9] Now take a world, w_1, where angles can be trisected with ruler and compass, but there are no Meinongian square circles; and take a world, w_2, with square circles in it, but in which angles behave wisely. According to some, w_1 and w_2 are two *different* ways the world could not be. This intuitively suggests that the kingdom of the absurd is not like Hegel's night, in which all cows are black. Such worlds may therefore be justifiable objects of philosophical analysis, given that a certain amount of logical structure can be recognized in them: we can reason about them and make discriminations about what goes on within them. The *degree* of structure to be assigned to them is the main point on which impossible worlds theorists diverge. No logical chaos, anyway, is expected. Impossible worlds are nowadays proposed by various authors as a natural extension of possible worlds theories,[10] having useful applications in the study of the notions of propositional content, intentional state, belief management, etc.[11] Rescher and Brandom's pioneering work has been quite influential in this direction.

6.2.3 Non-Adjunctive Worlds

Following the original notation of the essay, let us express the holding/not holding of a state of affairs in a world as follows:

$|P|_w = +$
$|P|_w = -$

---

[7] For this distinction, see Priest (1992): 292, (2001): 171-2.
[8] See Restall (1997) and (1999): 55-6.
[9] A claim some paraconsistent theorists endorse, anyway: see e.g. Mortensen (1989).
[10] For a general introduction, see e.g. Beall and van Fraassen (2003): 86ff. See also Pasniczek (1998).
[11] See the monographic no. 38(1997) of the Notre Dame Journal of Formal Logic, entirely devoted to impossible worlds.

("The state of affairs (described by) sentence *P* obtains/does not obtain in world *w*").[12] Now, as Rescher and Brandom claim, standard possible worlds are exhaustively described by maximal consistent sets of sentences, that is to say, they are constrained by the (LNC) and the (LEM): exactly one of the pair *P*, ¬*P* obtains in *w*.[13] Non-standard worlds are obtained by means of two recursive ontological operations having standard worlds as their base, and called *schematization* and *superposition*. I will symbolize them, respectively, as ∩ and ∪ (not to be confused with the ordinary set-theoretic operations). A schematic world $w_1 \cap w_2$ is a world in which all and only the states of affairs obtaining both in w_1 and in w_2 hold:

$$|P|_{w1 \cap w2} = + \Leftrightarrow |P|_{w1} = + \text{ and } |P|_{w2} = +.$$

An *inconsistent* world $w_1 \cup w_2$ is a world in which all and only the states of affairs obtaining either in w_1 or in w_2 hold:

$$|P|_{w1 \cup w2} = + \Leftrightarrow |P|_{w1} = + \text{ or } |P|_{w2} = +.$$

A schematic world can be, in general, ontologically "underdetermined": we can expect such worlds to host *vague* objects. The vagueness in question is not merely a so-called *de dicto* (semantic or epistemic) fact, but a *de re*, metaphysical phenomenon.[14] Our main interest is in inconsistent worlds obtained via superposition; they should be, dually, "overdetermined".[15] An inconsistent world is an impossible world for sure – ac-

---

[12] Therefore, let us take *P*, *Q*, ... , as propositional parameters (or 0-ary predicates) for a little bit.

[13] See Rescher and Brandom (1980): 3.

[14] "In the case of a schematic world, the situation is not just that *we* don't know whether *P* or its contradictory ~*P*, but that *the world itself is indeterminate* in this regard in its make-up. [...] The situation is one of *ontological underdetermination* – with regard to certain envisageable states of affairs the world is simply 'incomplete.'" (*Ibid*: 5).

[15] "The make-up of the [inconsistent] world embodies a synthesis or fusion of incompatible states of affairs. This can be thought of as taking a number of individually and separately altogether self-consistent worlds, and ramming them together [...]. The crucial facet of the stance we are taking here is that the obtaining status (+ or –) of *P* and that of ~*P* in a[n im]possible world are to be independent of one another. The cardinal doctrine of

tually, the authors prefer to keep calling these worlds "possible", and change the notion of possibility itself: any "feasible target of rational consideration and scrutiny" is possible, and inconsistent worlds definitely are.[16] However, Rescher and Brandom have not been followed by the subsequent literature, where, as we have seen, it is customary to talk just of impossible worlds.

Now, the assignment of +s and –s is not compositional – and the non-adjunctive features of the approach emerge:

> It is crucial to note, however, that while we may have it that $|P|_w = |\sim P|_w = +$ in some inconsistent world w, we shall certainly never have it that $|P\ \&\ \sim P|_w = +$. Our stance is that two mutually inconsistent states of affairs might well both be realized in a non-standard world, whereas a single *self inconsistent* state of affairs can never be realized. Contradictions can be realized distributively but not collectively: *self-contradiction* must be excluded. We shall always have $|P\ \&\ \sim P|_w = -.$[17]

You may notice that non-adjunctiveness affects the underlying *ontology*. Rescher and Brandom want to maintain Adjunction as a rule of inference:

$$\frac{\alpha,\ \beta}{\alpha \wedge \beta}\quad (\wedge I)$$

whereas (a half of) the corresponding truth-functional semantic clause for conjunction fails for inconsistent worlds. That is, the right-left direction of:

(S∧) $T_w(\ulcorner \alpha \wedge \beta \urcorner) \Leftrightarrow T_w(\ulcorner \alpha \urcorner)$ and $T_w(\ulcorner \beta \urcorner)$[18]

---

the analysis is that *the ontological status of P and that of ~P should be seen as strictly independent issues*" (*Ibid*: 6).
[16] See *Ibid*: 4 and 32.
[17] *Ibid*: 7.
[18] We will always adopt the subscript after the truth predicate when we are dealing with a(n im)possible world semantics; so "$T_w(\ulcorner \alpha \urcorner)$" says that α is true at world w.

("$\alpha \wedge \beta$ is true at world w iff α is true at that world and β is true at that world"). Therefore, "we have to do with an unorthodox *semantics*, not an unorthodox *logic*".[19] As a result, the theory only tolerates what the authors call *weak inconsistency*, that is, *distributive* true contradictions:

$T_w(\ulcorner \alpha \urcorner), T_w(\ulcorner \neg \alpha \urcorner),$

but from these such *strong* inconsistencies as:

$T_w(\ulcorner \alpha \wedge \neg \alpha \urcorner)$

never follow.[20] It is easy to understand in which sense, and to what extent, the approach is paraconsistent:

> Our present tactic [...] is to accept $P \& \sim P \vdash Q$, but to block the move from the two mutually contradictory theses P and $\sim P$ to their self-contradictory conjunction $P \& \sim P$. This blockage is not – to be sure – accomplished within *logic* itself (for we shall retain [($\wedge$I)]), but rather in the *semantical* setting of any world-relative or system-relative assertion (as per [(S$\wedge$)]).[21]

Rescher and Brandom are therefore forced to rework the notion of logical consequence. Precisely, anything follows as a consequence of a *collective* contradiction, i.e., $\{\alpha \wedge \neg\alpha\} \vDash \beta$, and logical consequence for single-premise sets of sentences is completely classical. Paraconsistency is obtained only by means of the deviant semantic behaviour of conjunction, i.e., because of the fact that, as we have just seen, $\{\alpha, \beta\} \nvDash \alpha \wedge \beta$.[22]

---

[19] *Ibid*: 18. "*The present approach* – as we have seen – *dispenses entirely with any need to modify the principles of classical logic*. Despite its provisions of a non-standard ontology and a non-standard semantics, nevertheless, at the crucial level of logical machinery, it requires no innovations or renovations whatsoever" (58).

[20] See *Ibid*: 24.

[21] *Ibid*: 18.

[22] See *Ibid*: 16-7. "This failure of the orthodox semantical adjunction principle [...] has the far-reaching consequence that any and every inference rule which involves the combining of distinct premises (i.e., any rule that deploys a series of premises distributively) must itself be treated in its light of these considerations [...]. All multi-premissed inferential principles must be construed in a collective rather than distributive manner" (19-20).

The book proposes, then, several applications of the theory just described, also with deep philosophical reflections. It is claimed that an ontology of (both vague and inconsistent) Meinongian objects can be based on it. In addition, a general non-adjunctive strategy for the solution of semantic and set-theoretic paradoxes, and a theory of identity and individuation for inconsistent objects, are described.[23]

6.3 Non-Adjunctive Troubles

The most common objection to traditional non-adjunctive approaches has to do with the decidedly non-standard behaviour of conjunction. Even though something has to work non-classically in any paraconsistent logical theory, one may doubt whether, within these frameworks, conjunction is *conjunction* (pay attention to the *italics*). One may claim that *conjunction* is nothing but the connective whose meaning is captured by the homophonic clause (S∧) above. The equivalence (S∧) consists in, it is claimed, is not open to discussion, because conjunction is the Boolean functor whose meaning is completely captured by its truth table. To be sure, it is not forbidden to call "conjunction" an operator with deviant truth conditions (it's a free country). Nevertheless, Jaskowski's discussive conjunction, •, defined by means of (DC), has no clear intuitive justification. So it appears that the Avoid Overload Condition stated in § 5.3.2 has been violated.[24]

A second point is that non-adjunctive theories tolerate contradictions only in a very restricted sense. Consider Jaskowski's system, which is connected to a completely classical modal approach, such as that of S_5. Are the "inconsistent discussive theses" actual *contradictions* admitted in D_2? What Jaskowksi calls "conjunctive law of overcompleteness", that is, the discussive counterpart of (PS_i) (the imported version of (PS)):

---

[23] See Chs. 9, 10, 14 and 16.

[24] "First, it is totally opaque why a modal functor such as [◊] should poke its nose into the meaning of ordinary extensional conjunction. [...] Secondly, this approach to conjunction totally destroys the normal relationships between conjunction, disjunction and negation. [...] It is worth saying again that some classical logical relations will have to go paraconsistently. However, the wholesale destruction of the relations normally taken to hold between conjunction, negation and disjunction clearly speaks against discussive conjunction" (Priest and Routley (1989c): 159-60). Similar considerations can be found in Bremer (2005): 42-3.

$(PS_{di})\ \alpha \wedge \neg\alpha \supset \beta$,

is a theorem of **D₂**. (PS_d) and (PS_{di}) are not inter-derivable, unlike their classical counterparts. The intuition is that when we accept two inconsistent claims (α, $\neg\alpha$) put forward by two *different* sources of information (witnesses, disputants, etc.) in a dialogical context, from the point of view of Jaskowski's "neutral referee" we only have that $\Diamond\alpha$ and $\Diamond\neg\alpha$. Given the definition (DI) of discussive implication, the rejection of (PS_d) follows. However, the system would become "overcomplete" if one of the assumed positions turned out to be intrinsically inconsistent. This would amount to accepting something that is or entails a *collective* contradiction – what the dialogic rule excludes. Since anything follows from a collective contradiction, the Weak Condition of Non-Triviality is satisfied only in a rather attenuated sense. As a matter of fact, it seems that we are still dealing with a kind of *parameterisation*: inconsistency is actually circumvented by means of a distinction of respects, namely, the different sources of information or participants in the debate. According to Priest and Routley:

> The main problem with the discursive approach is just that it does not take the [...] dialethic motivation (that there are true contradictions) seriously. Contradictions may be "true" but this amounts to no more than "true in different worlds". Moreover each possible world is as consistent as any classicist could wish: the approach is much too modally based to accommodate inconsistency satisfactorily.[25]

Quite similar considerations hold for Rescher and Brandom's theory: their semantics is inter-translatable with Jaskowski's.[26] They suggest that the instances of the Abstraction Principle producing the set-theoretic paradoxes be split into two halves. Therefore, e.g., for Russell's paradox, we will have $\forall x(x \in R \rightarrow x \notin x)$ and $\forall x(x \notin x \rightarrow x \in R)$, each one included in a separate sub-theory.[27] But:

> This strategy has only the appearance of paraconsistency. In essence it is just a revisionist classical position. For paring an inconsistent theory down to various consistent subtheories is a game that classical set theorists have been

---

[25] Priest and Routley (1989c): 162.
[26] As shown in Priest (2002c): 300.
[27] See Rescher and Brandom (1980): 36ff.

playing for eighty years. The classicist is quite happy with both the above fragments of set theory.[28]

6.4 Truth-Functionality, Sub-Valuations, and the Italics Argument

In spite of the above critiques, Jaskowski's discussive logic is still pursued and developed today, e.g., in Max Urchs' works.[29] On the other hand, Rescher and Brandom's non-standard worlds do seem to take the idea of metaphysical violations of the (LNC) more seriously than Jaskowski's approach. The authors have tried to justify the semantic anomaly of conjunction as mirroring an *actual* anomaly of the worlds: they are "inherently non-truth-functional", that is, there is no recursive way of giving the truth-in-w conditions of a molecular sentence in terms of the truth-in-w conditions of its atomic components. Now, one may see this fact as a good reason to proscribe such worlds. However, one may also appreciate the situation as "a recognition of the difficult facts of life outside the sphere of standardist simplicity".[30]

Such an embryonic motivation for rejecting Adjunction for metaphysical reasons has been recently revitalized and developed by Achille Varzi,[31] within the context of a broader campaign against the exclusively truth-functional conception of the connectives. A logic and its respective semantics, as Varzi observes, always presuppose an *ontology*. If a logical law or consequence has to hold in all circumstances, "a systematic account of what should count as a genuine circumstance (in the relevant sense) is part of what it takes to define a logic, i.e. a theory of logical validity".[32] The view that conjunction cannot but be truth-functional is supported by the fact that ordinary semantics integrates in such semantic clauses as (S∧) an *assumption* on worlds, that is, a constraint on admissible interpretations. Interpretations where "and" or "or" mean something different from the Boolean functors are given no chance, as it were, since the beginning. However, this comes *in disguise*, since "typically, as a matter of standard practice, the semantics of the logical operators is

---

[28] Priest and Routley (1989c): 161; see also Bremer (2005): 38-9.
[29] See Urchs (1995), (2002).
[30] Rescher and Brandom (1980): 20.
[31] See Varzi (1997), (2004).
[32] Varzi (2004): 97.

spelled out as being part and parcel of a recursive definition of truth". The meaning of the logical operators is "fixed *indirectly* through a recursive definition of the truth-value of the statements in which they occur. It is imposed *ab initio* upon the entire semantic machinery".[33]

We should notice that this has to do precisely with the first of the three clashes of intuitions mentioned in § 5.3.1. According to Varzi, the status classically assigned to conjunction and disjunction *presupposes* our stipulations on what counts as an admissible circumstance. And if we have a different and wider notion of admissible circumstance (such as the one of non-standard world in the Rescher-Brandom approach), we may change the status of our connectives. But now, it would not be fair to say that we resorted to the *change-of-subject* Quinean trick: we just got rid of a limited and restricted way of understanding the notion of circumstance, or world.[34] Kripke has one time argued that the failure of truth-functionality within supervaluational approaches is a sign that supervaluationism misses the very point of the meaning of disjunction. When you say "$\alpha \vee \beta$", I am perfectly entitled to ask: "Ok, which one then (if not both?)". But now the truth-functional thesis on the meaning of conjunction and disjunction begins to look as what we may call, adapting Tappenden and Varzi, the *Italics Argument*: "you claim that 'either α or β' holds, so *either α or β* (stamp the foot, bang the table) must hold!":[35]

> In a way, this sort of objection can be dismissed on the grounds of its unfair appeal to intuition. Change of semantics, change of subject – says the objection. Fair enough. But who got the semantics right in the first place?[36]

Now, such a wider notion of circumstance appears to be legitimate: admissible non-adjunctive circumstances are provided by the aforementioned cases of incompatible evidences in a trial; by corrupt databases; by tales and fictional worlds (for instance: in one of Conan Doyle's stories, Watson limps because of a war wound in the leg; in different stories, the wound is in Watson's shoulder and he does not limp. It does not follow, though, that Watson both limps and does not limp in any of Conan Doyle's writings). Such "impossible" situations, as non-adjunctivists ar-

---

[33] *Ibid*: 104.
[34] See *ibid*: 108.
[35] See Tappenden (1993); Varzi (2004): 104.
[36] Varzi (2004): 103.

gue, *can* be analyzed and show a certain amount of logical structure.[37] We will see in the following that an analogue defence is proposed within different paraconsistent approaches, in order to justify the deviant behaviour of *negation*.

Furthermore, there is a set of obvious dualities between the syntactic (LNC1) and the (LEM); between the semantic (LNC2), such as $\neg(T(\ulcorner\alpha\urcorner) \wedge T(\ulcorner\neg\alpha\urcorner))$, and the Principle of Bivalence, $T(\ulcorner\alpha\urcorner) \vee T(\ulcorner\neg\alpha\urcorner)$; between ontologically "overdetermined", inconsistent objects, and "underdetermined", vague objects; and so on.[38] This suggest the possibility of building a *sub-valuational* semantics, as a complementary approach with respect to supervaluationism: a semantics which makes $\alpha \wedge \neg\alpha$ false, even if both α and $\neg\alpha$ are true, just like supervaluational semantics makes $\alpha \vee \neg\alpha$ true, even if neither α nor $\neg\alpha$ are. Various authors, including Dominic Hyde, have developed sub-valuational paraconsistent semantics, also as a means to provide a treatment of vagueness. With the relevant exception of Williamson's and Sorensen's epistemic strategies, all the main approaches to vagueness require some under-determinacy of reference and/or the rejection of bivalence. Hyde has proposed a dual theory based upon over-determinacy and gluts of truth values. It also preserves all classical tautologies, just as it happens in the supervaluational approach.[39] In general, the philosophical intuition would be that, instead of satisfying neither a vague predicate nor its negation, a borderline object may satisfy both of them. Notice that this may *spread* inconsistency all over the empirical world, in a double sense. First, we admit that concrete particulars, the objects of everyday experience, may be inconsistent. Second, if borderline cases can be inconsistent, inconsistent objects are more or less everywhere, because our natural language is filled with vague predicates.[40]

---

[37] "If an impossibile world is one in which there are discrepancies of the sort illustrated by the Watson example – a world in which certain facts both do and do not obtain – then we can keep such worlds under logical control exactly as we can keep the Holmes stories under control" (Varzi (1997): 622-3).

[38] For an inquiry on some of these dualities, see Restall (2004).

[39] See Hyde (1997). See also Varzi (1997), (2000), and prospective applications suggested in Beall (2004a), pp. 10-11.

[40] Unger (1979a-b) has inferred that ordinary objects (including him, you and me) simply do not exist. This is not mandatory, of course, for a friend of inconsistency.

6.5 Which World are You Living In? – Part One

I will not enter into the details of sub-valuational semantics. We can now address, instead, a subject that often comes into play when we deal with the semantics of paraconsistent-related theories. The issue is a metaphysical one or, better, it deals with *the point at which we should draw a line* between ontology, on the one side, and epistemology and semantics, on the other. What is, in fact, the *metaphysical* status of the Rescher-Brandom non-standard worlds, or of Varzi's circumstances?

Let us begin with the former. In their book, Rescher and Brandom realize that their characterization of non-standard worlds is largely parasitic upon standard possible worlds: in order to understand what a non-standard, either schematic or inconsistent world is, we must (a) know what a standard (maximally consistent) possible world is, and (b) understand the recursive schematization and superposition operations, $\cap$ and $\cup$. Also the treatment of modal logic in Chapter 14 of their book proposes an adaptation to non-standard worlds of the binary accessibility relation of standard Kripke semantics, in which (a) from a non-standard world, only non-standard worlds of the same kind are accessible: for instance, only inconsistent worlds are accessible from inconsistent worlds; and (b) accessible worlds are ultimately "made" (via the recursive $\cap$ and $\cup$ operations) of standard worlds, which are accessible from the standard worlds we began with: for instance, from an inconsistent world $w_1 \cup w_2$, only inconsistent worlds made of worlds accessible from w_1 and w_2, are accessible. Such a modified accessibility relation is therefore completely dependent upon the regular accessibility between standard worlds.[41]

Now, everyone knows that the ontological status of *ordinary* possible worlds is itself controversial. Positions range from a realistic stance, such as David Lewis', taking possible worlds at face value as real entities, independent of language and thought;[42] to what one may call a conceptualist view, in which talk about possible worlds is considered as talk about ways in which we can conceive the world as different from what it is – hence comes Kripke's famous motto that worlds are stipulated, not discovered by powerful telescopes;[43] to a linguistic approach, which construes talk about possible worlds as talk of maximally consistent sets of

---

[41] See Rescher and Brandom (1980): 70.
[42] See Lewis (1973), Ch. 4; (1986).
[43] See Kripke (1972).

sentences.[44] *A fortiori*, one may cast doubts on the ontological status of worlds that appear to be conceptually dependent on ordinary possible ones, such as non-standard worlds. Rescher and Brandom propose, then, what they call the Parity Thesis:

> We are not concerned to argue that possible worlds should be considered as part of the furniture of the universe, only that there is nothing to choose between standard and non-standard possible worlds in this regard. [...] Recall that we do not want to argue that we should treat any possible world as real, only that the considerations which can be advanced in favour of so treating standard possible worlds apply equally to non-standard possible worlds.[45]

Various authors in the paraconsistent tradition, including Priest, have resumed the Parity Thesis.[46] In spite of their declared purpose, though, it seems that Rescher and Brandom do not manage to gain persuasive arguments in favour of Parity, mainly because non-standard, non-adjunctive worlds appear to be *representations* of worlds. I shall elucidate the point by exploiting some more intuitive dualities between non-adjunctive sub-valuations and inconsistency, on the one side, and supervaluational semantics and other semantics of vagueness, on the other.

6.5.1 The Russellian Fallacy

Supervaluationism, many-valued logics, etc., have demonstrated their usefulness in the treatment of vagueness. According to the vast majority of authors, though, there is no *de re* vagueness: vagueness can only belong to representations of the world, such as language and thought, not to

---

[44] See Hintikka (1969).

[45] Rescher and Brandom (1980): 64-5.

[46] "How one should understand [non-normal worlds] metaphysically is, of course, a contentious question, and I want nothing I say here to prejudice that issue. For the rest of this article, you (the reader) are at liberty to read in your favourite story about the nature of possible worlds" (Priest (1992): 292). "As far as I can see, any of the main theories concerning the nature of possible worlds can be applied equally to impossible worlds: they are existent nonactual entities; they are nonexistent objects; they are constructions out of properties and other universals; they are just certain sets of sentences" (Priest (1997c): 580). A definitely realist position on non-standard worlds has been endorsed by T. Yagisawa (1998), who called it *extended* modal realism. The idea is that exactly the same reasons that would lead us to accept Lewis' modal realism hold also for impossible worlds: in "There are ways other than the ways the world could be", the existential import of quantification should be taken as seriously as Lewis does for possible worlds.

the world itself.[47] Or, at least, this is the dominant position – and it is compatible with most theories of vagueness.[48] For instance, according to supervaluationism, vagueness arises from our failure to make decisions on the borders of such predicates as "tall", "bald"; "rich", "old", etc. To turn this phenomenon into a *de re* vagueness, according to most philosophers, is to commit what Russell called the *fallacy of verbalism* - "the fallacy that consists in mistaking the properties of words [or beliefs] for the properties of things".[49] It amounts to a confusion of the *ordo repraesentationis* with the *ordo essendi*. I shall henceforth call it the Russellian Fallacy.

Now, "dualizing" back to the question of inconsistency: it may be observed that the examples of non-standard worlds, or circumstances, proposed within the non-adjunctive approaches, actually seem to be *theories*, *cognitive* states, or *descriptions*. Jaskowski's discussive contexts are dialogical situations, with participants in a discussion expressing opposite standpoints; Varzi's paradigmatic circumstances crucially involve storytelling; other cases appeal to databases filled by different sources of information, which can end up inconsistent.

It is important to emphasize that, in a sense, such an observation is not a direct objection against the non-adjunctive approach. On the contrary, sub-valuational semantics can be as useful in the treatment of linguistic inconsistency as supervaluational semantics has been in the treatment of the corresponding vagueness. Nevertheless, the point is that Rescher and Brandom at the outset promised us a theory capable of exhibiting *ontological* violations of the (LNC), not merely epistemic or linguistic inconsistencies. The failure of Adjunction in epistemic, and especially inductive contexts, is a recognized fact; but Rescher and Brandom had initially claimed that "this principle must be given up for *alethic* contexts as well, once one's possible-world semantics envisages the prospect of inconsistent worlds".[50] The intuitive reading of non-standard worlds does not seem to reach the alethic level, though, and appears to be confined within

---

[47] For an introduction to the issues of vagueness, see Keefe and Smith (1996); Keefe (2000).

[48] It is fair to say that some, e.g., Tye (1990), Parsons and Woodruff (1995), support the view that the world itself can be vague, that is to say, there can be ontologically vague objects, sets, and/or states of affairs. The discussion on this point has focused on a famous argument against vague objects by Gareth Evans: see Evans (1978).

[49] Russell (1923): 62.

[50] Rescher and Brandom (1980): 18.

epistemic or, in a wide sense, representational contexts. The models of non-adjunctive theories (what the theory should capture, express, or represent) seem to be theories in their turn, representations. As Rescher and Brandom themselves observe, "the consistent standardist can thus respond that non-standard models are representations of representings (namely beliefs)"; therefore, "non-standard models and weakly inconsistent theories are just two different ways of representing beliefs".[51] This seems to be inconsistent (no pun intended) with the initial promise to exhibit "hard, *existential* inconsistency", not just "soft, *epistemological* inconsistency".[52] To quote Timothy Smiley's dismay at the end of his examination of the Rescher-Brandom approach: "what else can one do but ask for one's money back?".[53]

Now, one may propose to solve the problem by adopting some *idealistic* or broadly constructivist position, thereby partly collapsing the distinction between "epistemic contexts" and "alethic contexts". In a metaphysically idealistic framework, the work to be done in order to establish the ontological furniture of the world coincides with an analysis of our conceptual scheme (and is sometimes taken as *relative* to it, to our ways of worldmaking).[54] A conceptual scheme may well be inconsistent. This also appears to be the final conclusion of *The Logic of Inconsistency*, whose last Sections propose a retrieval of Peirce's "methodological realism". Methodological realism makes of consistency and completeness regulative ideas for our knowledge;[55] and it is accompanied by an ontological idealism, consisting in "the claim that the signs in this sense which make up our beliefs at a time are what actually exist".[56] According to the metaphysically realist classicist, real world is both consistent and complete, and inconsistent and/or incomplete circumstances actually consist in beliefs or representations in human minds. On the contrary, for the

---

[51] *Ibid.*: 128-9.

[52] *Ibid.*: 2.

[53] Smiley (1993): 19.

[54] On this point see, e.g., Goodman (1978), Sacks (1989), Sosa (1999), Talmy (2000) and, for an idealistic but non-relativistic position, Hirsch (1982).

[55] "One generally laudable characteristic of an inquiry is that over the long run it converges to a *consistent* world. That is, we have a bias towards wanting our inquiries to be progressive in the sense of weeding out inconsistent beliefs". Ditto for completeness; therefore, "the conjunction of these two conditions is the requirement that an inquiry converge to a standard, consistent *and* complete world" (Rescher and Brandom (1980): 111).

[56] *Ibid*: 124-5.

metaphysical idealist the world actually inhabited by humans *is* structured by their (occasionally inconsistent and/or incomplete) beliefs. We will see in subsequent Chapters that also other models for paraconsistent theories and logics, once they are given an intuitively plausible interpretation, look pretty much like epistemic structures, or systems of beliefs, etc. (this will become particularly relevant when we come to talk about relevant logic). Now, one does not need to be an admirer of Hegel (which I am) to consider metaphysical idealism as a respectable philosophical position. However, the evaluation of the claim, made by authors adhering to strong paraconsistency or dialetheism, of having exhibited *real* contradictions, has to be suspended, *unless* they consciously embrace such an idealistic position. I will come back to this issue at length in the following.

6.6 David Lewis' Fragmentation Strategy, and Preservationism

An indirect corroboration of the situation I just described comes from one of the still most convincing non-adjunctive proposals: the one put forward by David Lewis – not a strong paraconsistentist for sure – in his *Logic for Equivocators*. The failure of Adjunction is taken as a stratagem to handle inconsistencies in our belief corpus (or databases, textbooks, fictional works, etc.). This does not entail any metaphysical side effect endangering the (LNC). Lewis' fragmentation strategy is explicitly inspired by Jaskowski, Rescher and Brandom. Nevertheless, it begins by *replacing* the ordinary, "ontological" notion of *truth* with that of *truth-according-to-the-corpus*: "anything that is explicitly affirmed in the corpus is true-according-to-it".[57] A subset of a set of beliefs may include α, another subset, $\neg\alpha$. It does not follow, though, that the corpus as a whole includes their conjunction $\alpha \wedge \neg\alpha$. I may have believed that (1) Nassau Street ran east-west; that (2) the railroad in its proximity ran north-south; and that (3) the two were parallel. Therefore, each sentence was true-according-to-my-belief-corpus. However, from such a distributive inconsistency, no corresponding collective inconsistency followed:

> Now, what about the blatantly inconsistent conjunction of the three sentences? I say that it was not true according to my beliefs. My system of beliefs was broken into (overlapping) fragments. Different fragments came into action in different situations, and the whole system of beliefs never mani-

---

[57] Lewis (1982): 102.

fested itself at once. [...] I am inclined to think that when we are forced to tolerate inconsistencies in our beliefs, theories, stories, etc., we quarantine the inconsistencies entirely by fragmentation [...]. In other words, truth according to any single fragment is closed under unrestricted classical implication.[58]

Some classic works on the functioning of human memory may provide some support to Lewis' model. It seems that our mental hard disk is far too big a stock for us to perform an exhaustive research each time we aim at retrieving some information. Human memory works in a partitioned way.[59] Information is divided into largely separate sectors, and since it is difficult to compare different fragments, inter-sectoral inconsistencies are likely to pass unnoticed for a long time. To quote Roy Sorensen: "contradictions rarely parade as p & $\sim p$. Most are small, quiet, and colourless".[60]

This line is also followed by weakly paraconsistent theories that have come to be called *preservationist* logics – an approach started by P.K. Schotch and R.E. Jennings in the 80s, and developed mainly by Bryson Brown.[61] The main aim of preservationism is to provide accounts of consequence in which the property preserved by the consequence relation is not *truth* or, syntactically, consistency at all, but some other desirable property we may wish to conserve when we want to reason and draw inferences in situations of inconsistent information – situations of what Brown has called "*faute de mieux* reasoning". In Schotch and Jennings' non-adjunctive approach, the property to be preserved is the level or degree of inconsistency l of a given set S of sentences. This is measured by the number of subdivisions we have to introduce into S in order to restore consistency for the separate subsets, or to make them (classically) satisfiable. More precisely, a *partition* of S is a group of disjoint consistent subsets whose union is S. The level of S, $l(S)$, is now the least n such that S can be partitioned into n sets. The consequence relation at issue, preserving $l(S)$, is called *forcing*. In Bryson Brown's (2001) approach, what is dealt with is ambiguity. An inconsistent set S is assigned some level of ambiguity on the basis of the number of its sentential variables we should

---

[58] *Ibid*: 103-5.
[59] See Howe (1970); Katzley (1975); Lindsay and Norman (1977).
[60] Sorensen (2001): 155.
[61] See Schotch and Jennings (1980), (1989); Brown (1999), (2001); and Brown and Schotch (1999).

parameterise, or disambiguate, in order to render S consistent. The logic, then, is supposed to preserve the degree of ambiguity.

In both approaches, there is no question of having a semantics that allows sentences to be true together with their negation.[62] The same holds for the notion of *truth-according-to-the-corpus* proposed by Lewis. It is quite clear that it has (intentionally) nothing to do with a realist view of truth, according to which α is true iff it corresponds to reality. To say that α is true-according-to-corpus C is to say that α ∈ C, that is, it is a member of a set of sentences (representations, descriptions, etc.) in which (a fragment of) the corpus consists; or that the belief (that) α is an element of the structured collection of beliefs C in which (a fragment of) the corpus consists. We will see in the following that various semantics for paraconsistent logics have been accused of introducing an illicit (i.e., not explicitly acknowledged, unlike what happens in Lewis' approach) shift from *true* to *told true,* or, *believed true*, or *true-according-to-the-corpus*. Evidently, when "… according to the corpus" gets lost in the story, epistemic and semantic issues can be mixed.

---

[62] And "it is easily seen that the weakly paraconsistent logics here under review are designed to be user-friendly. They are developed with actual real-life reasoning in mind. They do this in such a way that requires us to see that what the competent reasoner often must not do is seek for consequences under a relation that guarantees truth-preservation" (Woods (2003): 112).

7 Positive-Plus Systems

> Will the real negation please stand up?
>
> J.M. Dunn, *Relevance Logic and Entailment*

7.1 Logics of Formal Inconsistency and C-Systems

Following Priest and Routley (1989c), I have gathered under the label of *positive-plus* systems an archipelago of formal systems mainly due to the Brazilian school of paraconsistent logic, initiated by Newton C.A. da Costa and developed by numerous co-workers.[1] Such systems are introduced in the overview by W.A. Carnielli and J. Marcos as a subgroup of the so-called *logics of formal inconsistency*, i.e., logics that express the notion of inconsistency within the formalized object language and differentiate it (to a certain extent, as we shall see) from that of contradictoriness.[2] Positive-plus systems bear this name because they are paraconsistent logics whose positive fragment preserves some standard (classical or intuitionist) positive logic essentially unchanged. The paraconsistent features of these systems is obtained by placing on top of the orthodox positive logic a profoundly modified treatment of negation – hence come, as I will try to show, the major difficulties for this approach.

The most prominent *positive-plus* systems are da Costa's hierarchies of C-calculi. The initial aim, in da Costa's words, was to build a logical framework that could underlie inconsistent theories in which "there are 'good' theorems, whose negations are not provable, and 'bad' ones, whose negations are also theorems".[3] And the C-systems have had numerous applications;[4] I will consider those having to do with set theory in

---

[1] See Carnielli, Coniglio and D'Ottaviano (2002).
[2] See Carnielli and Marcos (2002): 1.
[3] Da Costa (1974): 498.
[4] "There are certain cases in which we might think of studying directly an inconsistent theory. For example, a set theory containing Russell's class (the class of all classes which are not members of themselves) as an existing set, or a theory whose aim be the systematization of Meinong's theory of objects. Apparently, it would be as interesting to study the inconsistent systems as, for instance, the non-Euclidean geometries: we would obtain a better idea of the nature of certain paradoxes, could have a better insight on the connec-

§ 11. In this Chapter, I will concentrate only on the strictly logical core of the approach. Da Costa also produced the first quantified paraconsistent logic; but I will stick to the propositional fragments, for it is at this level that most of the interesting philosophical issues raise.

7.1.1 Bits of Positive-Plus Syntax

Let us begin with a little bit of proof-theory. An axiomatic base to start with in order to build the positive-plus systems is the following (sometimes labelled as $\mathbf{C_{min}}$, the "minimal paraconsistent logic"):[5]

(A1) $\alpha \to (\beta \to \alpha)$
(A2) $(\alpha \to \beta) \to ((\alpha \to (\beta \to \gamma)) \to (\alpha \to \gamma))$
(A3) $\alpha \to (\beta \to \alpha \wedge \beta)$
(A4) $\alpha \wedge \beta \to \alpha$
(A5) $\alpha \wedge \beta \to \beta$
(A6) $\alpha \to \alpha \vee \beta$
(A7) $\beta \to \alpha \vee \beta$
(A8) $(\alpha \to \gamma) \to ((\beta \to \gamma) \to (\alpha \vee \beta \to \gamma))$
(A9) $\alpha \vee (\alpha \to \beta)$
(A10) $\alpha \vee \neg \alpha$
(A11) $\neg\neg\alpha \to \alpha$

The only rule of inference is *modus ponens*. The positive fragment, as we can see, corresponds to the positive part of classical propositional logic. The basis is anti-intuitionist, though, since it comprises (A10) and (A11), that is, Strong Double Negation and the (LEM). $\mathbf{C_{min}}$ also includes the axiom corresponding to Adjunction, (A3); and (A1), which is a paradox of material conditional, and gets discharged, as we shall see, within relevance logics as a "fallacy of relevance". Adding *ex falso* in the form of (PS1), i.e., $\neg\alpha \to (\alpha \to \beta)$, would be sufficient to restore classical propositional logic.[6] If we remove (A9), we obtain da Costa's weakest origi-

---

tions amongst the various logical principles necessary to obtain determinate results, etc." (*Ibid*: 497-8).
  [5] Carnielli and Marcos (2002): 10.
  [6] See Bremer (2005): 110.

nal system, called $\mathbf{C_\omega}$. The Deduction Theorem holds for $\mathbf{C_{min}}$, therefore we can proceed to the other systems by adding inference rules (instead of axioms) to the base.

7.1.2 Climbing the C-Ladder

It is customary to introduce two unary operators, "°" and "•": $°α$ and $•α$ should be read, respectively, as "$α$ is consistent" and "$α$ is inconsistent". These notions can be taken within the different systems as primitive, characterized syntactically or semantically, and be independent of each other or inter-definable. Specifically, not in all positive-plus systems consistency and inconsistency of a given formula ($°α$, $•α$) are equivalent to *contradictoriness* and *non-contradictoriness* ($α \wedge \neg α$, $\neg(α \wedge \neg α)$). Following Carnielli and Marcos' presentation, we obtain the system **bC**, the *basic logic of (in)consistency*,[7] by adding to the base the rule:

$$\frac{°α, α, \neg α}{β} \quad \text{(bC)}$$

This tells us that if $α$ is assumed as consistent, yet both $α$ and its negation hold, anything follows. Tersely put, a *consistent contradiction* (!) detonates. Such a feature of this kind of logics is nicely expressed by saying that they are *gently explosive*.[8] One of the *desiderata* set forth by da Costa in his seminal paper was that a paraconsistent logic should retain "the most part of the schemata and rules of C_0 [that is, classical propositional logic] which do not interfere with the first conditions [on a good non-explosive logic]".[9] Therefore, he maintained that Explosion should *hold* within contexts that are assumed as consistent. This is the intuitive motivation behind (bC) and analogue rules that can be found in the positive-plus systems.

I will come to the issue of the weakness of da Costa negation in due course. Meanwhile, it should be noticed that, by means of it and the consistency operator, a strong negation is definable:

---

[7] See Carnielli and Marcos (2002): 36.

[8] "In any such a gently explosive logic, given a contradictory theory there is always something 'reasonable' – to wit, consistency – which one can add to it in order to guarantee that it will become trivial" (*Ibid*: 28).

[9] Da Costa (1974): 498.

(Df~) $\sim\alpha =_{df} \neg\alpha \wedge °\alpha$.

Strong negation embodies the idea that α is consistent; therefore, the coexistence of α and $\sim\alpha$ is itself explosive.[10] Contraposition:

$$(\alpha \rightarrow \beta) \rightarrow (\neg\beta \rightarrow \neg\alpha),$$

does *not* hold in **bC**. If it held, via (A1) we could obtain $\alpha \rightarrow (\neg\alpha \rightarrow \neg\beta)$ and we would straightforwardly find ourselves out of an interesting paraconsistency. There are reasons for wanting Contraposition to fail within a paraconsistent approach, anyway, quite independently of the strength of the involved conditional. Suppose that if something is a swan, it is white, and suppose that some strange animal turns out to be a swan and both white and not white. By Contraposition, it would follow that it is not a swan, i.e., it is false that it is a swan; and this seems unwarranted. There is no reason to assume that the concept *swan* behaves inconsistently merely because of its being instantiated by an inconsistent thing. It is in the spirit of the paraconsistent, i.e., non-detonating logical enterprise, to reject the idea that, just because an object is inconsistent with respect to one predicate, it has to be inconsistent with respect to others.[11] A conditional must satisfy *modus ponens*, that is, it must preserve truth forwards, but it need not preserve falsity backwards (notice that this has been argued, at least for some conditionals, also independently of the issue of paraconsistency).[12] Furthermore, restricted forms of contraposition with an extra consistency assumption can hold within the positive-plus systems,[13] for instance the rule:

$$\frac{°\beta, \alpha \rightarrow \beta}{\neg\beta \rightarrow \neg\alpha} \quad (Cont°)$$

This is a way to perform a *classical recapture*: by adding the appropriate consistency restrictions, classical laws and/or rules of inference can

---

[10] *Ibid*: p. 500. Despite being explosive, this strong negation is not equivalent to *classical* negation – it becomes such only when we reach the logic **Ci**. More on this point in Carnielli and Marcos (2002): 49-50.
[11] On this point, see da Costa and Wolf (1980).
[12] See e.g. Stalnaker (1968), Lewis (1973).
[13] See Carnielli and Marcos (2002): 39, for more detailed explanations.

POSITIVE-PLUS SYSTEMS / 155

be regained, provided we are in contexts that are assumed consistent. We move from **bC** to the systems **bbC** and **bbbC** by adding rules that capture dualities between ° and • (which are quite reasonable, since consistency and inconsistency should be polar opposites):

$$\frac{\neg \bullet \alpha}{\circ \alpha} \quad \text{(bbC1)}$$

$$\frac{\neg \circ \alpha}{\bullet \alpha} \quad \text{(bbC2)}$$

$$\frac{\bullet \alpha}{\neg \circ \alpha} \quad \text{(bbbC1)}$$

$$\frac{\circ \alpha}{\neg \bullet \alpha} \quad \text{(bbbC2)}$$

We get to the system **Ci** ("where contradiction and inconsistency meet",[14] i.e., they become equivalent) by adding a rule that characterizes •:

$$\frac{\bullet \alpha}{\alpha \wedge \neg \alpha} \quad \text{(Ci)}$$

and we reach **Cil** by also adding one that characterizes °:

$$\frac{\neg(\alpha \wedge \neg \alpha)}{\circ \alpha} \quad \text{(Cil)}$$

From now on, (in)consistency and (non-)contradictoriness are identified again. Here begin the well-known da Costa C-systems. They are n-ary hierarchies ($1 \leq n \leq \omega$) of sentential calculi C_n, and of first-order calculi without (C^*_n) and with ($C^=_n$) identity. As anticipated, I shall deal only with the sentential part (C_0, C^*_0 and $C^=_0$ are, respectively, the classical sentential, predicative, and predicative-with-identity calculi). C_1 is ob-

---

[14] *Ibid*: 44.

tained by adding three rules that regulate the way in which consistency propagates from sub-formulas to the molecular formula:[15]

$$\frac{°\alpha \wedge °\beta}{°(\alpha \wedge \beta)} \quad (°\wedge)$$

$$\frac{°\alpha \vee °\beta}{°(\alpha \vee \beta)} \quad (°\vee)$$

$$\frac{°\alpha \rightarrow °\beta}{°(\alpha \rightarrow \beta)} \quad (°\rightarrow)$$

This means that formulas composed only of consistent sub-formulas are themselves consistent. The different ways in which consistency and inconsistency propagate between composed formulas and their components produces the hierarchic variety. We can have weaker and weaker logics by having, for each C_n, more and more premises required in order to guarantee consistency.[16] Any calculus is strictly stronger than those who follow it, that is, C_{n+1} is weaker than C_n, and C_ω (corresponding to C_{min}) is the weakest in the hierarchy.[17] As Diego Marconi has remarked, "da Costa hierarchies are therefore, as it were, sensitive to the *degree* of contradictoriness of the universe of discourse".[18] The Brazilian school has produced for these calculi numerous syntactic (e.g., their presentation in Gentzen-like natural deduction style) and metalogical (e.g., interesting semantics with the respective completeness proofs) results.[19]

---

[15] See *Ibid*: 63. See also da Costa (1974): 499; da Costa and Marconi (1989): 9-10, for axiomatic presentations.

[16] "One ought not to take α^0 as expressing consistency unless it, itself, behaves consistently. This thought motivates the weaker da Costa system C_2, which is the same as C_1, except that α^0 is replaced everywhere by $\alpha^0 \wedge \alpha^{00}$. Of course, there is no reason to suppose that this expresses the consistency of α unless it, itself, behaves consistently. This thought motivates the da Costa system C_3 where α^0 is replaced everywhere by $\alpha^0 \wedge \alpha^{00} \wedge \alpha^{000}$. And so on for all the da Costa systems Ci, for finite non-zero i" (Priest (2002c): 306fn).

[17] Da Costa (1974): 501. See also da Costa and Marconi (1989): 10.

[18] Marconi (1979): 53.

[19] See e.g. Raggio (1968); da Costa and Alves (1976); for a classic survey, da Costa and Marconi (1989): 8-12.

7.2 Da Costa Negation and its Troubles

As I have anticipated, criticisms of these systems have focused mainly on the behaviour of negation. Let us follow a semantic approach now. Typically, the semantics of negation for most (thought, admittedly, not all) da Costa systems is given by the following clauses:

(S¬1) $F(\ulcorner\alpha\urcorner) \Rightarrow T(\ulcorner\neg\alpha\urcorner)$

(S¬2) $T(\ulcorner\neg\neg\alpha\urcorner) \Rightarrow T(\ulcorner\alpha\urcorner)$.

(S¬1) guarantees that at least one of α and $\neg\alpha$ is true, thereby validating (together with the standard clause for disjunction) axiom (A11), that is, the (LEM). (S¬2) does the job of validating (A10), i.e., Strong Double Negation, the intuitive justification being that if it is not the case that $\neg\alpha$, then, since via (S¬1) one among α and $\neg\alpha$ has to hold, it is the case that α. The semantics invalidates Explosion both in its collective and in its distributive form, that is:

$\{\alpha, \neg\alpha\} \nvDash \beta$
$\{\alpha \wedge \neg\alpha\} \nvDash \beta$.

The Weak Condition of Non-Triviality, therefore, is respected. Of course, the basic idea is that α and $\neg\alpha$ can both be true, whereas β is false. However, the semantics also turns on all the three main clashes of intuitions described in § 5.3.1. First, da Costa negation is not truth-functional but, at best, partially truth-functional: given that α is false, this suffices via (S¬1) to establish that $\neg\alpha$ is true; but given that α is true, $\neg\alpha$ may be true as well as false (because of this, the positive-plus approach has also been labelled as non-truth-functional). This entails that replacement of equivalents fails. It also entails that negation typically resists an algebraic treatment. These facts are sufficient to conjecture that positive-plus logics violate our Avoid Overkill Condition: they contravene a very strong intuition – the one according to which something is a negation only if it (truth-functionally) switches truth values: when applied to a true sentence, it produces a false one, and *vice versa*. According to Routley and Priest, such a failure of extensionality is sufficient to certify that da

Costa negation "is not our friendly neighborhood extensional negation, but a radically intensional functor of some sort".[20]

Moreover, within positive-plus systems (LNC1), that is, the syntactic (LNC), $\neg(\alpha \wedge \neg\alpha)$, *cannot be a theorem*. The negation of $\mathbf{C_{min}}$ is so weak that it does not have any negative formula as a theorem.[21] And if (LNC1) were a theorem, beginning with **Ci1**, and within all the da Costa systems $\mathbf{C}_n$, all formulas would be taken as consistent via the rule (Ci1). Therefore, given the rule (bC), all these systems would turn out to be simply (not just gently) explosive. Of course, there are quite intuitive reasons for not wanting (LNC1) as a paraconsistent theorem. This was, in fact, the first *desideratum* in da Costa's original list: the failure of $\neg(\alpha \wedge \neg\alpha)$ was taken as a condition of adequacy on any logic for inconsistent theories.[22] Since contradictions and local inconsistencies are expected, we may suppose that some instances of (LNC1) are *false*. Hence came also da Costa and Marconi's reply to the first semantic charge concerning the loss of truth-functionality for negation:

> The truth conditions for $\neg A$ are determined by the truth conditions for A and the truth conditions for $\neg(A \wedge \neg A)$. […] And it is quite plausible, from a paraconsistent standpoint, that the truth-value of a negative formula depends on whether or not the world is consistent "at point A", for paraconsistent semantics, contrary to classical semantics, does not assume that the world is consistent at every point: so that it must "check" in each single case, so to speak.[23]

Things are not so straightforward, though. Once we accept both α and $\neg\alpha$, therefore (since we retain Adjunction) $\alpha \wedge \neg\alpha$, there is no *a priori* bar on contradictions anymore. Therefore, there seems to be nothing special in accepting $\neg(\alpha \wedge \neg\alpha)$, too. It is just one more contradiction to take on board, so to speak. We will soon meet several paraconsistent logics (and, especially, logics suitable for *strongly* paraconsistent or dialetheic uses) in which (LNC1) *is* a theorem. More importantly, as Priest and Routley also observed:

---

[20] Priest and Routley (1989c): 164. Mortensen (1980) put similar considerations forth.
[21] See Bremer (2005): 110.
[22] See da Costa (1974): 498.
[23] Da Costa and Marconi (1989): 18. Batens (1980): 228-9 had already put forth a similar argument for the non-truth-functionality of negation.

The law of non-contradiction [taken as (LNC1)] has traditionally been seen as a central property, if not a defining characteristic, of negation. [...] That an account of negation violates the law of non-contradiction therefore provides *prima facie* evidence that the account is wrong. This is [a] piece of evidence that *da Costa negation is not negation*.[24]

You may recall the traditional definitions of contrariness and sub-contrariness given in § 1: α and β are contraries iff $\alpha \wedge \beta$ is logically false, sub-contraries iff $\alpha \vee \beta$ is logically true (and contradictories iff they are both contraries and sub-contraries). Now, it is sometimes claimed that, given the failure of the (LEM) within intuitionist logic, intuitionist negation would identify a contrary of α, not its contradictory. A "dual" claim can now be made for da Costa negation: the fact that in (mainstream) positive-plus systems $\neg(\alpha \wedge \neg\alpha)$ fails whereas $\alpha \vee \neg\alpha$ holds can be taken as showing that da Costa negation actually produces a sub-contrary, not the contradictory. Therefore, it is a pseudo-negation, since "negation is a contradiction forming functor, not a subcontrary forming functor".[25]

The point is that any theory of negation should be a theory of the relation of *contradictoriness*, i.e., a relation involving pairs of contradictories, whose linguistic expression may or may not involve an explicit "not" – for instance, "Plato is a man" and "Plato is not a man"; "Some Australian is a philosopher" and "No Australian is a philosopher". Now, contradictories are unique up to logical equivalence. This is not the case with sub-contraries: something may have more than one sub-contrary. This explains the *indeterminacy*, as it were, of the da Costa operator. In his anthology on the formalization of dialectics, Marconi has claimed that the da Costa negation of α expresses a "negation-set", each element of which would be somehow opposed to α. This would reflect, according to Marconi, "a certain indeterminacy of a dialectical theory", as if "dialectical negation did not 'choose' between the various elements" variously opposed to a given element: "a more generic, less univocally determined concept than that of the ontological counterpart of classical negation".[26] This can be hardly taken as a point on behalf of such a negation, though.

---

[24] Priest e Routley (1989c): 164-5.
[25] *Ibid*: 165.
[26] Marconi (1979): 53-4.

On the contrary, according to Priest and Routley it clearly shows that our Avoid Overkill Condition has been massively violated.[27]

Da Costa negation also lacks almost all the inferential properties commonly associated to the connective. Typical laws/rules involving negation that hold even in minimal logic, Contraposition and Double Negation Introduction, fail for most da Costa systems, unless we introduce the consistency restrictions (we have seen that there are reasons for rejecting Contraposition within a paraconsistent approach; but we shall meet fully paraconsistent logics that preserve it in the following). In addition, *reductio* and De Morgan Laws fail. Even those who maintain that inferential features are the source of meaning for logical constants, therefore, have good reasons to claim that da Costa negation has few resemblances with negation.[28]

Finally, consider the following result of translatability into the classical calculus (be it **CC**). By recursion on the complexity of α, it is easy to show that:

(TR) $\vdash_{CC} \alpha \Leftrightarrow \vdash_{Ci} {*}\alpha$,

---

[27] "[The fact that da Costa negation is a sub-contrary forming operator] explains why ~ [here the tilde is the da Costa negation] is not truth functional. For the truth value of a subcontrary of A is not determined by the truth value of A. ~ is not an extensional functor [...]. So ~A is a sub-contrary of A, but which? For although the contradictory of a statement is unique, it may have many sub-contraries. Which is ~A? [...] There are no other constraints on ~ to determine which sub-contrary functor it is. Hence the answer to this question must be radically indeterminate" (Priest and Routley (1989c): 165-6).

[28] According to Routley "the weakened negation systems lack all forms of contraposition though surely some are correct, and indeed there is little basis for regarding the so-called negations of these systems as genuine negations at all rather than, say, positive modal connectives, e.g. weird truth or necessity connectives" (Routley (1979a): 305). See also Mortensen (1980), and Lenzen (1996), where the point is discussed entirely from an inferential point of view. Lenzen distinguishes "*dispensable* principles which, though they are valid principles of classical negation, need not necessarily be satisfied by arbitrary other negations"; and "*indispensable* principles which a logic L always has to satisfy if its monadic operator ~ is to count as a genuine negation" (40). And failure of Contraposition is considered by itself a sufficient reason not to count an operator as a negation in any way (see 43ff.). Similar criticisms may hold for the system called **PI**, proposed by Diderik Batens in Batens (1980) as a basic extensional paraconsistent logic. **PI** is in fact a positive-plus system, too. Its negation has the same semantics as da Costa's, and Contraposition and *reductio* do not hold.

that is, a given formula α is a theorem of **CC** (an explosive logic) iff its *-transformed is a theorem of **Ci** (a paraconsistent, gently explosive logic). The recursive characterization of *-transformation is the following:

(1) If α is an atomic formula, *α = α;
(2) If # is a binary connective, *(α # β) = *α # *β;
(3) *¬α = ~*α[29]

*-transformation consists in systematically replacing da Costa negation with the strong negation. Now, as da Costa acknowledged, since strong negation "has all properties of the classical negation" in his C-systems, they are trivialized by "each formula of the type [α ∧ ~α]".[30] As Priest and Routley pointed out:

> Is the lack of a genuine negation operator in the C systems merely a matter of omission? The answer is a quick and simple 'No'. For if we were to add an operator, -, with the obvious conditions for negation,
>
> $v(-A) = 1$ iff $v(A) = 0$,
>
> it is easy to see that non-paraconsistency would be reinstated. For then $\{A \wedge -A\} \models_C B$. Thus the C systems achieve their paraconsistency only at the cost of dispensing with negation.[31]

The gently explosive features of these systems have also been criticized. Carnielli and Marcos claimed that the rejection of any form of explosion, as pursued in some relevant paraconsistent logics, or within the framework of Priest's logic of paradox, **LP** (with which we shall deal in the following Chapters) is unnecessary. It leaves us with no way to express within the formal language the distinction between *tolerable* and *intolerable* contradictions, which is accepted even by strong paraconsistentists:

> Under our present point of view, proposing a logic in which no single contradiction can ever have a harmful effect on their underlying theories is quite an

---

[29] See e.g. Carnielli and Marcos (2002): 51.
[30] Da Costa (1974): 500.
[31] Priest and Routley (1989c): 166.

extremist position, and may take us too far away from any classical form of reasoning. [...] So one may conjecture that consistency is exactly what a contradiction might be lacking to become explosive – if it was not explosive from the start. Roughly speaking, we are going to suppose that a 'consistent contradiction' is likely to explode, even if a 'regular' contradiction is not.[32]

That is to say, precisely when it is *not* the case that $°α$, the pair $α$, $¬α$ does not trivialize the calculus. Manuel Bremer has simply replied that if both $α$ and $¬α$ hold, and the notion of consistency has to retain bits of what we normally mean by it, then $°α$ is intuitively false. How could we have $α$, $¬α$ *and* keep claiming that $α$ is consistent?[33]

These difficulties seem to speak against the philosophical import of the Brazilian approach to paraconsistency (which is why it has been dealt with quickly in this Chapter). They do not entail the loss of any strictly logical, technical and applicative interest. C-systems are being constantly developed and adopted by various researchers in the paraconsistent area. With a few adjustments, they can also serve as the underlying logics of some paraconsistent set theories (I will say something on this in § 11). One of the most interesting aspects of the approach is the idea of expressing consistency and inconsistency within the object language via appropriate sentential operators. As we will see in the following Chapter, the insight can be exported into quite different paraconsistent logical approaches.

7.3 Adaptive Strategies

I will end this Chapter with a quick glance at the so-called *adaptive logics*, developed mainly by Diderik Batens and his co-workers.[34] What is at issue in the adaptive approach are the dynamic processes by means of which we produce and *take back* inferences in the presence of contradictions. Rational activity is a temporal process, and inferences are made from the body of information available at a certain time. Nevertheless, some part or other of it may later turn out to be inconsistent. Then we may have to withdraw some inferences, which presupposed the consistency of the data. This is a different situation from the one studied in non-

---

[32] Carnielli and Marcos (2002): 26-7.
[33] See Bremer (2005): 117.
[34] For a survey, see Batens (2000); see also Meheus (2003).

monotonic logics: we are not dealing with "external" changes resulting by adding further premises to a given set, but with "internal" dynamics: "within the reasoning on a conclusively given premise set the insight on logical relations within the premise set and its consequences increases".[35]

More than a logic, or a group of logics, the adaptive approach looks like a set of inferential *strategies*, consisting in the different ways in which inferences are withdrawn in front of newly discovered inconsistencies. For instance, we may use the rule of Disjunctive Syllogism to prove things which would otherwise be unavailable, or to abbreviate otherwise long and cumbersome proofs. However, such a rule can be correctly applied only to consistent premises, and since we usually we have no decision procedure guaranteeing consistency, we may later find that we have to take back the inferences depending on (DS). An adaptive structure is generally characterized by an *upper limit* and a *lower limit* logic. The upper limit logic is a strong one and may typically coincide with ordinary classical logic: within it, we allow ourselves the unrestricted use of all classical rules/principles, in order to draw more consequences in fewer steps. But classical logic, of course, is explosive, so it becomes useless when inconsistencies emerge. Here paraconsistency begins to play. The lower limit logic is typically a paraconsistent system in which some classical rules or principles do not hold. Sometimes **LP** is taken as the lower limit logic; but Batens' favourite system, **CLuN**, is a positive-plus logic with a weak and, like da Costa's, non-truth-functional negation.[36] We may notice, however, an important difference: in the Brazilian approach we have to know what is consistent and what is not *before* formalizing via the (in)consistency operators. In the adaptive approach, we reason under the hypothesis that a given formula, or set of formulas, behaves consistently; lacking an algorithm for consistency, we may have to withdraw our supposition. The whole process is naturally framed within a conception of logic as an "inferential engine", such as the one captured by a presentation in terms of natural deduction: we can refer to different stages in a proof, to what has been inferred *so far*, etc. We may, for instance, simply add to an ordinary natural deduction linear proof a column displaying the "consistency assumptions", that is, indexes for the formulas whose consistency is presupposed at each inferential step.[37]

---

[35] Bremer (2005): 93.
[36] See *Ibid*: 101-2; see also Batens (1986), (1989); Batens and De Clercq (2004).
[37] See Bremer (2005): Ch. 7, for details and some sample proofs.

The philosophical motivation for the adaptive approach has often been linked to a general propensity to privilege weak paraconsistency: adaptive logics are taken as providing techniques for reasoning in situations of occasionally inconsistent information, without admitting worldly inconsistency. Nevertheless, the approach seems to be neutral with respect to the issue of dialetheism; a kind of adaptive strategy for recapturing classical reasoning has been proposed within a broadly dialetheic context, too, as we shall see in the following Chapter. It is true, nevertheless, that Batens is not a dialetheist. In an influential paper called *Against Global Paraconsistency*, he argued that paraconsistent logic cannot be "universal": it cannot include classical logic as a particular case, or as a good approximation for consistent contexts (a position which is supported, on the contrary, by Routley, Priest and others, as we shall see at length in the following). Some of Batens' arguments have to do with the inferential and expressive weakness of paraconsistent negation, and with the claim that the metatheory of a paraconsistent theory ultimately cannot be itself paraconsistent. Global paraconsistency "does [not] actually permit an adequate description of consistent domains, e.g., the meta-theory of most logics, including paraconsistent logics".[38] I will return to these issues in § 14. Meanwhile, we may appreciate the fact that an adaptive approach can help us hold a middle way; the full strength of classical logic is available when possible and, in case things go wrong (i.e., in the presence of inconsistencies), we resort to the paraconsistent capacity of avoiding detonations:

> The contextual point of view is not affected by the shortcomings of global paraconsistency, is attractive for independent reasons, and does not require that any domain be in principle safeguarded against inconsistency. If this alternative is viable, as I believe, then we are able to stick out our necks and use the stronger, classical means to describe the domains that we believe to be consistent; if the belief turns out to be mistaken, we still may change our minds.[39]

---

[38] Batens (1990): 210.

[39] *Ibid*: 227-8. "Without adaptivity we had to reason using some [paraconsistent logic] in all contexts which we suppose to contain contradictions. Given that quite a lot of standard logic is missing (including contraposition, transitivity of identity, etc.) that is a severe restriction. We cannot capture a lot of (harmless) consequences in that field then. Philosophy as that area of universal talk about semantics and epistemology would have to use such a restricted logic. It is questionable how many of its theses and arguments could really (i.e. without hidden recourse to standard logic) be expressed. Adaptivity, on the

other hand, makes clear that reasoning from the present contradictions is rather the exception than the rule" (Bremer (2005): 95).

8 The Logic of Paradox

> Even to understand the word 'doublethink' involved the use of doublethink.
>
> George Orwell, *Nineteen Eighty-Four*

8.1 Overview

I have already mentioned Priest (1979), one of Priest's early writings on dialetheism. He proposed there a logical system called logic of paradox, and labelled as **LP**, which I will investigate in this Chapter. The system has been variously developed and discussed by Priest himself as well as by other scholars in the paraconsistent tradition, especially because of the intuitive appeal of its semantics. Therefore, I shall not describe it proof-theoretically, but via an approach in terms of truth tables and logical consequences. I will also deal with its first-order extension **LPQ**. Other paraconsistent theories we met were not limited to the sentential level, but included calculi and/or semantics for predicative languages (for instance, Rescher and Brandom's logic of inconsistency and da Costa's hierarchies). Most of the philosophical issues concerning paraconsistency emerge, though, at the sentential level: after all, the main concern is how to avoid such things as *ex falso* and the Disjunctive Syllogism, while at the same time keeping most of the remaining logical landscape. Nevertheless, in the development of inconsistent theories with an underlying paraconsistent logic it is vital to have the resources of predication and quantification. We happen to have to do with inconsistent objects, that is, objects both having and lacking certain properties: for instance, such inconsistent sets as R and V, or perhaps Meinongian objects, like the square circle. And we need a language capable of talking about them adequately. Such a language shall not have a standard semantics for sure, since within it the extension of a predicate and its complement are mutually exclusive (and exhaustive).

LP and **LPQ** have been taken by various authors as a good starting point for a paraconsistent treatment of extensional (i.e., different from the conditional) operators. However, their discussion will lead us straight into theoretical issues, and to an evaluation of some of Priest's philosophical ideas. As should be clear by now he is, together with Richard Routley, the author who has most thoroughly explored the philosophical spin-offs of strong paraconsistency. I will deal particularly with Priest's solution to the classical recapture issue.

8.2 True, False, True and False

8.2.1 The Original LP

LP is a three-valued logic, whose values are $P(\{1, 0\}) - \emptyset$: you obtain them by taking the power set of the classical values, and subtracting the empty set. This last move is connected with Priest's philosophical arguments against gaps, that is, sentences lacking truth value. As we have seen in the first part of the book, Priest believes that gaps do not provide a satisfactory way out of the semantic paradoxes. In *In Contradiction*, the rejection of truth value gaps is also motivated by means of a teleological account of truth: truth is what you aim at when you assert. This is a one-player game, and in a one-player game anything less than success is a failure; therefore, it seems wrong to admit truth value gaps, which would be something like a tie:[1] a tie is like kissing your sister.

Sentences are evaluated in **LP** via a function v whose output is one of the remaining values: (a) true, $\{1\}$; (b) false, $\{0\}$; (c) true *and* false, $\{1, 0\}$. The last is, of course, the peculiar value of the **LP** approach to paraconsistency: some sentences are both true and false, against (LNC2a). This truth value is also labelled as *paradoxical*.[2] In the paraconsistent literature, it is also claimed that true non-paradoxical sentences are *true only*, and false non-paradoxical sentences are *false only*. I shall talk of an evaluation *function*, following Priest's initial position. Criticisms put forth by Timothy Smiley and Anthony Everett in 1993 led paraconsistentists to realize that, after all, it is better to talk about an evaluation *relation* between sentences and truth values. The change was introduced in order to deal with some so-called *hypercontradictions* – particularly infectious contradictions that appear to be a sort of revenge liar challenging dialetheism. I will deal with them in the last part of the book, though. Meanwhile, I shall follow Priest's initial arrangement, for the sake of simplicity.

---

[1] See Priest (1987): Ch. 4. This criticism has been criticized in its turn by a well-known gapper, namely Parsons (1990).
[2] See Priest (1979): 226.

One of the enjoyable aspects of the semantics of **LP** is its treatment of negation. It is a choice negation, whose clauses – following Priest's early favourite notation – look like this:

(S¬1) $1 \in v(\neg\alpha) \Leftrightarrow 0 \in v(\alpha)$

(S¬2) $0 \in v(\neg\alpha) \Leftrightarrow 1 \in v(\alpha)$.

We should read "$1 \in v(\alpha)$" as "α is true under è v", "$0 \in v(\alpha)$" as "α is false under v".[3] Negation conforms to (Neg1), then, and it definitely seems not to violate the Avoid Overload Condition – in fact, it is explicitly designed with the aim of making as little changes as possible with respect to the classical framework. To be sure, this is a non-classical operator, since its clauses are not homophonic and do not respect (Neg2). The following basic intuition, though, is undoubtedly respected: negation is the operator that (truth-functionally) toggles truth and falsity. Unlike the positive-plus approach, where we have just two truth values but truth-functionality fails, truth-functionality is retained here.[4] We just add the small postscript that something can be both true and false. Bremer has therefore claimed that "**LP** negation is an extension of standard negation".[5] The tabular presentation usually given is the following:

α	¬α
1	0
0	1
1, 0	1, 0

The negation of a paradoxical (true and false) sentence is itself paradoxical. This is the standard tabular presentation one can find in most papers and books.[6] However, the notation is slightly equivocal because, for

---

[3] See Priest (1987): 75.

[4] "The properties of negation are neat and simple and no extra semantic postulates have to be added, as in da Costa's approach, to ensure bits of double negation. Moreover, there can be no doubt that the negation of this approach is negation. The semantics are recursive and extensional. Thus [such a negation] is not an intensional functor" (Priest and Routley (1989c): 169).

[5] Bremer (2005): 46.

[6] See e.g. Bremer (2005): 47ff.

instance, if α is *true only* its value should not be 1, but its singleton, {1}. As far as I know, little importance has been given to the ambiguity. The clauses for conjunction and disjunction are even more straightforward, and partly homophonic:

(S∧1) $1 \in v(\alpha \wedge \beta) \Leftrightarrow 1 \in v(\alpha)$ and $1 \in v(\beta)$

(S∧2) $0 \in v(\alpha \wedge \beta) \Leftrightarrow 0 \in v(\alpha)$ or $0 \in v(\beta)$

(S∨1) $1 \in v(\alpha \vee \beta) \Leftrightarrow 1 \in v(\alpha)$ or $1 \in v(\beta)$

(S∨2) $0 \in v(\alpha \vee \beta) \Leftrightarrow 0 \in v(\alpha)$ and $0 \in v(\beta)$.[7]

The first clause in each pair corresponds to the classical, homophonic one. The second clause is classically redundant, but of course things change when truth and falsity become partially independent. The corresponding matrices are the following:

α	β	α∧β	α∨β
1	1	1	1
1	0	0	1
1	1,0	1,0	1
0	1	0	1
0	0	0	0
0	1,0	0	1,0
1,0	1	1,0	1
1,0	0	0	1,0
1,0	1,0	1,0	1,0

At the outset, the conditional gets the ordinary definition:

(Df→) $\alpha \rightarrow \beta =_{df} \neg \alpha \vee \beta$.

LP matrices follow those of mainstream three-valued logics (they correspond, for instance, to the strong Kleene matrices of **K₃**),[8] but their in-

---

[7] See Priest (1987): 75.
[8] See Priest (2001): 120.

terpretation is changed. The third value is not, like in traditional three-valued logics, "indeterminate" (usually labelled as ½), but {1, 0} or "paradoxical". The important point is that designated values in **LP** are {1} *and* {1, 0}, therefore the value "paradoxical" is a designated one. The underlying intuition would be that a given formula gets a designated value when it is *at least* true, and it is a tautology when 1 appears among its values in all the lines of its truth table.⁹ This entails that whereas, for instance, within Łukasiewicz's three-valued logic $\alpha \vee \neg \alpha$ is not a tautology, in **LP** it is, since it always gets designated values. What we are most interested in is that the syntactic (LNC1) is a logical law, too:

α	$\neg$	$(\alpha$	$\wedge$	$\neg\alpha)$
1	1	1	0	0
0	1	0	0	1
1, 0	1,0	1,0	1,0	1,0

Instances of $\neg(\alpha \wedge \neg\alpha)$ can never be false (only), but either true (only), or both true and false. The fact that $\neg(\alpha \wedge \neg\alpha)$ is a **LP**-tautology conforms to the intuition we spoke of in § 7.2, to which Priest, Routley and other paraconsistentists attribute the greatest importance. The validity of (LNC1) (as well as that of the (LEM)) is an essential condition for being allowed to claim that negation works within a formal system as *real* negation, that is, something which expresses the relation of contradictoriness between sentences (propositions, or whatever truth-bearer you like).¹⁰ It is also clear in which sense the (LNC), nevertheless, *fails* within **LP**: some (C1)-contradictions, that is, some instances of the schema $\alpha \wedge \neg \alpha$, can be true, even though they can never be true *only*, but at best (at worst?) true and false.

The definitions of logical consequence and logical validity are straightforward. Where Γ is a set of formulas, we have:

$\Gamma \vDash \alpha \Leftrightarrow$ under any evaluation v, $(1 \in v(\beta)$ for all $\beta \in \Gamma \Rightarrow 1 \in v(\alpha))$

---

⁹ This idea first came to Asenjo (1966); see also Asenjo (1989).

¹⁰ "This fact about contradictories obviously gives immediately two of the traditional laws of negation, the law of excluded middle [...] and the law of non-contradiction [...]. Note that one may formulate the law of non-contradiction in a number of very different, and non-equivalent, ways. As I have formulated it here [i.e., $\neg \Diamond(\alpha \wedge \neg\alpha)$], the truth of the principle is quite compatible with dialetheism" (Priest (2006): 78).

⊨ α ⇔ under any evaluation v, $1 \in v(\alpha)$.

All classical-bivalent tautologies are tautologies of **LP**. Therefore, Priest claims that the whole **LP** *extends* classical, bivalent logic.[11] This can be made technically precise: a standard classical interpretation is isomorphic to a **LP**-interpretation in which no atomic formula or parameter gets the paradoxical value $\{1, 0\}$; therefore, classical evaluations are special cases of **LP**-evaluations. A paraconsistent logic should preserve classical logic whenever the latter is correct. Classical logic is just, in a sense, incomplete: it cannot survive in inconsistent environments, since it is explosive. Furthermore, **LP** truth tables give a decision procedure at the sentential level, so **LP** is complete and decidable.

On the other hand, logical consequence has undergone important changes with respect to its classical counterpart. Within **LP** we have that:

$\{\alpha, \neg \alpha\} \nvDash \beta$
$\{\alpha \wedge \neg \alpha\} \nvDash \beta$
$\{\alpha \vee \beta, \neg \alpha\} \nvDash \beta$.

Therefore, these (distributive and collective) versions of *ex falso*, and Disjunctive Syllogism (the usual suspect in the Lewis-Popper argument for *ex falso*), fail. Let us pay attention to how this happens: such consequences fail *precisely* in paradoxical contexts: situations in which α and, therefore, $\neg \alpha$, are paradoxical (both true and false), and β is false (only), provide the required counterexamples. We have already had a glance at this point, and we shall take it up again it in the following Chapter.

But there is something quite strange with **LP**, due to the fact that both $\{1\}$ and $\{1, 0\}$ are taken as designated values: we have a case of *universal satisfiability*. If each atomic formula or parameter gets the value $\{1, 0\}$, then every formula gets the value $\{1, 0\}$; as a consequence, since $\{1, 0\}$ is designated, any set of sentences is satisfiable within **LP**.[12] The interpretation that lets it happen is sometimes called the *trivial model*. At

---

[11] See Priest (1979): 228; Bremer (2005): 48. "In a very obvious sense, the semantics subsume those of classical logic. For classical logic is just the case where no parameter (and hence no formula) takes the dialetheic value $\{0, 1\}$. All that is wrong with classical semantics for the extensional connectives (and classical logic recognises no others) is that it 'forgets' this particular case" (Priest (1987): 76).

[12] See Priest (1987): 224; Beall and van Fraassen (2003): 124; Bremer (2005): 49.

first, this may appear as just a minor weirdness. Nevertheless, it has important philosophical side effects: in a sense, nothing is *ruled out* strictly *a priori*, i.e., on logical grounds alone, within a dialetheic framework based on such a logic. A trivial world is logically admitted. How do we know, then, that the *actual* world is not trivial? If the triviality of this world (and, therefore, *philosophical* trivialism) should be ruled out, it seems we have to obtain this by other means and, presumably, *a posteriori*. *A posteriori* conclusions may arguably be fallible. And this could spell troubles both for dialetheists and for those who aim at criticizing them. I will come to this issue in due course.

8.2.2 Fixing the Conditional

Most of the strictly logical problems for this version of **LP** come from the fact that its conditional is the standard material one. First, since all classical tautologies are preserved within **LP**, we find undesired laws, such as Contraction, i.e., $(\alpha \to (\alpha \to \beta)) \to (\alpha \to \beta)$. This makes **LP** unsuitable as an underlying logic for inconsistent semantics or set theories, which would be trivialized by the Curry paradox. For the same reason, also some versions of Pseudo-Scotus hold: for instance, (PS$_i$1), that is, $\alpha \wedge \neg\alpha \to \beta$. Consequently, **LP** as it stands fulfils neither the Strong, nor the Weak Condition of Non-Triviality. Conversely, several desirable logical consequences having nothing to do with Explosion fail; for instance:

$\{\alpha \to \beta, \beta \to \gamma\} \not\models \alpha \to \gamma$
$\{\alpha \to \beta, \alpha\} \not\models \beta$
$\{\alpha \to \beta, \neg\beta\} \not\models \neg\alpha$.[13]

Given the phase-displacement between tautologies and logical consequences, **LP** turns out to be too prodigal with the former, keeping such things as the horrible (PS$_i$1); and too restrictive with the latter, invalidating Transitivity (of the conditional), *modus tollens*, and violating the Modus Ponens Condition. As Priest himself notes, though, most of the rules corresponding to these logical non-consequences "are really variants on the disjunctive syllogism", resulting from "expressing '$\to$' in terms of

---

[13] See Priest (1979): 228.

'¬' and '∨'".[14] Therefore, we had better reject (Df→). So it seems that "**LP** as no acceptable conditional connective at all!".[15] On the other hand, given what we have seen in the previous Chapters, **LP** seems to score far better than the non-adjunctive and positive-plus approaches concerning the treatment of conjunction and negation. One may therefore assume **LP** as a basic paraconsistent arrangement for the other (different from the conditional) logical operators. Its conditional can then be replaced by something with the following truth table:

α	β	α → β
1	1	1
1	0	0
1	1, 0	0
0	1	1
0	0	1
0	1, 0	1
1, 0	1	1
1, 0	0	0
1, 0	1, 0	1, 0

The resulting system, called **RM₃**, fulfils the Modus Ponens Condition: when α → β and α get a designated value, so does β. Unlike **LP**, though, **RM₃** does not retain the standard tautologies. As usual, something classical has to go.[16] Alternatively, **LP** may be supplied with some relevant conditional. We should keep in mind this possibility, when we come to study the notion of entailment within relevant logics. In the first edition of *In Contradiction*, Priest favoured a strict conditional whose semantics was provided by means of a binary relation of accessibility *R* between worlds. Contraction was invalidated (and the Strong Condition of Non-Triviality respected) by allowing for *irreflexive* worlds, that is, worlds *w* such that it is not the case that *wRw*. Priest has recently changed his mind on the conditional, though,[17] and has adopted a standard relevant conditional whose semantics is framed in terms of a ternary accessibility

---

[14] *Ibid*: 232.
[15] Bremer (2005): 51.
[16] See Priest (2001): 122-3; Beall and van Fraassen (2003): 124-5.
[17] See Priest (1996), and (1987): 269-71.

relation and impossible worlds. I will talk of this kind of conditional in the next Chapter.

8.2.3 LPQ and Multi-Criterial Terms

We extend the language of **LP** into that of **LPQ** the usual way, that is, by introducing n-ary predicative constants, individual constants, variables and quantifiers, and possibly function symbols, with the appropriate syntactic rules defining well-formed terms and formulas. A **LPQ** model $\mathbb{M}$ is a pair, <D, v>, where D is a set of individuals, v is a function assigning denotations to the descriptive vocabulary. We also consider different assignments of values to the variables within an interpretation. Given a model $\mathbb{M}$, and relative to an assignment g, the denotations of individual constants and variables are individuals in D. So far, everything works the usual way. Interesting things begin to happen with the semantics of the predicates. Classically, any predicate receives only an extension, since being in it is both necessary and sufficient to avoid being in its (Boolean) complement. When one discharges the idea that the extension of a predicate and its complement are exhaustive, as in non-bivalent first-order semantics, predicates also need an *anti*-extension (or counter-extension, or negative extension). This happens when we give up the idea that truth and falsity are exclusive, too. Given an n-ary predicate, P^n, we have:

(S$_{\text{Pred}}$) $v(P^n) =$ <P$^+$, P$^-$>, with P$^+ \subseteq$ D^n, P$^- \subseteq$ D^n and P$^+ \cup$ P$^- =$ D^n.

This means that in **LPQ** predicates receive ordered pairs as their denotations: the positive and negative extension of P^n, P$^+$ and P$^-$, are subsets of the n-ary Cartesian product of D – therefore, sets of ordered n-ples; their union coincides with D^n. Intuitively, P$^+$ and P$^-$ are the sets of things satisfying, respectively, P^n and its negation. Paraconsistent authors have intuitive motivations for such a characterization. We may assume that the semantics of a predicate is specified by means of its criteria of application. Now, ordinary language hosts predicates with different, and occasionally *conflicting*, criteria of application:

> The idea is that there might be criteria the application of which may result in grouping an object within the extension of the predicate in question. There may also be other criteria the application of which may result in grouping an object within the anti-extension of the predicate in question.

Inasmuch as these criteria need not be the exact complements of each other it might be that some object is grouped in the extension *as well as* in the anti-extension of a predicate. This object thus has the property/feature corresponding to the predicate, and it lacks the property (or has a complementary property).[18]

Typically, criteria of application can be captured by meaning postulates, and these may contradict each other. Conflicting meaning postulates may be embedded in our linguistic practice, and difficult to detect and identify. However, they may also be explicitly settled by laws. Suppose, for instance, that some norm states that a marriage performed by the captain of a ship counts as a legal marriage only if the ship was in open water throughout the ceremony. It turns out, then, that some older and still holding body of laws had established that such a marriage is valid also if the ceremony has only begun with the ship in open water, but has ended with the ship in the port. Then someone may turn out to be both a married man and a bachelor (and, of course, nobody would infer from this that he is not a man anymore, or both a man and not a man, etc). Some strong paraconsistentists have therefore pointed at the existence of multi-criterial predicates as a possible case against the (LNC) (and inconsistent legal systems have also been taken by Priest and others as speaking on behalf of dialetheism).[19] For instance, according to J.C. Beall the extensions of our ordinary language predicates "are constrained not only by their role in our overall theories, but also by our 'intuitions' about them". If such intuitions are inconsistent, a good formal account of the situation may reflect this fact, instead of destroying it by means of some regimentation (e.g., via the usual parameterisation, or distinction of respects).[20] Notice, nevertheless, a shift that makes us smell once again the Russellian Fallacy, that is, the fallacy of verbalism. Given that *predicates* with conflicting criteria of application inhabit our language, Bremer's quote above straightforwardly jumps to the conclusion that there are objects violating the (LNC), that is, objects which are inconsistent with respect to the corresponding worldly *properties*. Why should a semantic inconsistency turn into a metaphysical inconsistency so swiftly? I will come to the issue of the fallacy of verbalism at length in the following

---

[18] Bremer (2005): 56.
[19] See Priest (1987): Ch. 13.
[20] See Beall (2004a): 10. See also Priest and Routley (1989d): 503-5.

Chapter, though. So let us now go back to the remaining technical details of **LPQ**.

The clauses for quantified formulas in the semantics of **LPQ** are the following:

(S∀1) $1 \in v(\forall x\alpha, g) \Leftrightarrow$ for all d ∈ D, $1 \in v(\alpha, g[x/d])$

(S∀2) $0 \in v(\forall x\alpha, g) \Leftrightarrow$ for some d ∈ D, $0 \in v(\alpha, g[x/d])$

(S∃1) $1 \in v(\exists x\alpha, g) \Leftrightarrow$ for some d ∈ D, $1 \in v(\alpha, g[x/d])$

(S∃2) $0 \in v(\exists x\alpha, g) \Leftrightarrow$ for all d ∈ D, $0 \in v(\alpha, g[x/d])$

where $v(\alpha, g)$ is the evaluation of a formula α relative to the assignment g of values to its variables, and $g[x/d]$ is the same assignment as g, except that it assigns to x the individual d.[21] **LP** and **LPQ** have also been developed in the shape of semantic trees, for which we have, in particular, completeness proofs for **LPQ**.[22]

8.3 LPm and the Classical Recapture in Priest's Approach

Considering the radicalism of the dialetheic position asking us to question the *principium firmissimum*, the novelties introduced by **LP** with respect to classical logic are not dramatic. Something has to go, nevertheless. Some logical consequences, with the corresponding rules of inference, are not acceptable anymore within **LP** (and, as we will see in the next Chapter, neither they are in most relevant logics). The most notable loss is that of Disjunctive Syllogism. *Ex falso* looks counter-intuitive to the neutral eye, too (after all, classical logicians themselves call it a *paradox*); but we commonly use (DS) and feel there is nothing wrong with it. I know that Achille Varzi is either in New York or in Italy; and I know he is not in New York; therefore, he is in Italy. If any consistent strategy addressing the logical paradoxes has the burden of explaining which premise in the derivation of the paradoxes has to be dismissed, and why it appeared so plausible to us, that only the derivation of a contradiction cast

---

[21] See Priest (1987): 77; Bremer (2005): 55-6.
[22] See e.g. Bloesch (1993); Bremer (2005): 52-9, who also provides some sample proofs via semantic tableaux.

doubts on it; reciprocally, paraconsistency has the burden of explaining why such inferential patterns as (DS) seem to us perfectly justified. Detailed elucidations concerning the classical recapture are also supposed to be provided: how are we to regain legitimate theorems of ordinary mathematics, whose proof crucially involves some step of (DS)?

You may notice that this is quite a different position with respect to, say, intuitionism. Typically, paraconsistency does *not* claim that there is something wrong with ordinary mathematics. It may claim that there is something deeply wrong with mainstream set theories, as we know from § 3. However, as a rule it accepts that classical mathematical reasoning is valid. Priest proposes to call *quasi-valid* an extensional inference which is classically valid, but dialetheically invalid.[23] And his solution to the problem of the classical recapture is, tersely put: counterexamples to such inferential schemas as (DS) occur only in inconsistent contexts. Therefore, we can use a quasi-valid inference, *provided* we are in the safe realm of the consistent.

Technically, it is not complicated to give a precise meaning to this proviso, especially since most paraconsistent logics "generalize" classical logic, in the sense that they coincide with it in all cases (i.e., all interpretations) that are consistent (the *difficult* issue, on the other hand, is once again a philosophical one, as we shall see in the following). Priest has tackled the technical side via the *minimally inconsistent* **LP** approach, **LPm**. The basic intuition comes from Batens' adaptive strategy. Given a set of premises (data, beliefs, etc.) from which we are supposed to extract consequences, "we can cash out the idea that the situation is no more inconsistent than we are forced to assume by restricting ourselves to those models of the information that are, in some sense, as consistent as possible, given the information".[24] **LPm** is non-monotonic, and consistency is a default assumption in the sense of non-monotonic logics. **LP** interpretations can be ordered with respect to their degree of inconsistency – for instance, given two models of **LP**, M_1 and M_2, M_2 is said to be strictly more inconsistent than M_1 ($M_1 < M_2$) iff more atomic sentences are made both true and false by M_2 than by M_1. Logical consequence is then defined with respect to models that are *minimally* inconsistent: a minimally inconsistent model of a set of formulas Γ is the least inconsistent model of Γ in the ordering, and α is a minimally inconsistent consequence of Γ

---

[23] See Priest (1987): 111-12.
[24] Priest (1987): 222.

($\Gamma \vDash_m \alpha$) iff every minimally inconsistent model of Γ is a model of α. **LPm**-consequence is the same as classical logical consequence in any consistent context, and "*LPm* is a more generous notion of consequence than *LP*, which allows for classical inferences – such as the disjunctive syllogism – provided inconsistency does not 'get in the way'".[25] Priest has also shown that, in the most relevant cases, if the **LP**-consequences of some set of formulas are non-trivial, so are their **LPm**-consequences (this fact has been called Reassurance). Further details can be found in Priest (1991), and (1987): Ch. 16.

8.3.1 Consistency as the Default Option

Disjunctive Syllogism is a quasi-valid rule of inference, which holds provided we are in consistent contexts. And consistency, according to Priest, is the default option. I have previously mentioned the debate among dialetheists on the possibility of admitting contradictions in the empirical world, therefore, concrete inconsistent objects – e.g., via a paraconsistent *de re* interpretation of the pervasive phenomenon of vagueness. Many feel this option implausible, and limit the possibility of true contradictions to such territories as semantics and set theory, therefore having at most abstract inconsistent objects, such as sets. However, even those who take seriously the idea of perceivable contradictions, as Priest does, claim that "dialetheias appear to occur in a quite limited number of domains: certain logico-mathematical contexts, certain legal and dialogical contexts [...] and maybe a few others".[26] Hence, we normally use and consent to quasi-valid arguments, and (DS) is employed both in ordinary mathematics and in everyday reasoning. Priest proposes, then, the following Methodological Maxim:

(M) Unless we have specific grounds for believing that the crucial contradictions in a piece of quasi-valid reasoning are dialetheias, we may accept the reasoning.[27]

---

[25] *Ibid*: 225.
[26] Priest (1987): 116.
[27] *Ibid.* See also Priest (1979): 235.

Since exceptions easily pass unnoticed, and given the ideology of consistency dominating standard logical and philosophical practice since Aristotle, we came to believe that (DS) is valid *simpliciter*, not quasi-valid.

A dialetheist and a classical logician, therefore, reason exactly in the same way in any (apparently) consistent environment. Following the Methodological Maxim (M), the dialetheist tolerates quasi-valid inferences and, typically, (DS). However, if in the context initially taken as consistent a deep-rooted contradiction follows from seemingly analytic principles, the classicist is taken to disaster by her explosive logic. On the contrary, the dialetheist can *choose*: if she believes the contradiction to be independently rejectable, she will have to adjust her theory in order to eliminate the inconsistency. If, on the other hand, she believes the contradiction to be true, she will just accept it. This time, an analogy with intuitionism does work: the intuitionist may allow herself the full force of classical logic and the (LEM), *provided* she is reasoning within finite (or at least decidable) domains. Similarly, classical logical consequence can be understood by the dialetheist in terms of quasi-validity within consistent environments, and the dialetheist can infer from her body of information just as much as the classicist. The latter "explodes" in inconsistent situations, though, whereas the former does not. Dialetheism turns out to be an extension of classical logic also in this sense.[28] In addition, as we have seen, such logical strategies as the adaptive approaches, or the non-monotonic **LPm**, show us how to handle the issue technically, by moving from standard logic to a weaker paraconsistent framework in case of need.

---

[28] See Priest (1987): 117-9. See also Priest (1978): 142-3: "provided a theory is consistent then we can use classical logic. But of course classical logicians are interested only in consistent theories anyway. Others are pointless. Thus the full strength of classical logic, wherever it can be used non-trivially, is taken over by paraconsistency. Its problem-solving power is preserved as required. Where its *extra* strength comes from is precisely in being able to function as the underlying logic of inconsistent theories. Disjunctive Syllogism cannot be used in such circumstances and the theory does not collapse into triviality. Thus we see that paraconsistent logic is related to classical logic in much the same way that, say, special relativity is related to Newtonian mechanics. Just as Newtonian mechanics is generally incorrect but usable for low velocities (i.e. agrees with special relativity for low velocities), so classical logic is incorrect generally but usable for consistent situations (i.e. it agrees with paraconsistency in consistent situations)".

8.4 Consistency and the Metalanguage

I claimed that a counterexample to:

$$\frac{\alpha \vee \beta,\ \neg \alpha}{\beta} \quad (DS)$$

is provided within **LP** by assuming that α is paradoxical, both true and false. One may therefore be tempted to think that (DS) is, as it were, an *enthymematically* valid inference rule. Since it is quasi-valid, that is, a good rule provided we do not find ourselves in the realm of the inconsistent, an inferential schema which adds to (DS) an extra premise to the effect that α is consistent, i.e., non-paradoxical, i.e., not a dialetheia, might do the trick. We may try to introduce some operator(s) expressing consistency in the formalized object language, of the kind employed within the logics of formal inconsistency. Hence comes the following extension of **LP(Q)**.

8.4.1 Semantic Operators in LP(Q)

The extension adopts six object language operators.[29] To understand them, you may just recall that the set of truth values in **LP** is $\{\{1\}, \{0\}, \{1, 0\}\}$. Now, consider the following tabular presentation:

α	$T\alpha$	$F\alpha$	$\Delta\alpha$	$\nabla\alpha$	$°\alpha$	$\bullet\alpha$
1	1	0	1	0	1	0
0	0	1	0	1	1	0
1, 0	1	1	0	0	0	1

The interpretation is the following: "$T\alpha$" says that α is true, and "$F\alpha$" says that α is false.[30] "$\Delta\alpha$" says that α is *true only*, i.e., true and not a dialetheia, and "$\nabla\alpha$" says that α is *false only*, i.e., false and not a

---

[29] And it is taken directly from Bremer (2005): 50.

[30] You may pay attention to the following fact: the truth and falsity *predicates*, *T* and *F*, I have been using all the time, apply to sentence names, so they would be constitutively metalinguistic (if Tarski was right on the metalanguage issue, which as we know is controversial, especially for a strong paraconsistentist). What we have introduced here, T and F, are monadic sentential operators added directly to the "object language": they *use* the relevant formula, do not *mention* it.

dialetheia. "°α" says that α is not paradoxical, that is, either true only or false only; "•α" says that α is paradoxical, that is, both true and false. To be sure, the abundance can be easily reduced, since the operators are inter-definable in numerous ways. The important point is that the semantics *of* all these operators is *bivalent*: when you put any of them in front of a formula, the result will never be something whose value is {1, 0}. The underlying justification for such a move seems to be that there should be a way of talking *about* the truth values (therefore, the (in)consistent status) of sentences, which is not itself inconsistent: when we claim that α has a certain consistent or inconsistent status, what we claim should not itself be inconsistent, but either true only, or false only.

8.4.2 Forcing Consistency

Now, by some Carnapian principle of tolerance we may consent to the introduction of almost any operator, provided that it does not appear to lead to triviality, and its definition is clear-cut enough (it's a free country). Nevertheless, Priest would probably disagree with the spirit of the whole initiative. Both in the '79 paper containing the first presentation of **LP**, and in subsequent writings, he has rejected for philosophical reasons the idea that *any* operator can *force* consistency. The plain reason for this is that when we say "α behaves consistently", our very claim can itself be inconsistent:

> The trouble here is that there is no a priori guarantee that 'A is not paradoxical' is not itself paradoxical. Thus 'A is not paradoxical ∧ A ∧ ¬A' may itself be true. And if it is, all of the premises of the enthymematic disjunctive syllogism are true whilst the conclusion is arbitrary. Nor it is difficult to show that attributions of paradoxicality may themselves be paradoxical. [...] To summarize so far: there is no statement that can be made that forces a formula to behave consistently. We can say 'A behaves consistently', but because our "metatheory" is liable to be inconsistent, this cannot *force* consistent behaviour. This is just one of the hard facts of paraconsistent life.[31]

To get the point, you may recall that the role of those sentential operators is to provide the formal ("object"-) language with means of expressing semantic facts we would also describe by "metalinguistic" sentences

---

[31] Priest (1989): 623-4.

in which semantic predicates are attached to sentence names. Now, a thorough dialetheist has independent philosophical reasons for rejecting the idea that some "metalinguistic" or "semantic" fragment of her theory is inconsistency-proof. One of the essential motivations for (strong) paraconsistency, as we know, has to do with the rejection of Tarski-like approaches trying to solve the various semantic paradoxes by means of a hierarchy, or of some rigid separation between object language and metalanguage. As Priest has claimed, "the beauty of the paraconsistent approach to logical paradoxes is that it finally renders the object-language/meta-language distinction unnecessary in any shape or form".[32] To be sure, we may also keep calling "metatheory" the part of the theory that includes the semantic notions, but (apart from the fact that the terminology might now sound misleading) this should be considered as a part or sub-theory of the whole dialetheic theory. For instance, according to Routley "of course we can retain the distinction between object-language and metalanguage, basic to metamathematics; but we are free to characterize 'true in English', for example, within English".[33] This has technical outcomes, too: Priest has conducted a campaign within the paraconsistent community, to support the idea that the metalogic of paraconsistent logics should be paraconsistent, too (in particular, it should use some non-classical, paraconsistent set theory,[34] like the ones we will meet in the following part of this book). Intuitionists' claim that their logic is the right one is philosophically uncertain unless they have an intuitionist metatheory, too. Similarly, there is no reason why the (strong) paraconsistent theorist should be allowed to pass unnoticed when, after claiming that the One True Logic is paraconsistent, she uses a classical metatheory to obtain her metalogical results.

Consequently, (DS) cannot be taken as an enthymematically valid inferential pattern. No premise capable of guaranteeing the consistency of the crucial α (by using or mentioning it) is available. The fundamental philosophical point is that, according to the dialetheist, contradictions *do not have a very different status from any other sentence:* in some cases, they can be true; they can be rationally accepted, and asserted. For instance, the idea that I am a fried egg is as unwarranted as the idea that I am a fried egg and not a fried egg, and much *less* warranted than the the-

---

[32] Priest (1984a): 161. See also Priest (1979): 220ff, and (1987): Chs, 1 and 9.
[33] Routley (1979a): 323.
[34] See Priest (1987): 258-9.

sis that the liar is both true and false. In other words, "the dialetheist, [...] considers the hypothesis of inconsistency or, better, local inconsistency (i.e. the truth of α ∧ ¬α for particular α) to be no different, in principle, from any other hypothesis".[35] Therefore, it has to be evaluated case by case, on its own merits and, of course, in a defeasible way:

> I am frequently asked for a criterion as to when contradictions are acceptable and when they are not. It would be nice if there were a substantial answer to this question – or even if one could give a partial answer, in the form of some algorithm to demonstrate that an area of discourse is contradiction free. But I doubt that this is possible. Nor is this a matter for surprise. Few would now seriously suppose that one can give an algorithm – or any other informative criterion – to determine when it is rational to accept something. There is no reason why the fact that something has a certain syntactic form – be it $p \land \neg p$ or anything else – should change this. One can determine the acceptability of any given contradiction, as of anything else, only on its individual merits.[36]

I believe that the lack of means to guarantee consistency in any part of the theory is one of the main problems – if not *the* problem – of dialetheism. This shall be dealt with in § 14. In that context, it will be noteworthy to discover that such a difficulty was anticipated by Aristotle, who expressed it under the label of ἔλεγχος: the elenctic argument against the rival of the βεβαιοτάτη ἀρχή. Meanwhile, let us face less dramatic issues. After all, classical logicians may agree that, when we reach a certain level of expressive power, there is no algorithm for consistency. One may even concede that we have to tackle contradictions on the basis of fallible heuristics, and nevertheless insist that we need something like a general principle in order to discriminate between acceptable and unacceptable contradictions, even though no algorithm is available. A proposal that has been advanced is the following.

As we have seen, the cases on behalf of strong paraconsistency are based upon the fact that some contradictions appear to follow from strongly *intuitive* principles, such as the (unrestricted) T-schema and the (unrestricted) Abstraction Principle of set theory. Intuitions, by Cohen's definition, are untutored beliefs based on no inference.[37] They have to do with the idea of *analyticity*, taken rather loosely as the idea of principles

---

[35] *Ibid*: 115.
[36] Priest (1998): 423.
[37] See Cohen (1986): 73ff.

included in the core of a single notion. The point may be made also by resorting to Stephen Schiffer's distinction between happy face and unhappy face solutions to the paradoxes.[38] In a happy face solution, we pick a member of the paradoxical set of sentences and we are able to refute it without *ad hocness*. In an unhappy face solution, we cannot gain consistency because we are rationally committed to accept each member, so that rejecting any of them would violate a conceptual, i.e., strictly analytical, norm. Now, it is of the very concept of truth, the dialetheist argues, that the truth predicate satisfies the unrestricted T-schema; and it is of the very concept of set that, given any condition P, there is a set, the set of all Ps, to which something belongs iff it is a P.[39] One may therefore distinguish between ordinary (rejectable) inconsistencies, and inconsistencies stemming from *the analysis of unitary concepts*. A dialetheist may propose, then, as a necessary condition on the truth of a contradiction, that it spreads in such an analytic way from some basic intuition on some basic notion. Analytic intuitions are, so to speak, existence-entailing: the contradictions of naïve set theory show that inconsistency lies in what it is to be a set. More generally, "if the analysis of an intuitive concept, of what it is to be an X, if the very idea of Xhood yields an inconsistency, then it lies in the very nature of Xhood that there are inconsistent Xs".[40] So, as Woods conjectures, a dialetheist may accuse her consistent rival of committing some sort of "*converse* begging the question" fallacy, when the latter rejects some such contradiction. She is refusing to stick to premises she *should* accept (and maybe, as in the case of Frege facing "the contradiction", premises that were accepted until the contradiction emerged), because they follow analytically from basic concepts she shows to fully share. All consistent solutions to the semantic and set-theoretic paradoxes are unhappy face solutions.

We will find such a "realistic" attitude again among strong paraconsistentists in the following. In my opinion, nevertheless, pleas for intuitions and analyticity are troublesome especially in this context. As we have seen since § 5.3.1, intuitions typically conflict with each other *precisely* when fundamental notions are at issue, and paraconsistency has proved a battleground for such clashes: "analytic intuitions are dialectically impo-

---

[38] See Schiffer (1998).

[39] At the beginning of *In Contradiction,* Priest conjectures that these principles "are *a priori* and, one might argue, analytic, specifying as they do (at least in part) the defining conditions of the notions concerned" (Priest (1987): 4).

[40] Woods (2003): 189.

tent when directed at the analytically other-minded".[41] In his most mature writings, Priest appears to have developed a cost-benefit strategy, defending the truth of some contradictions by stressing the broadly theoretical and epistemological advantages of the theories that include them. The dialetheic choice of accepting the logical paradoxes as sound arguments wins because of its simplicity, explanatory power, lack of *ad-hocness*, etc.[42] The best corroboration for dialetheias, in this perspective, is a justification of the dialetheic *research program*, and this depends in its turn on successes achieved by applying (strongly interpreted) paraconsistent logics. We will look at some applications in the third part of the book. Before reaching them, though, we have to meet the most promising ones among these logics.

---

[41] *Ibid*: 194.
[42] See e.g. Priest (2006): Part III.

9 Relevance Logic and Impossible Worlds Again

> The law that it is impossible to affirm and deny simultaneously the same predicate of the same subject is not expressly posited by any demonstration [...]. For grant a minor term of which it is true to predicate man – even if it be also true to predicate not-man of it – still grant simply that man is animal and not not-animal, and the conclusion follows: for it will still be true to say that Callias – even if it be also true that to say that not-Callias – is animal and not not-animal.
>
> Aristotle, *Posterior Analytics*

9.1 Overview

When someone claims that some statement is "irrelevant" within an argumentation, she usually means that it is of no utility in getting to the conclusion. At the core of the *relevance logic* research program lie two ideas related to this, a positive and a negative one. The positive idea is that the notion of relevance goes beyond the mere realm of pragmatics; it fully belongs to *logic*, and it can be supplied with a rigorously formal treatment. The negative idea is that the classical logical apparatus is guilty of irrelevance, that is, it retains as good patterns of reasoning arguments in which some premises are irrelevant with respect to the conclusions. Relevance logic is also called *relevant* logic (mainly in Australasia, it seems to me), to imply that it is the One True Logic – and I shall use both terms interchangeably in this Chapter. In the hierarchy proposed by Susan Haack to classify the various degrees of logical departure from the standard formalism, relevance logic would occupy for some of its aspects the seventh rank, that is, the level of the "challenges to the standard conception of the scope and aspirations of logic [...] often enough associated with challenges to classical metaconcepts".[1]

---

[1] Haack (1978): 155; see also 201.

Systems of relevance logic have been built since the pioneering works of A.R. Anderson and N.D. Belnap in the Sixties. These logics are also, in my opinion, the best, both most developed and most promising, paraconsistent approach around. The relevant tradition is nowadays well established; it has achieved an impressive amount of metalogical results, and several applications within mathematics, philosophy, and computer science. Because of this, I shall enter into more (also technical) details than I did for some of the paraconsistent approaches examined in the previous Chapters. Particularly the semantics provided for these formal systems raise numerous interesting questions on one of the most important philosophical issues dealt with in this book: how to provide a plausible interpretation for a logic underlying theories that violate the (LNC). This will lead us to deepen some of the considerations we have already met (e.g., in § 6 on non-adjunctive approaches), concerning the intuitive understanding of the models of a paraconsistent logic.

The literature on relevance is far too vast to provide even a partial overview of the subject: J.M. Dunn and Greg Restall recently mentioned a list containing over 3000 entries in papers and essays on relevant logic and related subjects.[2] Nevertheless, we have a rationale to delimit, albeit roughly, the area under discussion. This is provided by the fact that our main topic is the (LNC)-controversy. Now, Anderson and Belnap's original problem was *not* the one of building logical systems that could underlie inconsistent theories without leading them to explode. In particular, they were not prone at all to (what later came to be called) strong paraconsistency or dialetheism, i.e., to the admission of true contradictions, and/or inconsistent objects and/or states of affairs. The paraconsistent features of their systems came, as it were, as a by-product of their quest for a satisfying formalization of the notion of *relevant implication* that could avoid the paradoxes of the conditional. After having exposed the main general ideas, I shall therefore limit myself to those features of relevance that are relevant to paraconsistency (and leave out other aspects, such as the place of relevance within the wider framework of substructural logics). Even after restricting our field this way, it will be hard to do justice to the subject. Then again, we will learn quite soon that the question of relevance and the one of the violation of the (LNC) are probably closer relatives than Anderson and Belnap wished.

---

[2] See Dunn and Restall (2002): 1.

9.2 Relevant Implication

9.2.1 The Paradoxes of Material Conditional

It is no accident that, within classical logic itself, Pseudo-Scotus is called a *paradox* of material conditional. No classical logician blames the inference as incorrect. The trouble is with *how* you infer something, rather than in what you infer. *Ex falso* is often called *negative* paradox, because negation occurs in it. Its mate, the positive paradox, is sometimes called "*a fortiori* reasoning", and was captured by medieval logicians via the Latin motto *verum ex quolibet*:

(VEQ) $\alpha \to (\beta \to \alpha)$.

The schema tells us that, if a given α holds, then α is entailed by any formula β. (VEQ) is classically provable, e.g., as follows:

(1)	1	α	Ass
(2)	2	β	Ass
(3)	1, 2	$\alpha \wedge \beta$	1, 2, $\wedge$I
(4)	1, 2	α	3, $\wedge$E
(5)	1	$\beta \to \alpha$	2, 4, $\to$I
(6)		$\alpha \to (\beta \to \alpha)$	1, 5, $\to$I

The weird features of (VEQ) can be highlighted by providing it some content: the schema is instantiated, e.g., by the inference according to which "If the earth is round then, if pigs can fly, then the earth is round". Weirdness increases when we consider that, since β stands for any sentence, the schema is also instantiated by "If the earth is round then, if pigs can*not* fly, then the earth is round". Our sense of disease depends on our perceiving no correlation between the roundness of the earth and the fact that pigs can, or cannot, fly. Material conditional does not involve any causal or conceptual connection between its antecedent and its consequent. It expresses a mere truth function, according to which it is not the case that the antecedent holds and the consequent does not. On the contrary, to accept a statement of the form "If..., then..." we are intuitively inclined to require that some content connection holds between its antecedent and its consequent. However, as we have seen at the beginning of this part of the book, the nuisance cannot be solved by merely switching

190 / How to Sell a Contradiction

to Lewis' intensional implication, which is equally subject to the paradoxes of strict implication.[3] It is not even an issue reducible to the standard treatments of counterfactual conditionals.[4] According to relevance logicians, the very idea of entailment is essentially relational; it has to express some connection between antecedent and consequent, or between premises and conclusions, that cannot be shrunk to truth-functional or (standard) modal notions.

9.2.2 Implication and the Conditional (Boil Down to the Same)

Relevance logicians usually talk in a rather indistinct fashion of *implication* and of the *conditional*. In this Chapter, I am following them with the terminology. But even though "paradoxes of material implication" is a traditional expression, Quine has taught us to pay heed to the distinction between "If..., then..." and "... implies...". The former is a connective, so sentences should replace the dots in it; the latter is a verb, so nouns are supposed to surround it. It is well known that, according to Quine, intensional logic was conceived in sin – or, better, in two original sins: essentialism, and the use/mention confusion. The latter took place when, aiming at expressing the necessity of the connection between antecedent and consequent, Lewis placed it within the object language by coining his strict implication, instead of pushing it up to the metalanguage, as provability or deducibility.[5] Relevance logic has been accused of being merged in the same confusion between levels.

It has nevertheless been observed[6] that what is at issue in the relevance logic enterprise is a more general notion of logical entailment: something that, hierarchic distinctions notwithstanding, should hold for *all* logical levels articulating inferences. That the issue is a broader one is manifested by the fact that, within the classical perspective itself, the (pseudo-)connection captured by material conditional is mirrored in the metalinguistic sign of theoremhood (or syntactic consequence), ⊢, for any logic in which the Deduction (Meta)Theorem holds (this is also the role played in our natural deduction calculus by Conditional Proof, (→I)). A

---

[3] For a classic relevant criticism of strict implication, see Meyer (1974).
[4] See e.g. Stalnaker (1968), (1984): Ch. 7; Lewis (1973).
[5] See Quine (1976): Chs. 13 and 14.
[6] See Dunn (1986): 120-2; Dunn and Restall (2002): 3-5.

renewal of the former will therefore require a renewal of the latter. As Anderson and Belnap have claimed against the Quinean distinction:

> It is philosophically respectable to "confuse" implication or entailment with the conditional, and indeed philosophically suspect to harp on the dangers of such a "confusion". (The suspicion is that such harpists are plucking a metaphysical tune on merely grammatical strings.)[7]

9.2.3 "Washing Dirty Money Through Mexico"

The initial diagnosis of the paradoxes of material conditional is that they are "fallacies of relevance", lacking a content connection between antecedent and consequent, or between premises and conclusions. "Be relevant" is, famously, one of Grice's conversational maxims – the Maxim of Relation.[8] As I have anticipated, though, the relevant approach is characterized by its refusing to confine such a fallacy in the dominion of pragmatics. I could not improve Bremer's plain remark according to which "it seems implausible that we use logic to derive some conclusions *and then* revise this set of conclusions to retract all Irrelevant deductions".[9] According to Dunn, "the theory of Grice makes much use of a basically unanalyzed notion of relevance. One of Grice's chief conversational rules is 'Be relevant' but he does not say much about just what this means".[10] Relevance logic may be taken as a formal analysis of the notion.

Let us say something on the mainstream notion of *relevance*, which does not seem to be quite univocal. Systems of relevant logic have been initially developed via axiomatic presentations (we will see a few examples soon). However, they are equally well described in terms of natural deduction. The notation in this case also keeps a record of the assumptions on which formulas depend (so Anderson and Belnap (1975) have claimed that natural deduction provides a proof-theoretic motivation for relevance). The dependence can be expressed by indexing the formulas employed by the various inferential steps: once each formula assumed within the proof has been provided with a distinct numeral, this can be passed on through the various applications of the inference rules. Now,

---

[7] Anderson and Belnap (1975): 473.
[8] See Grice (1975).
[9] Bremer (2005): 37.
[10] Dunn (1986): 124.

let us begin with a negative case. Consider again the first four lines of the derivation of (VEQ) above:

(1) 1 α Ass
(2) 2 β Ass
(3) 1, 2 $\alpha \wedge \beta$ 1, 2, $\wedge$I
(4) 1, 2 α 3, $\wedge$E

At line (4), we are back with the α we started with. But now, it is indexed "1, 2", as is shown by the assumption column. This is a typical case of irrelevant manoeuvre: β has not been *actually used* to obtain α at line (4). α was there since the beginning, and the combination of Conjunction Introduction and Elimination only aims at giving the illusion that α finally depends on β. Dunn has claimed that such irrelevant tricks of classical derivations are comparable to "washing dirty money through Mexico".[11] The example is striking. The next task is to understand what "actually used" positively means.

9.2.4 The Variable Sharing Property

From the proof-theoretic point of view, the various systems of relevance logic are built around the idea of avoiding the paradoxes, (VEQ) and (PS), in all of their shapes. Different arrangements are available (we shall look at some of them soon), but the philosophical point behind them has to do with the connection between antecedent and consequent. Exactly how such a connection should be made explicit is not quite straightforward, and one may say that the very concept of relevance is not univocal within the relevant theoretic framework. A first notion of relevance is the "content-overlap" one. A minimal or necessary requirement, the so-called Weak Relevance Condition or Variable Sharing Property should capture it:

(VSP) If $\alpha \to \beta$ is a theorem, then α and β share at least one sentential variable.[12]

---

[11] *Ibid*: 141.

[12] "A formal condition for 'common meaning content' becomes almost obvious once we note that commonality of meaning in propositional logic is carried by commonality of propositional variables. So we propose as a *necessary*, but by no means sufficient, condi-

(VSP) aims at expressing the intuitive idea that the link between the antecedent and the consequent of logically valid entailments has to based upon some "analyticity" or commonality of meaning – which is lacking if we stick to the classical material conditional.[13] Let us notice that, on the basis of (VSP), neither positive-plus systems nor the basic **LP** are relevant. Therefore, relevance logics are a sub-group of paraconsistent logics, in the sense that the requirements to be satisfied for a logic to be relevant are stricter than the fulfilment of the Weak Condition of Non-Triviality.

Relevantists have claimed that a reform of entailment goes hand in hand with a reform of deduction. Let us look at the (standard) Deduction (Meta)Theorem again:

(DT) $\alpha_1, \ldots, \alpha_n \vdash \beta \Leftrightarrow \alpha_1, \ldots, \alpha_{n-1} \vdash \alpha_n \to \beta$.

Starting with:

(K) $\alpha, \beta \vdash \alpha$

which holds merely by the ordinary definition of $\vdash$, two applications of (DT) to (K) lead us directly to the positive paradox, (VEQ):

$\vdash \alpha \to (\beta \to \alpha)$.

In a natural deduction framework, the same problem holds for any system having Conditional Proof, ($\to$I), which simulates precisely (the left-right half of) the (DT). Again, the problem is that in (K) β is *irrelevant*. A second and perhaps partly different notion of relevance, which may be called "actual-use" relevance, comes into play here. A deduction:

$\alpha_1, \ldots, \alpha_n \vdash \beta$

is relevant with respect to a given assumption α_i ($1 \leq i \leq n$), as the relevant story goes, only if α_i has been actually used in order to obtain β.

---

tion for the relevance of A to B in the pure calculus of entailment, that A and B must share a variable" (Anderson and Belnap (1975): 32-3).

[13] See Dunn (1986) 145-6; Anderson and Belnap (1975): Sect. 5.1.2 and, for a proof of (VSP), Sect. 22.1.3.

"Actually used" can be made precise, in the simplest case, as follows: we have a chain of inferential steps – say, (E→) or the *modus ponens* – beginning with α_i and getting to β, with some marker appearing at the line in which α_i is assumed, passing on each time at least one of the premises of the *modus* is marked, and reaching β. (DT), then, has to be reshaped so that it allows us to "discharge" the assumption α_j only if the deduction of β is relevant with respect to α_j. A deduction is relevant *simpliciter* iff it is relevant with respect to all the assumptions. The same proviso should be respected in the (re)formulation of (I→) in natural deduction.[14]

9.3 Relevance and Contradiction

Richard Routley took the supposition that there are no contradictions in the world as a sort of religious faith. A classicist believes in the Consistency of the Whole; a relevance logician suspends her judgement; a dialetheist, just like an atheist, discards the hypothesis.[15] Routley's philosophical career evolved into this kind of atheism,[16] whereas, in this perspective, the relevantist would be a natural agnostic on the (LNC). This has been taken by some as entailing that relevance logics are best suited for weakly paraconsistent purposes, e.g., for the formalization of patterns of (relevant) deduction in the presence of inconsistent data or information, with no need to take a stance on the worldly validity of the (LNC). The relevant stance is often introduced as requiring, for an inference to be correct, *classical* truth-preservation *and* something more (namely, some content connection between premises and conclusion, etc.). Nevertheless, some of the main assumptions of the whole research program of relevance appear to gain real plausibility only if they are linked to a fairly strong paraconsistency. Let us see why.

Presenting the so-called independent argument for *ex falso*, I have mentioned Popper's claim that the proof should be understood and accepted by every thinking man. Similarly, C.I. Lewis' challenge to those

---

[14] For detailed explanations and some proofs, see Anderson and Belnap (1975): 17ff; Dunn (1986): 133-9.

[15] See Routley (1979a): 301.

[16] As for how this happened, I cannot but refer you to the wonderful fictional (?) story told in Priest's *Sylvan's Box* (Priest (1997c)) on how Routley changed his mind at the beginning of the 80s, after finding an actual impossibile object. It seems that *Sylvan's Box* played a role in somehow changing David Lewis' mind on paraconsistency (see Lewis (2004): 176).

who may question his proof was meant as a rhetorical move ("pick the inference you want to reject!"). However, as it often happens in philosophy, someone's proof of a counter-intuitive conclusion is someone else's refutation of one of the premises.[17] Anderson and Belnap claimed that the Lewis-Popper proof is "self-evidently preposterous", and that "it is immediately obvious where the fallacious step occurs".[18] When the referee of a paper submitted by them to the "Journal of Symbolic Logic" in the late 50s pointed out that Disjunctive Syllogism failed in their calculus, Anderson and Belnap's bold rejoinder was simply to declare that things had to go that way.[19] I expressed (DS) as an inference rule:

$$\frac{\alpha \vee \beta, \neg \alpha}{\beta} \quad (DS)$$

Accepting Double Negation, which holds in most mainstream relevant systems (and replacement), (DS) is equivalent to Ackermann's so-called "γ–rule":

$$\frac{\neg \alpha \vee \beta, \alpha}{\beta} \quad (\gamma)$$

As we have seen in § 5.2, this is nothing but *modus ponens* for the material conditional defined by (Df→). The admissibility of γ was the first in the list of the "open problems" for relevance logic raised by Anderson in his influential (1963) paper. I will cut a long story short and say that, cumbersome provisos apart, many relevantists' solution amounts either to a rejection or to some form of strong limitation of both (γ) and (DS), especially when one comes to the applications of relevant logics (such as the relevant arithmetic **R#** which we shall meet in § 12).[20] Within the mainstream relevant systems, one can prove:

---

[17] Or, famously, "my *modus ponens* is your *modus tollens*". This is what John Woods called "Philosophy's Most Difficult Problem" (see Woods (2003): 14).

[18] Anderson and Belnap (1975): 165.

[19] The paper was published. See Anderson and Belnap (1959): 302.

[20] See Routley and Routley (1972); Dunn (1986): 151ff; Read (1988): 31ff. One way of regaining (DS) for some relevant systems is to make it work like the rule of Necessitation in standard modal logic: when both $\alpha \vee \beta$ and $\neg \alpha$ are theorems, then β is a theorem, too (see Anderson and Belnap (1975): 299).

$$\alpha \wedge \neg\alpha \to \alpha \wedge (\neg\alpha \vee \beta),$$

from which, if the aforementioned rules held, (PS$_i$1) (i.e., the imported *ex falso*), would follow by transitivity. We also know, on the other hand, that (DS) is invalid in such paraconsistent logics as **LP**. Now, let us go back to the argument against the Lewis-Popper proof we met in § 5.2. Recall that the proof of (PS$_i$1) was the following:

(1)	1	$\alpha \wedge \neg\alpha$	Ass
(2)	1	α	1, $\wedge$E
(3)	1	$\neg\alpha$	1, $\wedge$E
(4)	1	$\alpha \vee \beta$	2, $\vee$I
(5)	1	β	3, 4, DS
(6)		$(\alpha \wedge \neg\alpha) \to \beta$	1, 5, $\to$I

Let us follow our *bona fide* intuitions: if one assumes a formula, this amounts to the hypothesis that it is true and, if one assumes its negation, this amounts to the hypothesis that it is false (after all, by (Neg1) truth is falsity of negation). Then, at step (3) α is taken as false. According to step (4), at least one of α and β has to be true. (DS) gives us β, at line (5). But at line (2) α was taken as *true*. This mirrors the plain fact that we are reasoning in an inconsistent context (we began by assuming $\alpha \wedge \neg\alpha$ – that is, semantically, α is taken as both true and false). Now this is not a *milieu* in which (DS) can work appropriately. The other rules are not under discussion here: ($\vee$I) and ($\wedge$E) mirror plain truth-functional features of conjunction and disjunction. However, (DS) gives us β at line, (5) only because, to begin with, have both $\neg\alpha$ and α among our assumptions, obtaining $\alpha \vee \beta$ from the latter. Now in inconsistent contexts, that is, under the hypothesis that $\neg\alpha$ and α are both true, $\neg\alpha$ and $\alpha \vee \beta$ are, too, whatever is β and, particularly, even if β is false. Therefore, (DS) is just a bad rule of inference, since it is not truth-preserving: it may lead us from true premises to a false conclusion.[21]

In § 5, I claimed that this is the master argument against (DS). Now, essential to the argument is precisely the assumption of *inconsistencies*. This way is not open, though, to a relevant logician who does not admit

---

[21] See Dunn (1986) 152-3; Priest (1989): 622-3.

countermodels to the (LNC) (whatever the intuitive reading of the countermodels is supposed to be). Such a relevantist can only claim that what is wrong with (DS) and (γ) has to do exclusively with considerations of *relevance*. Anderson and Belnap did try to defend the idea that the object-language counterparts of (DS) and (γ), namely:

$$\neg\alpha \wedge (\alpha \vee \beta) \to \beta$$

$$\alpha \wedge (\neg\alpha \vee \beta) \to \beta$$

are "nothing less than [...] fallac[ies] of relevance",[22] and should be questioned because of *this*. Such an idea is less than convincing, though. It is not explained by (VSP): the Weak Relevance Condition is violated by (PS), but it is fulfilled by (DS) and (γ) – and, indeed, by all the steps of the Lewis-Popper proof. The fact that the fallacy of relevance supposedly committed by (DS) and (γ) cannot be characterized in terms of (VSP) may itself be interpreted as a confirmation of the idea that sharing a variable is not *sufficient* for relevance.[23] However, didn't Anderson and Belnap tell us that "content-overlap" relevance was "*carried by* commonality of propositional variables"? Supporters of strong paraconsistency have constantly stressed this point:

> The antecedent and consequent of th[ese] principle[s] satisfy Anderson and Belnap's own condition for relevance; that is, they have a variable in common. Admittedly, this was never intended as a sufficient condition, merely a necessary one. However, A ∧ (¬A ∨ B) looks, for all the world, as if it is relevant to B. [...] Anderson and Belnap are quite right that [(DS)] is formally invalid (though of course certain of its substitution instances may be valid). However, the explanation of this is provided by paraconsistency. [(DS)] is not a fallacy of relevance: it just plain fails to be truth preserving. If A and ¬A are true, then the premises are true, whatever B is. Of course, Anderson and Belnap do not believe in true contradictions and so cannot take this line. This makes their rejection of [(DS)] philosophically unstable.[24]

---

[22] Anderson and Belnap (1975): 165; see also Sect. 25.
[23] Some even claim that (VSP) alone is of virtually no help whatsoever in fixing the valid rules: see Diaz (1981): 17.
[24] Priest (1989): 622. For a nice discussion of the whole situation, see Woods (2003): 58-68.

Here we have the decisive shift, leading us from relevance to paraconsistency. According to dialetheists, such rules as (DS) and (γ) fail precisely in inconsistent contexts. In Priest's jargon, they are defeasible *quasi-valid* inference patterns, acceptable insofar as we are in the realm of consistency. Therefore, the issue of relevance and the admission of violations of the (LNC) appear to be, at least to some extent, two sides of the same coin.

9.4 Bits of Relevant Syntax

9.4.1 The Mainstream Systems

We can start our fast axiomatic overview of relevance with a basic system, **B**, of which the most popular systems are extensions. The axiom schemas for **B** are:

(A1) $\alpha \to \alpha$
(A2) $\alpha \to \alpha \vee \beta$
(A3) $\beta \to \alpha \vee \beta$
(A4) $(\alpha \to \gamma) \wedge (\beta \to \gamma) \to (\alpha \vee \beta \to \gamma)$
(A5) $\alpha \wedge \beta \to \alpha$
(A6) $\alpha \wedge \beta \to \beta$
(A7) $(\alpha \to \beta) \wedge (\alpha \to \gamma) \to (\alpha \to \beta \wedge \gamma)$
(A8) $\alpha \wedge (\beta \vee \gamma) \to (\alpha \wedge \beta) \vee (\alpha \wedge \gamma)$
(A9) $\neg\neg\alpha \to \alpha$

The rules of inference are:

$$\frac{\alpha \to \beta, \ \alpha}{\beta} \quad (R1)$$

$$\frac{\alpha, \ \beta}{\alpha \wedge \beta} \quad (R2)$$

$$\frac{\alpha \to \neg\beta}{\beta \to \neg\alpha} \quad (R3)$$

$$\frac{\alpha \to \beta}{(\gamma \to \alpha) \to (\gamma \to \beta)} \quad (R4)$$

$$\frac{\alpha \to \beta}{(\beta \to \gamma) \to (\alpha \to \gamma)} \quad (R5)$$

Therefore, we have *modus ponens*, Adjunction, a Contraposition, and forms of Concatenation and Transitivity. (R4) is sometimes called Affixing, and (R5) Suffixing. By adding new axioms, and sometimes dropping the corresponding rules, we obtain some of the mainstream relevant systems:

(A10) $(\alpha \to \neg\beta) \to (\beta \to \neg\alpha)$
(A11) $(\alpha \to \beta) \to ((\gamma \to \alpha) \to (\gamma \to \beta))$
(A12) $(\alpha \to \beta) \to ((\beta \to \gamma) \to (\alpha \to \gamma))$
(A13) $(\alpha \to ((\alpha \to \beta) \to \beta)$
(A14) $(\alpha \to (\alpha \to \beta)) \to (\alpha \to \beta)$.

Leaving some intermediate systems aside, by adding (A10), (A11) or (A12), (A13) and (A14), and dropping (R3), (R4) and (R5), we obtain the system **R** of relevant implication, initially developed by Belnap. We get **T** (the *ticket entailment* system) from **R** by dropping (A13), and adding a version of *reductio*:

(A15) $(\alpha \to \neg\alpha) \to \neg\alpha$.[25]

The system **E** of entailment can be obtained from **T** by defining necessity as:

$\Box\alpha =_{df} (\alpha \to \alpha) \to \alpha$,
and adding the axiom schemas:

(A16) $(\alpha \to \beta) \to (((\alpha \to \beta) \to \gamma) \to \gamma)$
(A17) $\Box\alpha \wedge \Box\beta \to \Box(\alpha \wedge \beta)$.

(A17) is required for the inductive proof of the Necessitation rule, and it can be avoided by adding Necessitation as primitive:[26]

---

[25] See Bremer (2005): 73-4.
[26] See Dunn (1986): 129.

$$\frac{\vdash \alpha}{\vdash \Box\alpha} \quad \text{(Nec)}$$

E is both a relevant and a modal logic; minor modifications apart, it differs from **R** mainly by adding necessity to the relevant implication formalized within **R**. Dunn has therefore claimed that, even though **E** was Anderson and Belnap's favourite system, it is **R** that should be taken as the paradigmatic relevance logic.[27]

A collateral system, **RM**, is obtained by adding to **R** the Mingle axiom schema:

(A18) $\alpha \to (\alpha \to \alpha)$.

The underlying idea of Mingle would be that of expressing the entailment from a given formula to a tautology in accordance with the (VSP) principle. But within **RM** one can obtain theorems violating the (VSP), such as $\neg(\alpha \to \alpha) \to (\beta \to \beta)$, so it is a "semi-relevant" logic.[28]

Works on relevance logics often adopt and study an important subsystem of these systems, *first degree entailment* (**FDE**), also called "logic of tautological entailment". A *zero degree* formula includes only the "extensional" connectives of disjunction, conjunction and negation. A *first degree* formula has the form $\alpha \to \beta$ and both α and β are zero degree formulas. So first degree formulas do not contain nested implications. The first degree fragments of the main systems **R** and **E** are identical, and there is an intuitive motivation for considering such fragment(s) separately, namely that if someone is inclined to read entailment as a metalinguistic relation, then the iteration of $\to$ becomes difficult to understand.[29] A shortcoming of the system is that "it does not allow to model nested entailments, whereas many intuitive logical truths for the conditional (Permutation, Transitivity etc.) are just that".[30]

---

[27] *Ibid*: 119.
[28] For details, see Anderson and Belnap (1975): Sect. 29.3.
[29] See Dunn (1986): 146.
[30] Bremer (2005): 65.

9.4.2 Dialectical Logics by Routley and Meyer, DL and DK

T, **R** and **E**, though, are too strong for paraconsistent purposes. We know that Anderson and Belnap did not mean to take issue with the (LNC). And **T**, **R** and **E** all include (A14), which is nothing but the axiom form of Contraction-Absorption. Therefore, the systems do not fulfil the Strong Condition of Non-Triviality: we cannot build inconsistent set theories or semantic theories on top of them, since they would be trivialized by some version or other of the Curry paradox.[31] We have to resort to weaker relevant systems, which, with a terminology due (as far as I know) to Ross Brady, have come to be called *depth* relevant logics.[32]

As typical examples of relevant logics satisfying the Strong Condition of Non-Triviality, let us consider the *dialectical logics*, **DL** and **DK**, developed by Routley and R.K. Meyer in some papers constituting cornerstones of classical paraconsistency.[33] The axiom schemas for **DL** (a "minimal dialectical logic")[34] are:

(A1) $\alpha \to \alpha$
(A2) $(\alpha \to \beta) \wedge (\beta \to \gamma) \to (\alpha \to \gamma)$
(A3) $\alpha \wedge \beta \to \alpha$
(A4) $\alpha \wedge \beta \to \beta$
(A5) $(\alpha \to \beta) \wedge (\alpha \to \gamma) \to (\alpha \to \beta \wedge \gamma)$
(A6) $\alpha \wedge (\beta \vee \gamma) \to (\alpha \wedge \beta) \vee (\alpha \wedge \gamma)$ (with $\alpha \vee \beta =_{df} \neg(\neg\alpha \wedge \neg\beta)$)
(A7) $\neg\neg\alpha \to \alpha$
(A8) $(\alpha \to \neg\beta) \to (\beta \to \neg\alpha)$
(A9) $(\alpha \to \neg\alpha) \to \neg\alpha$ (or equivalently: $(\alpha \to \beta) \to \neg(\alpha \wedge \neg\beta)$)
(A10) $P \wedge \neg P$

The rules are:

$$\frac{\alpha \to \beta, \ \alpha}{\beta} \quad (R1)$$

---

[31] Slaney (1989) has shown that **R** faces a version of the Curry paradox also without Contraction.
[32] See Brady (1984). Sometimes they are called *importive* relevant logics (see Routley (1979a): 306).
[33] See Routley and Meyer (1976); Routley (1979a), and (1979b): Appendix.
[34] See Routley and Meyer (1976).

$$\frac{\alpha, \beta}{\alpha \wedge \beta} \quad (R2)$$

$$\frac{\alpha \rightarrow \beta, \gamma \rightarrow \delta}{(\beta \rightarrow \gamma) \rightarrow (\alpha \rightarrow \delta)} \quad (R3)$$

We obtain **DK** by modifying (A6) into $\alpha \wedge (\beta \vee \gamma) \rightarrow (\alpha \wedge \beta) \vee \gamma$, dropping (A9), and adding the (LEM):

(A11) $\alpha \vee \neg\alpha$.

First-order extensions, **DLQ** and **DKQ**, have the following additional schemas for quantification:

(A12) $\forall x \alpha[x] \rightarrow \alpha[x/t]$
(A13) $\forall x(\alpha \rightarrow \beta[x]) \rightarrow (\alpha \rightarrow \forall x \beta[x])$
(A14) $\forall x(\alpha \vee \beta[x]) \rightarrow (\alpha \vee \forall x \beta[x])$
(A15) $\forall x(\beta[x] \rightarrow \alpha) \rightarrow (\exists x \beta[x] \rightarrow \alpha)$

(x is not free in α in (A13)-(A15)), and the rule:

$$\frac{\alpha}{\forall x \alpha} \quad (R4)^{35}$$

A strange feature of **DL(Q)-DK(Q)** is (A10). This is not a schema, but is meant to be a determinate contradiction by giving a fixed interpretation to the propositional parameter P. The aim of (A10) is to guarantee the inconsistency of the real world, or the actuality of at least one contradiction.[36] Some paraconsistent authors have criticized this move, though.[37]

---

[35] See Bremer (2005): 142. Slightly different presentations can be found in various works by Routley, but modifications are usually of minor importance: for instance, in Routley (1979a) the existential quantifier is defined; in Routley (1979b), **DK** lacks (A10).

[36] "$[P \wedge \neg P]$ is accordingly a real contradiction, and real contradictions hold in the actual case. [...] It is a virtual corollary of the dialectical fact, incorporated into DKQ, that reality is inconsistent, that freedom from contradiction does not provide an acceptable necessary condition for rationality or on rational belief or rational inquiry" (Routley (1979a): 312).

Dialectical relevant logics will resurface in the third part of this book, devoted to the applications of paraconsistency, because Routley and Brady have based some paraconsistent set theories on them. More than via their proof-theoretic presentation, though, the importance of Routley and Meyer's dialectical logics emerges by means of their underlying semantics. According to Routley, one of the main cases for these systems "derives from features of the semantics for relevant dialectical logics, in particular the way in which the semantics encompasses inconsistent and incomplete worlds".[38] Impossible worlds are coming on stage again.

9.5 Relevant Semantics

Precisely because they are the most developed and discussed paraconsistent logical theories, some of the intrinsic difficulties of paraconsistency emerge here in a clear fashion. The "problem of semantics" discussed in the following is not a purely technical one – that is to say, it is not the mere problem of finding an abstract structure with respect to which one can prove the adequacy of a given logical calculus. It is a philosophical question: the one of the intuitive interpretation of the formal structure itself. I have already made reference to the issue of intuitive interpretation in the discussion of non-adjunctive approaches. However, the point can be made more precise now, by exploiting a distinction that has gained currency in the literature: the one between *pure* and *applied* semantics.

9.5.1 Pure Semantics, Applied Semantics

This kind of terminology, as far as I know, has probably been introduced by Plantinga.[39] It is somehow misleading, since it seems to deal with the applications or the uses of a logic, whereas this is not what is at issue at all. Nevertheless, it has become a standard one by being adopted in various textbooks of philosophy of logic, such as those by Kirwan (1978) and

---

[37] For instance, according to Batens "that some set of formulas is correctly considered a logic presupposes, among other things, that it is closed under substitution for propositional variables; this is, as Anderson and Belnap (1975, 462) say, 'what makes it a logic'. However, if this is correct, it is hard to see how [(A10)] may be considered a theorem of logic" (Batens (1980): 223).
[38] Routley (1979a): 305.
[39] See Plantinga (1974): 126ff.

Haack (1978). Dummett talked of "merely algebraic notion of logical consequence", as opposed to "semantic notion of logical consequence properly so called":

> Semantic notions are framed in terms of concepts which are taken to have a direct relation to the use which is made of the sentences in a language [...]. It is for this reason that the semantic definition of the valuation of a formula [...] is thought of as giving the meanings of the logical constants. Corresponding algebraic notions define a valuation as a purely mathematical object which has no intrinsic connection with the uses of sentences.[40]

The history of the semantics for intuitionistic logic provides, in my opinion, a good example of the difference. The early intuitionistic semantics – e.g., the topological interpretations by Tarski and others[41] – may be reasonably qualified as pure semantics. As Dummett says, they "were developed before any connection was made between them and the intended meanings of the intuitionistic logical constants". Therefore, even though intuitionistic calculus is complete with respect to these semantics, "no one would think of this as in any sense giving the meanings of the intuitionistic logical constants".[42] The situation seems to have changed with the development of Kripke semantics for intuitionistic logic,[43] since in this case we appear to have an independent intuitive interpretation: intuitionistic worlds should represent cognitive situations, or stages of knowledge, or of information, increasing over time along different paths, on the basis of the conception of mathematics as a mental activity. At each time t, the mathematician has different possibilities to increase his knowledge in the transition to time t_1. This can be framed formally, for instance, by considering the accessibility relation as a partial order (reflexivity and transitivity hold), and by assuming variable but never decreasing domains, that is to say, if w_2 is accessible from w_1, then the domain of w_1 is a subset of the one of w_2.[44] Therefore, accessibility has an intuitive meaning, being thought of as representing possible progresses of knowledge. We have completeness proofs for intuitionistic logic with respect to such semantics, providing counter-models to the classical laws

---

[40] Dummett (1973a): 204.
[41] See van Dalen (1986): 243ff; Beall and van Fraassen (2003): 96ff.
[42] Dummett (1973a): 205.
[43] See Kripke (1963a).
[44] See van Dalen (1986): 246ff.

rejected by intuitionists – paradigmatically, Strong Double Negation and Excluded Middle. However, the technical details are not of great significance here. The important point is the underlying *heuristics*: such semantic structures provide, in my opinion, an intuitively justified representation of the mental activity of the mathematician, as an idealized subject *constructing* mathematical objects through her proofs.[45]

As a result, we seem to have moved from a pure to an applied semantics. Such a shift, according to Susan Haack, is essential for a formal system to qualify as a *logic* of some kind or other.[46] The general point is that, given *any* purely proof-theoretically described logic of a certain kind, it is always possible to get a cheap algebraic model for it: just build a sound and complete semantics from the Lindenbaum algebra of the logic itself. Some authors also make the stronger claim that, until the passage is made, no determinate meaning has been assigned to the logical vocabulary of a formal system. According to B.J. Copeland, for instance, any logical semantics follows these two stages in its development: the first step is the mere construction of the formal semantic apparatus; the second step is the interpretation of the mathematical formalism. And, at this point, logic cannot but defer to some extent to metaphysics, given that we need "an *explanation of the nature of the objects* which constitute the range of this [evaluation] function, of any indices occurring in the domain of the function, of any operations or relations on these indices, and so on".[47]

The distinction between pure and applied semantics is of great importance in order to evaluate some of the most discussed and famous seman-

---

[45] This does not mean that Dummett would agree with me on this point, or approve the Kripke semantics for intuitionistic logic. He claimed in several works that the correct semantics for intuitionism should not be phrased in terms of possible worlds at all, and the meaning of the intuitionistic logical vocabulary should be entirely captured inferentially, that is, via introduction/elimination rules of inference in the style of Gentzen.

[46] "The identification of a system as a system of logic requires appeal to its (intended?) interpretation. To identify a system as a sentence calculus one does not only need to know the axioms/rules and their formal interpretation by means of matrices; one also needs to know that the values are to represent truth and falsity, the letters 'p', 'q', etc. to represent sentences, ' - ' negation, ' & ' conjunction, ' v ' disjunction, and so forth. [...] I have already urged that the pure semantics, by itself, is not sufficient; to justify the claim of a formal system to be a modal logic (sentence logic) some intuitive account of the formal semantics, connecting that set-theoretical construction with the ideas of necessity and possibility (truth and falsity) seems essential" (Haack (1978): 30 and 189).

[47] Copeland (1986): 479; my italics.

tics for paraconsistent logics. During the Eighties and Nineties, authors like Copeland, van Benthem, Smiley, and others claimed that the semantics for relevance logics – possibly, the most developed paraconsistent approach – did not reach the status of applied semantics. Therefore, providing an intuitive reading and a not purely technical assessment for them was an open problem. It seems that the situation has changed nowadays. As we shall see, several authors have offered intuitive, somewhat convincing interpretations of some relevant semantics. However, the point is that precisely *these* interpretations give us an idea of the *sense in which* a logical calculus that tolerates contradictions can have a model, or a set of circumstances, satisfying it. Let us see why.

9.5.2 Four Corners and the American Plan

We have both algebraic and model-theoretic semantics for relevance logics. A very simple semantics for **FDE** has been proposed by Dunn (1976) and Belnap (1977), and developed by many authors. It is a many-valued semantics, whose key idea is to assume as truth values $P(\{1, 0\})$, that is to say, the power set of the classical set: $\{\{1\}, \{0\}, \{1, 0\}, \varnothing\}$. This is nearly the same strategy as the one of **LP**, but now we retain the empty set – so we have a four-valued truth-functional semantics with both gaps and gluts. Of course, the intuitive reading of the values is: true only, false only, both true and false, neither true nor false. This gives us the Four Corners of Truth described by classical Indian logician Sanjaya.[48]

The semantic clauses for the extensional connectives are straightforward, and the same as those of **LP**:

(S¬1) $1 \in v(\neg\alpha) \Leftrightarrow 0 \in v(\alpha)$

(S¬2) $0 \in v(\neg\alpha) \Leftrightarrow 1 \in v(\alpha)$

(S∧1) $1 \in v(\alpha \wedge \beta) \Leftrightarrow 1 \in v(\alpha)$ and $1 \in v(\beta)$

(S∧2) $0 \in v(\alpha \wedge \beta) \Leftrightarrow 0 \in v(\alpha)$ or $0 \in v(\beta)$

(S∨1) $1 \in v(\alpha \vee \beta) \Leftrightarrow 1 \in v(\alpha)$ or $1 \in v(\beta)$

---

[48] See Raju (1954).

(S∨2) $0 \in v(\alpha \vee \beta) \Leftrightarrow 0 \in v(\alpha)$ and $0 \in v(\beta)$.[49]

FDE is complete and decidable with respect to the semantics. In particular, negation works as an order-inverting map, that is, a glut in-glut out, gap in-gap out functor toggling 1 and 0:

α	$\neg\alpha$
1	0
0	1
1, 0	1, 0
$\varnothing$	$\varnothing$

The approach has been called *American Plan*, since it has gained a lot of currency with relevant logicians in the United States. The designated values are {1} and {1, 0}. Just like within **LP**, *ex falso*, expressed e.g. in the form of (PS$_i$1), $\alpha \wedge \neg\alpha \to \beta$ can be refuted by assigning to β a non-designated value ({0}, or $\varnothing$), and to α (therefore, to $\neg\alpha$, and $\alpha \wedge \neg\alpha$) the designated {1, 0}: contradictions can sometimes be both true and false, although they can never be {1}, true only.

9.5.3 The Australian Plan and Impossible Worlds

Much more controversial is the (im)possible worlds perspective initially due to Routley and Meyer, and developed by various relevant logicians mainly in Australia, therefore called *Australian Plan*.[50] It is a very flexible semantics for full-fledged relevant systems (with nested entailments), presented in different versions.[51] It has often been described as a paraconsistent *extension* of ordinary possible worlds semantics, including non-normal or impossibile worlds. For our purposes, the following characterization will do. A *Routley-Meyer (im)possible worlds structure* is a quintuple <W, N, o, *, R>, where W is a set of "situations (or worlds)";[52] N is

---

[49] See e.g. Dunn (1986): 192.
[50] As for such labels, see e.g. Meyer and Martin (1986); Restall (1995).
[51] See Routley and Routley (1972) for the initial intuition; Routley and Meyer (1973) and (1976); Routley (1979a). A simplified variant has been proposed in Routley and Priest (1992) for the basic relevant system **B**. I will not deal with it here, though.
[52] Routley (1979a): 309.

a subset of W including normal or "theorem-regular" situations, that is to say, worlds in which all the theorems of the logic hold; o is an element of N, representing the real world; * is a unary operation on W, i.e., a function from worlds to worlds (sometimes called "involution"); R is a ternary accessibility relation on W. The most controversial part of the story has to do with the intuitive reading of R and *, which are used in the semantic clauses for negation and the relevant conditional. Let us begin with the conditional.

Ordinary semantics for modal logics adopts a binary accessibility relation on worlds and, as is well known, we obtain models for different modal systems by varying the formal properties of the relation. This is no doubt the most celebrated result of modern modal logics. Now, a typical semantic clause for a *strict* implication is something like:

(S⥽) $T_w(\ulcorner \alpha \prec \beta \urcorner) \Leftrightarrow \forall w_1(R(w, w_1) \Rightarrow (T_{w1}(\ulcorner \alpha \urcorner) \Rightarrow T_{w1}(\ulcorner \beta \urcorner)))$,

that is to say: $\alpha \prec \beta$ is true at world w, iff at all worlds w_1 accessible from w in which α is true, β is true. Such a clause will not do for a relevant implication: it makes of $\alpha \to (\beta \to \beta)$ a logical truth, but this is a fallacy of relevance violating the (VSP).[53] So a(n im)possible worlds semantics for relevant logics cannot in general work with a binary accessibility relation (even though a restricted binary accessibility can be regained, as we will soon see). We can adopt, instead, the following clause making use of the ternary relation of the Routley-Meyer model:

(S→) $T_w(\ulcorner \alpha \to \beta \urcorner) \Leftrightarrow \forall w_1 w_2(R(w, w_1, w_2)$ and $T_{w1}(\ulcorner \alpha \urcorner) \Rightarrow T_{w2}(\ulcorner \beta \urcorner))$.[54]

(S→) is still close to the standard clause for strict implication, except that the worlds of the antecedent and the consequent, w_1 and w_2, have been *distinguished*. This is the decisive move: the new clause does not validate $\alpha \to (\beta \to \beta)$ anymore. Non-normal worlds have begun to play. Via the ternary relation, it is possible to build models for different relevant logics, in a way somehow analogue to standard modal logics: beginning with the basic system **B**, we obtain models for the stronger logics,

---

[53] See Bremer (2005): 66.
[54] See Routley and Meyer (1976); Dunn (1986): 201.

such as **T**, **R** and **E** by adding formal conditions on R.[55] Technicalities can be left aside, though. A first concern is, how do we *read* the clause (S→) and, specifically, such things as "$R(w, w_1, w_2)$"? A little bit of help from the literature:

> In interpreting $Rxyz$ perhaps the best reading is to say that the combination of the pieces of information x and y (not necessarily the union) is a piece of information in z [...]. On this reading $Rxyz$ can be regarded as saying that x and y are compatible according to z, or some such thing.[56]

> An entailment [α → β] is true at some world if this world sees an accessibility between *two other worlds* such that if [α] is true at the first of these worlds [β] is true at the other.[57]

We can also define by means of R a restricted binary accessibility (be it ∠) between worlds:

(Df∠) $w \angle w_1 =_{df} \exists w_2 (w_2 \in N$ and $R(w_2, w, w_1))$,

that is, binary accessibility holds between two worlds iff "some *normal* world 'sees' the accessing".[58]

9.5.4 The Routley Star

Before discussing the plausibility of it all, let us consider the famous Routley Star. This provides the semantics for negation, and negation, as we now know, is the main operator at issue when the (LNC) is under discussion. Given a world w, the "involution" operation produces a world w^* (on whose nature I will say more in the following). The clause for negation in this framework is something like:

(S¬) $T_w(\ulcorner \neg \alpha \urcorner) \Leftrightarrow$ Not $T_{w^*}(\ulcorner \alpha \urcorner)$.

---

[55] See e.g. .Dunn (1986): 208-9; Bremer (2005): 73-4.
[56] Dunn (1986): 200.
[57] Bremer (2005): 67.
[58] *Ibid.* See also Routley (1979a): 310.

The standard clause for classical negation maintains that $\neg\alpha$ is true (at w) iff α is not true (at w, i.e., in that very world) – this comes from our (Neg2). The Routley-Meyer negation is characterized by saying that $\neg\alpha$ is true at w iff α is not true at w^*.[59] Therefore, it is an "intensional" connective: to evaluate a negated sentence at w, one has to go and look at what is going on at w^*.

This provides a further counter-example to (PS): we just have to consider the case in which α is true at w, β is not true at w and α is not true at w^*: both α and $\neg\alpha$, then, are true at w while β is not, and detonation is avoided. Notice that *ex falso* is neutralized without admitting a truth value glut, as in the Belnap-Dunn semantics, or in similar paraconsistent many-valued logics, such as **LP** (although intuitive formal dualities and correlations between the American and the Australian Plan have been extensively explored).[60] With a few more conditions, we can validate various theorems of relevant logics, and make the relevant negation have certain intuitive inferential features, which make it look like a real negation. For instance, the following clause on R and *, sometimes called Inversion:

$R(w, w_1, w_2) \Rightarrow R(w, w_2^*, w_1^*)$,

is central to the validation of Contraposition; and the assumption that involution is of period two:

$w^{**} = w$,

is used in the validation of Double Negation.[61] The negation characterized by such clauses is usually called by relevantists *De Morgan negation*, since also De Morgan laws hold for it; therefore, it has almost all the inferential properties of standard negation (except that of being explosive, of course). Therefore, it seems that our Avoid Overload Condition is satisfied. On the other hand, the mere fact that extra conditions on * have to

---

[59] See Routley and Meyer (1976); Routley (1979a): 311. Sometimes the clause is spelled saying that $\neg\alpha$ is true in w iff α is *false* in w^* (see e.g. Copeland (1979): 402; Restall (1995): 144). In this case, though, the distinction between untruth and falsity is of no importance.

[60] See e.g. Restall (1995).

[61] See e.g. Routley (1979a): 310; Dunn (1986): 207; Restall (1995): 144-5.

be added in order to validate the desired logical laws may be questioned. More importantly, one may wonder, why should * poke its nose into the clause for negation? What kind of world is w^* with respect to w? Here is some more help from the literature:

> The operation * is a reversal operation which takes a situation into its reverse and hence incompleteness in a situation into inconsistency, and inconsistency into incompleteness, i.e., the reverse of a situation [w] where both A and ~A hold is a situation [w^*] where neither A nor ~A hold.[62]

> One way of thinking of [w] and [w^*] is to regard them as 'mirror images' one of another reversing 'in' and 'out'. Where one is inconsistent (containing both A and ¬A), the other is incomplete (lacking both A and ¬A), and *vice versa* (when [w] = [w^*], [w] is both consistent and complete and we have a situation appropriate to classical logic). Viewed this way the Routleys' negation clause makes sense, but *it does require some anterior intuitions about inconsistent and incomplete set-ups.*[63]

9.6 Ultralogic

One of the early uses of the Routley-Meyer semantics has been that of John Heinz: he exploited dialectical logic to build a theory of fictional objects, such as those we find in tales and stories.[64] The idea is also at the core of Routley's monumental retrieval of Meinongian ontology in *Exploring Meinong's Jungle and Beyond*.[65] Of course, Meinongianism has to do with various kinds of non-existent objects, but fictional ones provide, as it were, the paradigmatic case. Truth conditions for sentences talking of fictional objects are to a large extent stipulated, that is, they often depend on the author's say-so. Inconsistencies are expected, they may hold within a tale but, obviously, not everything happens in the story and characters act in it in largely comprehensible ways. The discourse on fictional objects appears to be naturally tied to an impossible worlds semantics: objects are specified by a certain set of (occasionally inconsistent) conditions, and have the relevant properties at (occasionally impossible) worlds that realise the characterization. Therefore, Heintz recommends

---

[62] Routley (1979a): 309.
[63] Dunn (1986): 191; my italics.
[64] Cf. Heintz (1979).
[65] Routley (1979b).

impossible worlds semantics to accommodate them. Of course, one may suppose that some worlds are inconsistent without including the actual one among them. However, Heintz is also sympathetic to the Routley-Meyer hypothesis that consistency may not be ascertainable in principle for the *actual* world either. Let us inspect further.

According to Routley and Meyer, a dialectical logical framework properly *subsumes* classical logic as a particular case. This should not be understood in the strictly logical sense, i.e., a formal system properly including another iff the former has all the theorems of the latter, plus something more. Following a metaphor paraconsistent authors are used to, the situation corresponds to the one of special relativity including classical Newtonian dynamics as a useful approximation for low velocities. We are not dealing with velocity, though, but with contradiction. Philosophers learn how classical negation works in their first class of logic, just as engineers learn Newtonian dynamics in their first class of mechanics. Nevertheless, Newtonian dynamics is, strictly speaking, incorrect and, as the paraconsistent story goes, classical negation is, strictly speaking, incorrect too. Classical logic simply *assumes* that, for each w, $w = w^*$, that is to say, that all worlds are maximally consistent. Such an assumption lies beneath the standard semantic clause of exclusion negation, as a constraint on admissible *interpretations*: precisely one between α and $\neg\alpha$ has to be true in any w. But to assume the "classical negation principle", as Routley and Meyer call it, is just to embed in the semantic clause a condition of admissibility which is imposed on *worlds* – an assumption of consistency and maximality which is too strong to bear.[66] Now, let us listen to the justification of De Morgan negation provided by the Routleys in their influential paper:

> The new negation is neither queer nor weak because it behaves exactly like classical negation over consistent and complete set-ups. Since its behaviour diverges from that of classical negation only over a class of situations not admitted by modal logic, it is not so much the negation introduced by our new rule [(S¬)] which is different as the class of situations that the rule entitles us to consider.[67]

---

[66] See Routley and Meyer (1976).
[67] Routley and Routley (1972): 338.

Therefore, the key point concerns, again, the underlying *ontology*. As we see, the rejoinder to the Quinean *change-of-subject* charge for negation is analogue to the one provided by Rescher, Brandom and Varzi for conjunction: we (as paraconsistentists) have just broadened our horizons beyond the classical, narrow-minded way of understanding *worlds*. Therefore, the position that defends classical-exclusion negation as the One True Negation, claiming that a connective that does not play the game of the ordinary semantic clause is not *negation*, now looks like another case of the Italics Argument.

Routley has consequently presented his relevant logic **DKQ** as an "ultralogic", a possible candidate to the status of universal logic, "one which is applicable in every situation whether realised or not, possible or not". And "a universal logic is like a universal key, which opens, if rightly operated, all links. It provides a canon for reasoning in every situation, including illogical, inconsistent and paradoxical ones".[68] Specifically, embracing a dialectical-paraconsistent logic is a more rational position than sticking to classical logic if it turns out that the issue of the consistency of the world cannot be settled once and for all. And, according to Routley and Meyer's challenge, the question cannot actually be settled. Of course, there is quite a lot of consistency in our world. This can be empirically established: I can ascertain that sentence α – say, "Routley is in Belnap's room at the University of Pittsburgh on May 2, 1974" is false, so the world is consistent with respect to α. However, to affirm the *absolute* consistency of the world is to make a metaphysical, empirically unverifiable thesis, since strictly universal claims can be falsified by a single counterexample. Even though we know that the world is *locally* consistent, just as it is locally Euclidean, the belief in its absolute consistency is a metaphysical faith.[69] This echoes Łukasiewicz:

> *We can never say with complete certainty that real objects contain no contradictions.* Man did not create the world, and he cannot penetrate all his mysteries; he is not even master of his own conceptual creations.[70]

---

[68] Routley (1979b): 893.
[69] See Routley (1979a): 301-2.
[70] Łukasiewicz (1910b): 62.

9.7 Relevant Troubles

Let us now highlight some *inconvenientia* of relevance logics. First, it has been often remarked that the syntactic tricks used by relevant logicians in order to avoid the paradoxes of material and strict implication are largely *ad hoc*: no independent justification for the dismissal of various intuitive principles of ordinary logic is provided, except that, by admitting them, various kinds of irrelevance follow.[71] Furthermore, a serious shortcoming of full-fledged relevant systems is that they are not decidable at the sentential level. This is a peculiarity of the approach: since the 60s, decidability results (some of them, rather cumbersome) were provided for weaker systems and for fragments of **R** and **E**. Between 1982 and 1984, though, Alasdair Urquhart proved that the full **T**, **R** and **E** are undecidable.[72] This may not deflect the logicians' interest, but makes them look rather bizarre from the philosophical point of view. The computational complexity of these systems also poses technical problems when one comes to the issue of the classical recapture: there is no general method to guarantee that a classically valid mathematical proof has its relevant counterpart.[73] It is interesting, nevertheless, that a crucial principle in Urquhart's negative strategy is Contraction, $(\alpha \rightarrow (\alpha \rightarrow \beta)) \rightarrow (\alpha \rightarrow \beta)$ (or some equivalent principle, such as Assertion, $\alpha \wedge (\alpha \rightarrow \beta) \rightarrow \beta$), i.e., something that must go if relevance has to be suitable for paraconsistent purposes.

9.7.1 Strong and Weak Assertions

The philosophically relevant issues concerning relevance, though, come exactly at the level of the intuitive reading of the semantics. By appealing to the distinction introduced above, several authors have claimed that relevant semantics remain *pure*, not reaching the status of applied semantics. In particular, algebraic semantics have often been disqualified as "disguised syntax", since the operators defined in the algebra seem to have a meaning just through their mirroring the connectives of the syntax

---

[71] This point has been made at length in Diaz (1981).

[72] See Urquhart (1984).

[73] Therefore, "what recent work on Anderson and Belnap systems makes plain is that systems such as R and their satellite systems cannot satisfy their own procedural motivation, namely, to facilitate the finding of proofs" (Woods (2003): 75).

of calculus.[74] As van Benthem observed in his discussion of the relevant perspective, "there is no end to syntactical creativity", but the real point is whether the calculus has an intuitive semantic interpretation.[75]

But something analogous may happen also with the Routley-Meyer (im)possible worlds semantics, which has been criticized, again, by van Benthem, Smiley, Copeland, and others. Its attractive features are due to its looking like a development of Kripke semantics, and this fact has probably allowed the former to take advantage of the respectability of the latter. However, such resemblance, as the detractors' story goes, is essentially deceptive. The three-place relation R of the model is not similar to the accessibility or alternativeness relations due to Kripke, Hintikka and others, exactly because of the aforesaid reason: the latter do have (at least, according to most logicians) an intuitive meaning, whereas the former does not.[76] The same lack of independent characterization affects the monadic operation *, which should give meaning to the relevant negation.[77] It is not clear, therefore, which is the "intended" meaning assigned to logical symbols by such structures.

Routley and Meyer attempted to make * more palatable, by attaching to it an intuitive meaning in terms of the notion of *assertion*, and by distinguishing between a *weak* and a *strong* assertion of a sentence. The weak way to assert α is to omit the assertion of its negation. What is weakly asserted in w^* is precisely what is (strongly) asserted in w, and *vice versa*. In this sense, De Morgan-relevant negation should reduce classical negation to a particular case: under normal circumstances, that is

---

[74] Looking at the algebraic models used in completeness proofs for relevant logics, Dunn admits that "the above kind of soundness and completeness result is really quite trivial (though not unimportant), once at least the logic has had its axioms chopped so that they look like the algebraic postulates merely written in a different notation" (Dunn (1986): 187).

[75] See van Benthem (1979): 340.

[76] Which is sometimes acknowledged by relevant logicians themselves: see Anderson, Belnap and Dunn (1982): 163-4.

[77] "By itself this 'star rule' is merely a device for preserving a recursive treatment of the connectives [...] and it does nothing to explain their tilde until supplemented by an explanation of [w^*]" (Smiley (1993): 17-8). "I will not say very much here about what intuitive sense (if any) can be attached to the Routleys' use of the [*]-operator in their valuational clause for negation. Indeed this question has had surprisingly little extended discussion in the literature [...]. The Routleys' [1972] paper more or less just springs it on the reader, which led me in Dunn [1976] to describe the switching of [w] with [w^*] as 'a feat of prestidigitation'" (Dunn (1986): 190-1).

to say, when we assert exactly what we do not deny, w and w^* coincide and we have a situation appropriate to classical negation.[78]

Copeland and others found the distinction between weak and strong assertion simply unintelligible. First, the claim that De Morgan-relevant negation is an extension of classical negation because it includes both the classical case (i.e., $w = w^*$) and inconsistent-incomplete situations, seems to be an unwarranted one. In order to understand the interpretation given to a connective by a semantic structure we should look at the behaviour of that connective in *all* the worlds/circumstances of the model, not only in a subset of them. Second, now the problem is that the notion of weak assertion seems to have *itself* been introduced purely *ad hoc*, and its connection with the notion of assertion as usually understood appears obscure:

> In exactly what sense of "affirm" could a babe-in-arms be said to have [weakly] affirmed Heisenberg's Uncertainty Principle simply because it has omitted to assert the negation of the Principle? Talk of strong affirmation and weak affirmation remains mere jargon until it is explained how these supposed activities mesh in with the remainder of the network of psychological and behavioural transactions between humans and sentences.[79]

Lacking further explanation on the nature of w^*, the star operation seems to be a formal trick. In addition, the structural conditions introduced above (that * is of period two, etc.) appear to be *ad hoc*, in the sense that their unique motivation is to validate the theorems giving to De Morgan negation the desired inferential features. But "if the only constraint on * is that the resulting theory should validate the right set of sentences, then we are indeed in the presence of merely formal model theory".[80]

The early relevant approach to paraconsistency has been, therefore, a very controversial one. Although technically feasible, the strategy has been considered by many as probably lacking philosophical importance: "judgment has to be suspended until the proposed formal semantics has received any kind of satisfactory explanation. (At the present stage, the

---

[78] See Routley and Meyer (1973): 202.
[79] Copeland (1979): 409.
[80] *Ibid*: 410.

sense in which contradictions may be said to be 'true' remains a purely formal one.)".[81]

9.8 Which World are You Living in? – Part Two

9.8.1 Infons and Beyond

Since the time van Benthem and Copeland wrote their reviews, much work has been done in the field and the situation has changed. Several authors have provided not only further technical developments and results, but also plausible intuitive interpretations for the non-standard semantic structures described above. Nevertheless, the point is that, when we consider *these* interpretations, we are in a similar condition to the one we found ourselves in when we discussed non-standard worlds and/or circumstances of the non-adjunctive approaches. Recall that the most intuitive reading of such structures was in terms of theories, databases, cognitive frameworks, systems of beliefs, etc. Consequently, the claim that such worlds realize "ontic" inconsistencies is liable of committing the Russellian Fallacy: the fallacy that consists in attributing to the world features (specifically, inconsistency) that pertain only to our representations of it – *unless* we assume the idealistic-constructivist perspective according to which our actual world and its furniture are framed by our theories and conceptual schemes, thereby collapsing to some extent the distinction between *ordo cognoscendi* and *ordo essendi*.

Plausible interpretations of relevant semantics seem to fit into this pattern. It is true that various authors, including Priest, have given **FDE** a strongly paraconsistent or dialetheic interpretation. In *Doubt Truth to be a Liar*, a dialetheic correspondence theory of truth naturally follows exactly from the four-valued semantics for **FDE**, so that "the logic of the world would be First Degree Entailment".[82] However, first, this is not how the American Plan was intended by its creators Belnap and Dunn, who gave it a strictly epistemic reading. According to Dunn, the four-valued semantics seeks to depict the fact that "one can have inconsistent and or incomplete assumptions, information, beliefs, etc., and this is what

---

[81] Van Benthem (1979): 333; see also 343: "can any, technical or philosophical, interest be attached to the proposed semantics? Precisely this question cannot be answered until the authors have told us much more about the intuitions behind their 'worlds', 'ordering relation' and especially their 'reversal operation'".

[82] Priest (2006): 52.

we are trying to model to see what follows from them in an interesting (relevant!) way". So "all this talk of something's being both true and false or neither is to be understood epistemically and not ontologically".[83]

Secondly, and more significantly, Bryson Brown (1999a), (2001) has developed interpretations both of **FDE** and of Priest's **LP** that fully capture their consequence relation from the point of view of preservationist logics. As we have formerly seen, preservationism has a strictly epistemic reading in terms of reasoning in situations of inconsistent information. This means that the **FDE**-consequence relation can be read "without adopting a semantics that allows truth or any other 'designated' [...] values to be assigned to inconsistent sets of sentences or to contradictions".[84] According to Brown, this shows that a strongly dialetheic reading of **LP**, or of **FDE**, is not mandatory at all: we can interpret the consequence relations of **LP** and **FDE** as preserving some (classically perfectly acceptable) semantic properties, rather than preserving dialetheic truth values. And this:

> Blocks the slippery slope argument in Priest [(2000b)] that aims to drive us from the second level of paraconsistency (at which we accept the existence of interesting or valuable inconsistent theories that can be modelled with the help of paraconsistent consequence relations) to the third (at which we accept the possibility that some such theories are *true*).[85]

That is, it blocks the "slippery slope" from weak paraconsistency to dialetheism. We are then driven back to our initial issue: is dialetheism backed up by the dialetheist's interpretation of her favourite paraconsistent logics? In recent years, a plausible reading of the Routley-Meyer semantics for De Morgan negation and the relevant conditional has been proposed in terms of *information flow*, within the framework of situation semantics.[86] This approach is quite new but, informally, the basic idea is

---

[83] Dunn (1986): 192-3; see also Dunn (1976). The epistemic reading was also implicitly subscribed by David Lewis, in Lewis (1982): 102.

[84] Brown (2004): 141.

[85] Brown (2004): 146. See also Brown (1999): 498-9: "At least some paraconsistent logics that have been developed using dialetheic valuations can be reinterpreted in preservationist terms – in fact, I conjecture that they all can. [...] The philosophical question of the tenability of dialetheism is, in my own view, open, as is the closely related question of the tenability of dialetheic interpretations of various paraconsistent logics".

[86] See Barwise and Perry (1983); Mares (1996); Bremer and Cohnitz (2004).

to model circuits hosting regulated flows of information – in Manuel Bremer's words:

> Information flows in *distributed systems* (like a circuit that connects a switch with a bulb). These systems can be considered *channels* along which we reason. The laws and rules operative in such a system are *constraints* that define a proper working system and allow to reason along these constraints (like knowing that the switch was pressed if the bulb is burning). Talking in situation semantics language we can say that information about one *situation* is derived from another situation by some channel. The basic building blocks of information are *infons*, which resemble Russellian propositions.[87]

In particular, within this framework we can have an intuitive characterization of the Routley Star. Greg Restall has convincingly defended such a characterization, particularly in a 1999 essay whose subtitle deserves being mentioned: "How I Stopped Worrying and Learned to Love the Routley Star".

Two pieces of information or infons, as Restall argues, can be compatible or incompatible with each other, thereby bringing in a certain partial compatibility or incompatibility between worlds containing them. We can therefore introduce a binary compatibility relation C between worlds and characterize negation via such a relation:

(S¬) $T_w(\ulcorner \neg\alpha \urcorner) \Leftrightarrow \forall w_1(C(w, w_1) \Rightarrow \text{Not } T_{w_1}(\ulcorner \alpha \urcorner))$,

that is to say: $\neg\alpha$ is true at w iff α is not true at all compatible worlds. We can read "$C(w, w_1)$" as saying that nothing given by w is rejected by w_1, so the set of worlds compatible with w may represent what is not excluded by what is true at w.[88] And "this provides us with an *applied* semantics for negation".[89] In this context, and given some formal conditions on C, the "reverse twin" of w, w^*, is the most informative, or comprehensive, among the worlds w_1 for which $C(w, w_1)$ holds. And "the Routley star is a simplification of our compatibility clause for negation when we assume that C is symmetric, directed and convergent"; therefore, "given that the compatibility semantics makes sense and is an applied semantics,

---

[87] Bremer (2005): 68-9.
[88] See Restall (1999): 61-2.
[89] *Ibid*: 61; my italics.

it follows that its simple retelling, involving the Routley star, also makes sense, and it too is an applied semantics".[90]

9.8.2 The Russellian Fallacy Again

For the sake of the argument, let us suppose that the information flow representation is sufficiently transparent (which is doubted by some paraconsistent authors, including Priest).[91] Now the question is, what is an *infon*?[92] If worlds are considered as make-ups of infons (not simple sets, but something with more arrangement imposed on them, or even structures that act themselves as *conduits* for information in some way: this does not make a difference to the point); and if an infon somehow is information *on* something; then worlds are structures of information. Therefore, truth-in-the-world, once again, does not seem to be at issue here, by definition. In this case, worlds retain the right to be inconsistent and/or incomplete, since it is out of question that information can be inconsistent and/or incomplete. Again, we are dealing with theories, beliefs, etc., that is to say, *representations*.

Routley and Meyer, in fact, used to move back and forth on this point in their works. In the 1972 paper the Routleys introduced worlds as "setups", characterizing them quite syntactically as sets of *sentences*.[93] However, in the Routley-Meyer 1973 paper, set-ups became what is *described by* a set of sentences.[94] Such slips obviously attract the charge of Russellian Fallacy. In John Woods' words: "this is dialectic rhetoric *par excellence*. It takes us from doxastic inconsistency to ontic inconsistency", that is to say, "from the fact that beliefs are sometimes inconsistent to the possibility that inconsistent beliefs are sometimes true, and so to the possibility that sometimes objects or states of affairs are inconsistent".[95]

On the other hand, the idea of an intuitive reading of relevant (im)possible worlds in terms of information was unwillingly provided by an aforementioned critic of relevant logics. We saw how, according to

---

[90] *Ibid*: 63; see also Bremer (2005): 72-3.
[91] See Priest (2001): 197-8.
[92] A popular response to this follows Barwise and Perry: an infon is an ordered *n*-ple <$R, o_1, ..., o_n, b$>, where R is a *n*-ary relation, $o_1, ..., o_n$ are objects, and b is a *sign-bit*: 0 or 1.
[93] See Routley and Routley (1972): 335.
[94] See Routley and Meyer (1973): 200-1.
[95] Woods (2003): 89-90.

Copeland, to grasp the meaning assigned to the logical vocabulary by a formal semantics we need an independent characterization of the evaluation function and of the ontological structure that provide a model for the calculus. Nevertheless, this is not enough yet: the core task of the semantic apparatus is to represent a semantic property Φ, such that understanding the condition under which a sentence containing one of the relevant logical symbols has that property is sufficient to understand the meaning of the symbol. Applied semantics is finally reached when we are able to read semantic clauses as "statement[s] of necessary and sufficient conditions for the possession, by sentences of the forms involved, of an antecedently understood property Φ".[96] The case *par excellence* of such a semantic property Φ, of course, is that of being true. Having an independent grasp of what "true" means in the context, we can understand what conjunction means, via the clause: "$\alpha \wedge \beta$ is true iff...", specifying a necessary and sufficient condition for a sentence containing it as its main connective to have that property. But if the worlds of the Routley-Meyer semantics are characterized as sets of sentences, beliefs of an individual, collections of hypotheses, *infons*, etc.; then the property of being true in a world actually means that "the sentence A is a *member of* the set of sentences j, this set representing the beliefs of a given individual, or a given mathematical or physical theory".[97] Now, membership of a set of beliefs or a theory is a property quite different from *truth*: the predicate "is a member of belief set w" does not in general satisfy the Tarskian schema, whereas, as we have previously seen, many (including me) consider the T-schema as a minimal constraint on a truth predicate.

In this sense, as David Lewis remarked, relevance logic looks like a "logic for ambiguity". The idea that a database can include inconsistencies without being trivialized by them is reasonable (we have also seen the fragmentation strategy proposed by Lewis himself), but then the standard approach to this simply adopts parameterisations. A sentence that appears to be both true and false is not given a *de re* interpretation, as describing an inconsistent state of affairs *in rerum natura*, but a *de dicto* one: it is taken as an ambiguous sentence, true in some disambiguations and false in some others. Ambiguities should be settled before formalizing. Now, who needs a logic for ambiguity?

---

[96] Copeland (1986): 480.
[97] *Ibid*: 486-7.

222 / How to Sell a Contradiction

> I reply: pessimists.
> We teach logic students to beware of fallacies of equivocation. It would not do, for instance, to accept the premise $A \lor B$ because it is true on one disambiguation of A, accept the premise $\sim A$ because it is true on another disambiguation of A, and then draw the conclusion B. After all, B might be unambiguously false. The recommended remedy is to make sure that everything is fully disambiguated before one applies the methods of logic.
> The pessimist might well complain that this remedy is a counsel of perfection, unattainable in practice. [...] Ambiguity is everywhere. There is no unambiguous language for us to use in disambiguation the ambiguous language. So never, or hardly ever, do we disambiguate anything fully. So we cannot escape fallacies of equivocation by disambiguating everything. Let us rather escape them by weakening our logic so that it tolerates ambiguity; and this we can do, it turns out, by adopting some of the strictures of the relevantists.[98]

9.9 Strong Paraconsistency as an Idealism

The most plausible intuitive reading of the Routley-Meyer semantics, therefore, suggests that for relevance logics may hold what seems to hold for the various non-adjunctive approaches: these paraconsistent theories are quite legitimate (actually, very interesting and useful) insofar as we pursue the task of building models for inconsistency-management in our cognitive luggage. As a matter of fact, several important applications of paraconsistent logics in the topics of cognitive management and belief revision are currently being developed.[99] It may not be hazardous to predict that these applications will soon constitute a serious alternative to the standard Alchourrón-Gärdenfors-Makinson theories of belief change. But these perspectives, to begin with, seem to legitimate only a *weak* paraconsistent approach; not a strong or dialetheic one, promising genuine counterexamples to the (LNC). Contradictions are "true"; once an interpretation of the semantics has been produced, though, the intuitive meaning of "true" seems to have turned in our hands into something like "included in a theory or a system of beliefs". Therefore, the dialetheist, or supporter of strong paraconsistency, faces the Russellian dilemma. On the one side: if she claims to have exhibited *true*, i.e., *real*, contradictions, she is liable of committing the fallacy of verbalism. On the other side: if

---

[98] Lewis (1982): 107-8.
[99] See e.g. Tanaka (1998a), (2005).

she bites the bullet and takes the shift from "true (of the world)" to "included in a theory" at face value, she appears to commit herself to some equivalence between *worlds* and *theories*, thereby taking the road of some constructivism or idealism: the view according to which the world is made of theories, or conceptual schemes.

It is noteworthy that Meyer and Martin seem to embrace straightforwardly the second horn of the dilemma, in a 1986 paper in which they conduct a defence of the Australian Plan that has since become a classic. Contrasting the usual and, as we have seen, *epistemic* reading of the Belnap-Dunn American Plan, Meyer and Martin stress the "ontic" features of the inconsistencies admitted within the Routley-Meyer (im)possible worlds semantics.[100] Nevertheless, at the same time they describe an inconsistent state of affairs as one in which "a given theory could both assert and deny A" – and, in this case, "the obvious thing to say is that the theory is just mixed up about A".[101] Just like in the Peircean perspective embraced by Rescher and Brandom, consistency and completeness are a regulative ideal – what our theories should be inclined to, in the long run. Hence comes the usual objection: there is a trade-off between worlds and corresponding long run theories. It is the world, not our favourite theories, that confers truth and falsity upon sentences.

The reply is worth listening, for it consists exactly in embracing a form of idealism, and *retorting* it against the classical logician:

> The World to which a given Logic is linked tends to be a lot like the Logic which is linked to it. When one investigates the techniques that are employed in proofs of completeness, one regularly runs into something not far from our Theory of Theories. [...] The semantic Universes in which our modelling goes on are suspiciously like formal theories themselves, endowed with what are supposed to be the right sort of properties. [...] This process of imbuing what one thought was syntactic with semantical significance can become quite bald, as it does in proofs of Classical first-order completeness in the style of Henkin 1949. To refute a given non-theorem A, one builds a "maximally consistent" theory T without A; on a few further manipulation, the theory T becomes the model in which A is refuted. [...] To that extent, Classical logic is self-justified, not independently justified because it has been referred to the World. For the World to which it has been referred has been made in its own image. So one doesn't really need worlds (or models, or whatever) to

---

[100] See Meyer and Martin (1986): 319.
[101] *Ibid*: 311.

tell semantical stories. In the clearest sense, they are just imaginative copies of preferred theories.[102]

Talking about non-adjunctive approaches, I have claimed that metaphysical idealism may be considered a quite respectable position – and by subscribing to a consciously idealistic position dialetheists may avoid the risk of the Russellian Fallacy. Furthermore, once an idealistic stance has been adopted the very *distinction* between weak and strong paraconsistency seems to fade. As we have seen in § 5.6, the distinction was put forward by supporters of paraconsistency as a difference between (a) the claim that there can be inconsistent but interesting and non-trivial theories; and (b) the claim that there can be true contradictions, thus – given a minimally realist stance on "true" – inconsistent objects and/or states of affairs (those that makes the contradictions true). However, such a difference cannot easily be maintained, when we somehow make the distinction between *theories* and *states of affairs* collapse. And this might be a *welcome* result for a strong paraconsistentist, supporting in a different, albeit circuitous, way the "slippery slope" from weak paraconsistency to dialetheism.

9.10 Brady's Logic of Entailment, DJ^dQ

I will end this Chapter with some information on the system of weak relevant logic proposed by Ross Brady,[103] and called DJ^dQ. This will be useful in the following, when we shall talk about paraconsistent set theories. I kept this logic for the conclusion because, despite its being clearly relevant in spirit, its semantics differs from those of the mainstream relevant logics (specifically, it adopts neither the ternary accessibility relation, nor the Routley Star).

9.10.1 Bits of Syntax

As for the syntax, it can be given axiomatically by the following schemas (although also normalized natural deduction presentations are available):

---

[102] *Ibid*: 324-6.
[103] See Brady (1996), (2000).

(A1) $\alpha \to \alpha$
(A2) $\alpha \wedge \beta \to \alpha$
(A3) $\alpha \wedge \beta \to \beta$
(A4) $(\alpha \to \beta) \wedge (\alpha \to \gamma) \to (\alpha \to \beta \wedge \gamma)$
(A5) $\alpha \to \alpha \vee \beta$
(A6) $\beta \to \alpha \vee \beta$
(A7) $(\alpha \to \gamma) \wedge (\beta \to \gamma) \to (\alpha \vee \beta \to \gamma)$
(A8) $\alpha \wedge (\beta \vee \gamma) \to (\alpha \wedge \beta) \vee (\alpha \wedge \gamma)$
(A9) $\neg\neg\alpha \to \alpha$
(A10) $(\alpha \to \neg\beta) \to (\beta \to \neg\alpha)$
(A11) $(\alpha \to \beta) \wedge (\beta \to \gamma) \to (\alpha \to \gamma)$
(A12) $\forall x \alpha[x] \to \alpha[x/t]$
(A13) $\alpha[x/t] \to \exists x \alpha[x]$
(A14) $\forall x(\alpha \vee \beta[x]) \to (\alpha \vee \forall x \beta[x])$
(A15) $\forall x(\alpha \to \beta[x]) \to (\alpha \to \forall x \beta[x])$
(A16) $\forall x(\beta[x] \to \alpha) \to (\exists x \beta[x] \to \alpha)$
(A17) $\alpha \wedge \exists x \beta[x] \to \exists x(\beta[x] \wedge \alpha)$

(x is not free in α in (A14)-(A17)). The rules of inference are:

$$\frac{\alpha \to \beta, \ \alpha}{\beta} \quad (R1)$$

$$\frac{\alpha, \ \beta}{\alpha \wedge \beta} \quad (R2)$$

$$\frac{\alpha \to \beta, \ \gamma \to \delta}{(\beta \to \gamma) \to (\alpha \to \delta)} \quad (R3)$$

$$\frac{\alpha}{\forall x \alpha} \quad (R4)$$

We also have two meta-rules:

(MR1) If $\alpha \Rightarrow \beta$, then $\alpha \vee \gamma \Rightarrow \beta \vee \gamma$
(MR2) If $\alpha \Rightarrow \beta$, then $\exists x \alpha \Rightarrow \exists x \beta$

(to be applied with minor provisos concerning the interaction with (R4)). **DJ^{d}Q** has quite a weak negation (and Brady believes there are good reasons for its being so):[104] neither the (LEM) nor *reductio* hold. Due to such a weakness, the logic respects both the Weak and the Strong Conditions of Non-Triviality, being a depth relevant logic.

9.10.2 Meaning Containment Semantics

As for the semantics, Brady's main idea is to replace the traditional relevant notion of entailment as content connection with a concept of entailment as analytic meaning or content *containment*: a sentence entails another iff the content of the latter is analytically included in the content of the former. More generally, the meaning or content of both logical and non-logical words is taken as having a decisive role in determining the validity of arguments. The content of a set x of sentences, $C(x)$, is characterized by Brady as the analytic closure of "the set of all sentences which can be analytically established from that set, using the properties of relations and terms of sentences in x".[105] The content of a sentence α, $C(\alpha)$, is the content of its singleton. The *range* of a sentence α, $R(\alpha)$, is, dually, the set of all sentences from which α follows analytically. The semantics has an algebraic, rather than truth-functional, approach: logical operators are interpreted (apart from minor provisos) as particular algebraic operations on contents. Entailment can be characterized as meaning containment, that is to say, $\alpha \rightarrow \beta$ iff $C(\beta) \subseteq C(\alpha)$.[106]

An advantage of the approach, according to Brady, is that it focuses on a single concept, which makes a difference with strong relevant logics using diverse stratagems to avoid irrelevant classical inferences. Of course, negation cannot have the usual algebraic treatment as Boolean complementation, because then $C(\alpha \wedge \neg\alpha)$ would be the universal content and, given (S→), this would make a contradiction entail everything. The content of the negation of a sentence is then taken as the set of sentences which are not members of the range of α, that is, $C(\neg\alpha) =_{df} \{\beta \mid \beta \notin R(\alpha)\}$; intuitively, as Bremer puts the point, it is the set of "all facts the

---

[104] See Brady (1996): 164ff. I will come to this in § 11, when I will talk of the theory of classes built by Brady on his logic.

[105] *Ibid*: 161.

[106] See Bremer (2005): 76.

absence of which made [α] possible".[107] **DJ^{d}Q** is complete with respect to the meaning containment semantics.[108]

The basic philosophical issue with the whole framework depends on the way we are supposed to assign a content to a sentence, i.e., to fix what can be analytically established from it. Such notions (or those aspects of them that can survive the Quinean attack to analyticity) are usually captured by something like meaning postulates. Nevertheless, as Bremer remarked, a meaning postulate is, itself, an entailment: it typically includes a conditional, that is, something whose meaning should be captured by (S→) in its turn; and this seems to produce some conceptual circularity in the theory.[109]

---

[107] *Ibid.* See Brady (2000): 118-9.
[108] See Brady (1996): 169-70.
[109] See Bremer (2005): 76.

Part III.
The Uses

10 Semantically Closed Language

> Although the meaning of semantic concepts as they are used in everyday language seems to be rather clear and understandable, still all attempts to characterize this meaning in a general and exact way miscarried.
>
> Alfred Tarski, *The Semantic Conception of Truth*

10.1 Applications

If you challenge the most venerable principle in the history of Western philosophy, you are supposed to concentrate on the foundational aspect of your discourse before moving on to the applications. Consequently, the work on the uses of paraconsistent logics and semantics is still in its beginnings in some fields. On the other hand, an innovative and progressive research program in its early phases often evolves in a turbulent manner. Therefore, it is likely that in a few years some of the results to be exposed in this part of the book will look a bit outdated. I do not have room enough to examine all the proposed applications. One of them, for instance, has to do with a retrieval of the pristine formulation of the infinitesimal calculus. According to Priest and Routley "infinitesimals had to be genuine inconsistent objects", since the calculus had to assume that "an infinitesimal was both zero and non-zero". The authors proposed, then, a formalization based on the second-order theory of reals, with functions specified via λ-abstraction.[1] More recently, though, Priest has favoured a chunk-and-permeate mechanism, which does not require treating infinitesimals as inconsistent objects.[2] Other applications involve quantum mechanics,[3] deontic and epistemic logic, the metaphysics of change and becoming,[4] and, of course, belief revision[5] and the manage-

---

[1] See Priest and Routley (1989b): 374-6.
[2] See Brown and Priest (2004).
[3] See Priest and Routley (1989b): 377-89.
[4] See e.g. Priest (1982), (1987): Chs. 11 and 12; Tanaka (1998b).
[5] See e.g. Tanaka (1998a); Priest (2006): Ch. 8.

ment of databases. One of the most intensively studied areas has to do with the combination of inconsistency and vagueness, suggested since the origins of paraconsistency by the dualities between (LNC1) and the (LEM), (LNC2) and the Principle of Bivalence, etc., which we have already met. Hence came the sub-valuational semantics, such as Hyde's and Varzi's, that I mentioned in § 6; and the relevant impossible worlds semantics for fictional entities we had a quick glance at in § 9.6.

In this part of the book, I shall consider only three philosophically essential applications of paraconsistency. The first two correspond to the two main cases on behalf of contradictions discussed in the first part: (1) a formal semantics for a semantically closed language capturing the relevant features of ordinary language; and (2) set theory with unrestricted Comprehension, mirroring our intuitive and "Cantorian" conception of set. The third one (3), to be dealt with in the final Chapter of this part, concerns inconsistent arithmetics.

10.2 *Desiderata*

The first essential case against the (LNC) comes, as we know, from intuitive, universal semantics. We may therefore want to have a formalized semantics modelling a semantically closed language: a language within which, as we are supposed to be able to do in ordinary English, we are capable of saying anything linguistically expressible, without submitting to the limitations imposed by the various consistent solutions of semantic paradoxes we met in § 2. Such semantics is expected to *include* the paradoxes, against (LNC2). Following Woods (2003), we can phrase the *desiderata* of an intuitive inconsistent semantic theory for a semantically closed language as follows: (1) the theory is supposed to preserve all our most rooted intuitions on the notion of truth – to begin with, the unrestricted T-schema. (2) it has to be formulated in (a fragment of) the very same language as that for which the truth predicate is characterized and the semantics is given. And (3) since this produces the paradoxes, the underlying logic must be paraconsistent, and must prevent the theory from being trivialized by the Curry paradox – so it should fulfil the Strong Condition of Non-Triviality.[6]

Given a suitable logic, it seems we only have to designate a one-place predicate as the truth predicate (or a two-place predicate as the satisfac-

---

[6] See Woods (2003): 173ff.

tion predicate, defining then the former via the latter); give recursive truth/satisfaction conditions, formulated in the same language for which the notion of truth/satisfaction is being defined; and show that the definition respects the Tarski Convention T, that is, all instances of the T-schema can be inferred the usual way. The most interesting theory of this kind so far is the one proposed in Chapter 9 of Priest (1987), to which I now turn. As we will see, the carrying out of the task is slightly more complicated than its announcement, and it produces some unexpected outcomes.

10.3 Priest's Semantically Closed Theory

The theory substantially uses **LPQ** as its underlying logic, but with some modifications. First, **LPQ** has to be supplied with a relevant conditional (which has been changed in the passage from the first to the second edition of *In Contradiction*, but allowing the construction of the semantically closed framework to go through in the same way). Second, Priest proposes two variants of the theory, the first of them respecting what he calls the Exclusion Principle. This is nothing but the left-right half of our (Neg2), that is:

(Exc) $T(\ulcorner\neg\alpha\urcorner) \to \neg T(\ulcorner\alpha\urcorner)$.

As we shall see, in this variant (Exc) is directly embodied in the semantic clause for negation. On the other hand, we know that (Neg2) is controversial. In particular, it makes any internal (C2b)-contradiction, $T(\ulcorner\alpha\urcorner) \wedge T(\ulcorner\neg\alpha\urcorner)$, entail an external (C2c)-contradiction, $T(\ulcorner\alpha\urcorner) \wedge \neg T(\ulcorner\alpha\urcorner)$, which might be an unwelcome result even for a dialetheist (on why this is so, I shall say more in the final § 14 of the book). Hence comes the second variant of the theory, going through without (Exc). To make it work in this second shape, we also have to do without Contraposition; so the modification of the status of negation entails that the relevant conditional become a non-contraposible one.[7]

---

[7] See Priest (1987): 126.

10.3.1 The Language of the Theory

The language L of the theory has the usual first-order logical vocabulary with connectives, quantifiers, and identity. We assume conjunction, negation, conditional and universal quantifier as primitive. Disjunction, biconditional and existential quantifier are defined the usual way.

Officially, variables are indexed by the natural numbers (whose set is N) and constants by numbers in a subset K of N – but the indexing can be usually omitted for practical purposes. So the set of variables is $\{v_n \mid n \in N\}$, and that of constants is $\{c_k \mid k \in K\}$. Among the things constituting the domain of the variables of L are numbers, sequences of objects, and linguistic expressions too. The reason is easily understood: we want to quantify both on linguistic and on worldly entities, and to be able to talk about relations between the former and the latter, such as the satisfaction relation. Priest sometimes uses distinct kinds of syntactic variables (for instance, $s, s_1, \ldots, s_n$ as variables for sequences of objects), but he claims this is a merely mnemonic device: officially, the logic is single-sorted,[8] which makes the language gain in (philosophically appreciable) uniformity. The descriptive vocabulary, as was to be expected, is rich; it includes semantic and arithmetic expressions. Here it is, together with the intuitive interpretations associated by Priest.

"*Sat*" denotes the satisfaction relation, which of course can hold between sequences of objects and formulas, so it is taken as a set of ordered pairs. In order to have a notion of satisfaction generalized to unary predicates, we assume that an object o coincides with its unary sequence, $<o>$.

We have, then, three one-place predicates: "*Nat*" denotes the set N of natural numbers; "*Form*" denotes the set of well-formed formulas of L; and "*Term*" denotes the set of singular terms of L. We also have the following functors. "*Succ*" denotes, as usual, the unary successor operation (and, in this context, if applied to things other than numbers, the operations produces that very thing as an output: $Succ(t)$, if $t \notin N$, is just t). "*Const*" denotes a unary operation mapping each index k to the respective constant c_k and, when applied to things not in K, it gives a dummy/fixed constant as its output. "*Var*" denotes a unary operation that does the same job for variables, mapping each natural number n to the respective variable v_n, and mapping non-numbers to a dummy/fixed variable.[9] "+" de-

---

[8] See *Ibid.*
[9] See also Bremer (2005): 128.

notes a binary operation of concatenation: given two sequences, x and y, it produces a longer sequence $x + y$, which is precisely their concatenation. "*App*" denotes a binary operation such that, intuitively, $App(t_1, t_2)$ is the value of the function (denoted by) t_1 at the argument (denoted by) t_2. "*Subst*" denotes a ternary function such that $Subst(t_1, t_2, t_3)$ is the same function as (that denoted by) t_1, except that its value at the argument (denoted by) t_2 is (denoted by) t_3 (for the sake of brevity, we will have $t_1(t_2)$ for $App(t_1, t_2)$, and $t_1(t_2/t_3)$ for $Subst(t_1, t_2, t_3)$).[10] "*Den*" denotes a binary operation such that, intuitively, if t is a term of L and s is a sequence, $Den(s, t)$ is the denotation of t when its free variables have obtained as denotations objects of s.

Following Priest's presentation, the name of a linguistic expression is built by underlining it, so e.g. $\underline{\alpha}$ is the name of α. The name of zero is $\underline{0}$. For each singular term of L, then, we specify its name as follows:

$\underline{v}_n$ is $Var(\underline{n})$, for $n \in N$

$\underline{c}_k$ is $Const(\underline{k})$, for $k \in K$.

Given all this, plus names of the logical and auxiliary symbols, the predicates, and functors, we can build the name of any expression of L by means of the concatenation operation: given a sequence of expressions, $e = e_1... e_n$, its name is $\underline{e} = \underline{e}_1 + ... + \underline{e}_n$. Therefore, L reproduces those features of English we wanted to capture with our formalization: we can talk within L of any expression of L, and attribute semantic properties to it.

10.3.2 Axioms and Details

Once L has been introduced together with the informal interpretation of its non-logical terms, we can move ahead to the axioms of the theory. These are divided into four groups: axioms of group 1 are called *mathematical*, since they discipline the arithmetic concepts; those of group 2, *syntactic*; those of groups 3 and 4, *semantic*, recursively characterizing, respectively, denotation and satisfaction.

---

[10] For details and slight complications due to the single-sortedness of the theory, see Priest (1987): 127.

Group 1

(1) (a) $Succ(x) \neq \underline{0}$
    (b) $Succ(x) = Succ(y) \rightarrow x = y$
    (c) $Nat(\underline{0})$
    (d) $Nat(x) \rightarrow Nat(Succ(x))$

(2) $a = App(s(x/a), x)$
(3) $x \neq y \rightarrow App(s, y) = App((s(x/a), y)$

(1a-d) are a fragment of Peano arithmetic (zero is a natural number, it is not a successor, etc.). But in this context variables do not range only over numbers; so, for instance, (1b) is true also of things that are not numbers – and, in this case, it is unimportant, since if $x \notin N$, $Succ(x) = x$. Priest claims that all the axioms of group 1 "would be provable in the more general context of (dialetheic) set theory/number theory".[11]

Group 2

(1) $x + (y + z) = (x + y) + z$

(2) (a) $Term(Var(x))$
    (b) $Term(Const(x))$
    (c) $Term(x_1) \wedge \ldots \wedge Term(x_n) \rightarrow Term(f + \underline{(} + x_1 + \ldots + x_n + \underline{)})$, for every n-ary functor f.
(3) (a) $Term(x_1) \wedge \ldots \wedge Term(x_n) \rightarrow Form(\underline{P} + \underline{(} + x_1 + \ldots + x_n + \underline{)})$, for every n-ary predicate P.
    (b) $Form(x) \wedge Form(y) \rightarrow Form(\underline{(} + x + \underline{\wedge} + y + \underline{)}) \wedge Form(\underline{(} + x + \underline{\rightarrow} + y + \underline{)}) \wedge Form(\underline{\neg} + x)$
    (c) $Form(x) \rightarrow Form(\underline{\forall} + Var(y) + \underline{(} + x + \underline{)})$

(1) is associativity for the concatenation operation. Sub-group (2) regulates singular terms: constants and variables are terms, and (2c) tells us that we obtain a singular term via the concatenation of an n-ary functor with n terms (I have omitted for brevity the names of commas in the concatenation). Sub-group (3) regulates atomic and molecular formulas: (3a)

---

[11] *Ibidem*, p. 161.

claims that we obtain a(n atomic) formula via the concatenation of an n-ary predicate with n terms; (3b-c) spell out how to obtain conditionals, conjunctions, negated and quantified formulas. The whole machinery expresses within L the things we usually claim in an informal or semi-formal metalanguage, when we present the syntactic clauses for well-formed terms and formulas in an ordinary formal language.

Group 3

(1) $Nat(x) \rightarrow Den(s, Var(x)) = s(x)$
(2) $Den(s, Const(\underline{k})) = c_k$ for $k \in K$
(3) $Term(x_1) \wedge \ldots \wedge Term(x_n) \rightarrow Den(s, \underline{f} + \underline{(} + x_1 + \ldots + x_n + \underline{)}) = f(Den(s, x_1), \ldots, Den(s, x_n))$, for every n-ary functor f.

These axioms characterize denotation for the terms of L.

Group 4

(1) $Term(x_1) \wedge \ldots \wedge Term(x_n) \rightarrow (Sat(s, \underline{P} + \underline{(} + x_1 + \ldots + x_n + \underline{)}) \leftrightarrow P(Den(s, x_1), \ldots, Den(s, x_n)))$, for every n-ary predicate P.
(2) $Form(x) \wedge Form(y) \rightarrow (Sat(s, \underline{(} + x + \underline{\wedge} + y + \underline{)}) \leftrightarrow Sat(s, x) \wedge Sat(s, y))$
(3) $Form(x) \wedge Form(y) \rightarrow (Sat(s, \underline{(} + x + \underline{\rightarrow} + y + \underline{)}) \leftrightarrow (Sat(s, x) \rightarrow Sat(s, y)))$
(4) $Form(x) \rightarrow (Sat(s, \underline{\neg} + x) \leftrightarrow \neg Sat(s, x))$
(5) $Form(x) \wedge Nat(y) \rightarrow (Sat(s, \underline{\forall} + Var(y) + \underline{(} + x + \underline{)}) \leftrightarrow \forall z Sat(s(y/z), x))$

These axioms characterize recursively satisfaction for formulas and, of course, are the most important ones. (1) claims that a sequence s satisfies an atomic formula iff the n-ary predicate of the formula applies to the objects of s denoted by terms $x_1, \ldots, x_n$. (2)-(5) generalize to molecular formulas obtained by connecting and quantifying. Let us pay attention to axiom no. 4. This expresses the semantics of a negation that conforms to the Exclusion Principle (Exc): it claims that a sequence s satisfies a formula $\neg \alpha$ iff s does not satisfy α.

We can now define the truth predicate for L the usual, Tarskian way:

$T(x) =_{df} \forall s(Sat(s, x))$.

A true formula is a formula which is satisfied by all sequences. Now one can prove within the theory any instance of the T-schema for L, that is to say:

(TH) For every closed formula α of L, ⊢ $T(\underline{α}) \leftrightarrow α$.

Therefore, Tarski's Convention T is fulfilled. Priest proves (TH) by means of a couple of auxiliary lemmas on terms and formulas, in the technical appendix of the Chapter of *In Contradiction*.[12] The proof uses the Exclusion Principle, but Priest claims we can also do without it, provided we introduce the following changes. We must (a) replace the conditional with a non-contraposible one; (b) introduce a new binary relation *Antisat* of *anti*-satisfaction, which "is to falsity what satisfaction is to truth";[13] (c) replace axiom (4) of group 4, that is, the clause for negation, with the following:

(4b) $Form(x) \rightarrow (Sat(s, \neg + x) \leftrightarrow Antisat(s, x))$;

and finally (d) add a bunch of axioms recursively characterizing anti-satisfaction in a way analogue to that in which the (modified) initial axioms of group 4 characterize satisfaction. Changes (a)-(d) give us the second version of the theory, as anticipated above.[14]

10.4 Does it Work?

Priest claimed that, by means of such a theory, "dialetheism [...] solves a fundamental problem of semantics",[15] if (a) we stick to the basic point of ordinary semantics, according to which the notion of truth/satisfaction is at the core of a theory of meaning, and (b) we take the idea that ordinary language is semantically closed seriously:

> Of, course, semanticists do not really believe that the semantics of English are expressible only in another language. At least, I have not noticed classes

---

[12] Cf. *Ibid*: 136ff.
[13] *Ibid*: 129.
[14] See *Ibid*: 129-30 and 138-9.
[15] *Ibid*: 133.

of Hindi, Urdu and Mandarin swelled by the ranks of semanticists keen to see whether these languages contain the key to the ineffable.[16]

On the other hand, Priest's theory has some strange features – or, conversely, one may claim that it seems to *lack* strange features, which should nevertheless be expected within a paraconsistent theory. For instance, it is not known whether the two versions of the theory (with/without (Exc)) are inconsistent. This appears to be strange: didn't we expect a theory containing and capable of defining its own truth predicate to be inconsistent? As John Woods has wondered, "is not the agnosticism of [Priest's theory] with regard to its own inconsistency inconsistent with this fact?"[17] According to Priest, the consistency of the two variants is due to "purely accidental reasons" depending on the characterization of the notion of satisfaction. Explicitly adding as an "extra" premise a formal liar would be sufficient to obtain the desired contradiction. Another way would be to augment the arithmetical axioms so that it is possible to prove the Fixed Point Lemma,[18] with methods close to the ones I mentioned in § 2.1.3 and § 4.2.[19]

Another problem is whether the two variants of the theory are non-trivial. As we shall see in the following Chapter, the equivalent of the issue of consistency proofs for paraconsistent theories are non-triviality proofs: we want to be guaranteed that the theory cannot prove anything. The modified logic **LPQ** underlying the construction fulfils the Strong Condition of Non-Triviality, so standard roads to triviality are closed. However, a real non-triviality proof for the formal semantics is still to be found. In 2003, Woods ended his assessment of the issue by claiming that it is not the case that "as things stand, [Priest's semantics] has been able to make good on the fundamental dialethic insight that there are provably inconsistent theories that are provably nontrivial".[20]

---

[16] *Ibid*: 134.
[17] Woods (2003): 181.
[18] Priest (1987): 130.
[19] For some inconsistencies resulting in the theory via this kind of addition, see Bremer (2005): 134-6.
[20] Woods (2003): 181.

11 Paraconsistent Set Theory and Metalogic

> There are more things in heaven and earth, Horatio, than are dreamt of in your philosophy.
>
> William Shakespeare, *Hamlet*

11.1 Desiderata

Strong paraconsistentists consider set-theoretic as well as semantic paradoxes as what they should look to anyone who does not wear the classicist's glasses: sound arguments. Such a perspective is viewed as capable of recovering naïve set theory, taken as the intuitive and analytic characterization of sethood:

> As a matter of contingent history, opinion has clustered around the position that because *ex falso* is true the Russell Paradox is bad enough to require the replacement of the old set theory with something new and different enough to prevent paradox from reobtruding. Here is a position in which when a theory T collides with classical logic we change T; we do not change logic. It is well to note, however, that in principle the reverse strategy is also available: *Retain* T and *change* logic.[1]

The ontological cost of the retaining is clear: inconsistent objects are admitted in the domain of sets, against (LNC3). What about the benefit? The rationale for a paraconsistent set theory is to avoid the theoretical difficulties that face mainstream consistent set theories: the *ad hoc* nature of the limitations to the Abstraction Principle; the lack of a universal set, and the consequent violation of the Domain Principle, that threatens the intelligibility of the very notion of set. Besides having an intrinsic interest, however, set theory also has an *instrumental* function with respect to mathematics. A paraconsistent set theory, therefore, is supposed to make available a solid foundation for mathematical practice: it should provide a cure to the ineffectiveness of the cumulative hierarchy for the set-

---

[1] Woods (2003): 8.

theoretic foundation of category theory, and particularly of the operations on large categories. It should justify the idea that mathematicians are allowed to work without paying heed to the restrictions imposed by the ordinary axiomatic set theories. Inconsistent sets are not so troublesome, when Hell does not break loose because of them.

We will meet soon a case of paraconsistent set theory that fails with respect to these *desiderata*, and then a couple of perspectives that seem to fare better. The paraconsistent arithmetics that can be built within this framework shall be dealt with in the next Chapter. First of all, though, I shall say something concerning the metalogical situation produced by these theories, precisely because of their inconsistency.

11.2 Gödel, Second Half

So-called limitative theorems, such as Gödel's, Rosser's and Church's, certainly are the most studied and celebrated metalogical results (which may be due, among other things, to the philosophical fascination of the idea of a *prohibition* they bring with themselves). However, the influence of classical logic is such, that these acquisitions are often exposed without mentioning a prerequisite of applicability they depend on, namely, the *consistency* of the theory for which the limitation is at issue. This is embodied, on the contrary, within Gödel's First Theorem, whose formulation in § 4.2 involved the consistency (and ω-consistency) of **PA**. Gödel was initially cautious on the possibility of generalizing his incompleteness results, but in a note added in '63 to his 1931 paper he claimed:

> In consequence of later advances, in particular of the fact that due to A. M. Turing's work a precise and unquestionably adequate definition of the general notion of formal system can now be given, a completely general version of Theorems VI and XI is now possible. That is, it can be proved rigorously that [Theorem VI:] in *every* consistent formal system that contains a certain amount of finitary number theory there exist undecidable arithmetic propositions and that, moreover, [Theorem XI:] the consistency of any such system cannot be proved in the system.[2]

"Theorem VI" is the First Incompleteness Theorem, which I have introduced in 4.2, and in which the undecidable γ is at issue. "Theorem XI"

---

[2] Gödel (1931): 616.

is Gödel's Second Incompleteness Theorem. This is the only other limitative theorem we will meet in this book[3] – and I will explain it in a quicker and more informal way than I did for the First Theorem. We only need to get an intuitive idea of what "in every *consistent* formal system ... the consistency ... cannot be proved in the system" means.

11.2.1 "Consistency... Must Be Taken in Faith"

As we may recall, the first half of the First Theorem claims that, if **PA** is consistent, then γ cannot be proved in it. Can we express the idea *within* **PA**? In order to do it, we need a "consistency statement" (be it *Cons*) for **PA**. We can characterize it here by means of our provability predicate:

$Cons =_{df} \neg \exists x Prf(x, \ulcorner \bot \urcorner)$,

Where the *falsum* constant stands for any inconsistency. We can now express within **PA** the fact that if **PA** is consistent, then γ cannot be proved in it, thus:

(1) $\neg \exists x Prf(x, \ulcorner \bot \urcorner) \to \neg \exists x Prf(x, \ulcorner \gamma \urcorner)$.

But the consequent of (1), $\neg \exists x Prf(x, \ulcorner \gamma \urcorner)$, is nothing but γ, that is, $\neg Th(\ulcorner \gamma \urcorner)$. Therefore (1) is equivalent to:

(2) $Cons \to \gamma$

and it can be shown that (2) (therefore, (1)) is provable within **PA**. Gödel's Second Theorem, now, simply goes as follows:

(3) If **PA** is consistent, then $\nvdash_{PA} Cons$.

For suppose we can prove within **PA** the antecedent of (1), or (2), i.e., suppose that the formula expressing the consistency of **PA** is a theorem *of*

---

[3] For expositions and proofs of the rest of the family, that is, Church's Theorem, Löb's Theorem, etc., one can check any good textbook of intermediate or advanced logic, e.g., Boolos, Burgess and Jeffrey (2002).

PA. Then via (→E) or *modus ponens* we could also prove γ. We would have, that is, such a simple proof:

(1) $Cons \to \gamma$ Theorem of **PA**
(2) $Cons$ Theorem of **PA**?
(3)   γ   1, 2, →E

However, this is ruled out by the First Theorem, which has claimed that, given consistency, γ is not provable in **PA**. The Second Theorem, (3), tells us that, if **PA** is consistent, then the sentence of its formal language L expressing the consistency of **PA** cannot be a theorem of **PA**: the theory cannot prove *its own* consistency.

Such a conclusion has had, possibly, an even bigger philosophical impact than the First Theorem. Hilbert believed that no *ignorabimus* had to be admitted in mathematics: that "we can provide a rigorous and completely satisfying foundation for the notion of number, and in fact by a method that I would call *axiomatic*",[4] and that by means of it one could provide arithmetics with finitary consistency proofs. However, the Second Theorem shows that no consistent formal system capable of expressing elementary arithmetics is, as it were, "self-contained" in this respect: its consistency cannot be proved within the very system. Michael Dunn has claimed that "to exaggerate perhaps only a little, the consistency of systems like Peano (even Robinson) arithmetic must be taken in faith".[5]

11.2.2 Escape from the Undecidable

Now, a paraconsistent set theory, or arithmetic, does not fulfil *precisely* the consistency requisite; which suggests that such theories could somehow emancipate themselves from Gödel's Theorems, and maybe from other limitative results afflicting their consistent cousins based upon an explosive logic. To be sure, consistency proofs are not at issue, since we would be dealing with (negation-)inconsistent theories. What the theory may hopefully prove, though, is its own non-triviality – which in these

---

[4] Hilbert (1904): 131.

[5] Dunn (1986): 161. Actually, since Gentzen's work we do have consistency proofs for formal systems expressing arithmetics. The proofs in question, though, share the feature of employing more powerful deductive tools than those available within the systems for which consistency is proved, and in any case, they cannot be carried out within the systems.

contexts is more often called absolute consistency. Paraconsistent authors have begun to show that this is the case since the 70s, by building inconsistent but non-trivial theories, whose non-triviality proof can be represented within the very theories and that, in this sense, somehow avoid Gödel's Second Theorem. Their inconsistency allows them to escape also from Gödel's *First* Theorem, and from Church's undecidability result: they are, that is, demonstrably *complete* and *decidable*.[6]

Could paraconsistentists refrain from boasting of the cheerful situation?[7] They sometimes draw philosophical consequences, too. One of them is the possibility of envisaging a retrieval of Hilbert's programme,[8] which was deemed dead by scholars[9] precisely because of the Gödelian incompleteness results. Another one is an idea we met towards the end of the first part of the book. You may remember that, according to Routley and Priest, by acknowledging that the theory that captures our intuitive notion of mathematical proof is inconsistent we can explain how we *learn* arithmetic – that is, via a fully recursive procedure:

> We appear to obtain our grasp of arithmetic by learning a set of basic and effective procedures for counting, adding, etc.; in other words, by knowledge encoded in a decidable set of axioms. If this is right, then arithmetic truth would seem to be just what is determined by these procedures.[10]

Therefore, we have a package of great expectations and promises, and we can now begin to take into account some paraconsistent set theories.

---

[6] For a quick review, see Bremer (2005): Ch. 13.

[7] "The Limitative Theorems of classical metamathematics are usually regarded as, at best, disappointing, at worst posing nasty philosophical problems. With the exception of the Löwenheim-Skolem Theorem [which we shall mention soon] [...] paraconsistent arithmetic is free from all these Theorems, and so problems. The fact that a theory solves problems that beset its rivals is well recognised as speaking strongly in its favour" (Priest (1994): 342).

[8] See Priest and Routley (1989d): 527-8. Priest seems to have become more and more cautious on this, though: see Priest (1994): 345.

[9] With a few exceptions, e.g., Detlefsen (1979).

[10] Priest (1994): 343.

11.3 Variations on Quine

The paper in which Quine criticized the "unnatural and inconvenient consequences" of Russell's theory of types was called *New Foundations for Mathematical Logic*. In the very same paper, Quine proposed a system for set theory that, therefore, came to be labelled as **NF**. The salient feature of the system consists in replacing the Russellian hierarchy with a procedure named *stratification*, which produces a sort of "type without type". A given formula α is *stratified* iff all its variables can be indexed by natural numbers, so that in all subformulas of α of the form $x \in y$, if x gets the number n, y gets the number $n + 1$. Non-stratified formulas (such as $x \in x$), which were ruled out as meaningless by the theory of types, are admitted in **NF** as well-formed. As it were, we are allowed to *say* such things as "$x \notin x$". We reshape the Abstraction Principle, though, by claiming that:

(Abs$_{NF}$) If $\alpha[x]$ is a stratified formula, then $\exists y \forall x (x \in y \leftrightarrow \alpha[x])$

(y is not free... , etc.). Abstraction can therefore be performed only on stratified formulas: we can talk about the Russellian condition *not a member of itself*, but we cannot dispense Existence to the corresponding set producing "the contradiction".[11] This excludes the unnatural multiplication of all the relevant notions. Above all, *if* a set x exists, then its "real", intuitive complement also exists, i.e., there is a set of all the things not in x.

One of the relevant model-theoretic characteristics of **NF** is that it does not have a model in a segment of the cumulative hierarchy. Those who admit only the (well-founded) sets in the hierarchy have seen this fact as an anomaly, and it is fair to say that most classical mathematicians and logicians do not pick **NF** as their favourite set theory. Hao Wang observed that "no integration of the membership predicate (with the right identity relation) compatible with the axioms of NF could make well-orderings of both the less-than relation among ordinals and that among finite cardinals".[12] The early version of the system **ML** – a natural extension of **NF** including also classes – proposed by Quine in *Mathematical*

---

[11] See Quine (1937); see also Fraenkel, Bar-Hillel and Levy (1973): 161ff.
[12] Wang (1986): 640.

Logic[13] was proved inconsistent (one can derive in it the Burali-Forti paradox), which according to some is a sign of the fact "that something is wrong with NF",[14] too. But certainly, a classical *contra* (such as being inconsistent) is often a paraconsistent *pro*. In particular, an advantage of **NF** is, unsurprisingly, that it has the universal set, V: just apply (Abs$_{NF}$) to $x = x$ (and Cantor's paradox does not follow, because Cantor's Theorem does not hold in **NF**). Therefore, if our main concern with standard set theories was the lack of a universal set, **NF** seems to be designed for us.

The Brazilian school had therefore proposed paraconsistent set theories obtained by combining the logical hierarchies of da Costa's C-systems with the set-theoretic ideas of **NF**. The theory built on top of the calculus $C^=_1$ is called **NF$_1$** and, corresponding to the da Costa hierarchies, we can also have **NF$_2$**, ... , **NF$_\omega$**. **NF$_1$** is inconsistent (e.g., Russell's paradox can be proved in it) but non-trivial.[15] Its non-triviality proof apparently cannot be represented within it, though – which might be taken as a deficiency with respect to the paraconsistent promises we mentioned above. Da Costa showed that **NF$_1$** could be translated into **NF**, the original Quinean system (and the non-triviality of the former entailed the consistency of the latter). The translation, though, was but the *-transformation we met in § 7.2, that is, a replacement of all occurrences of da Costa negation with the (fully explosive) strong negation. Therefore, the same criticisms as those advanced in that § against the strategy still hold.

In addition, the reformulation of the Abstraction Principle within this kind of theories depends heavily on the trick with negation. The Quinean formulation of (Abs$_{NF}$) required $\alpha[x]$ to be stratified, so non-stratified conditions did not generate any set. Within the initial **NF$_1$**, Abstraction can be performed also on non-stratified formulas, *provided* $\alpha[x]$ does not include occurrences of the strong negation. Furthermore, it must not include occurrences of the conditional: given its standard behaviour within the positive-plus logics (e.g. its validating Contraction-Absorption), it would trivialize the theory via a variant of the Curry paradox.[16] However,

---

[13] See Quine (1951).

[14] Woods (2003): 167. Quine was rescued from contradiction by Wang himself, who adjusted **ML** into a more cumbersome but consistent theory (see Wang (1950)).

[15] See da Costa (1974): 507; see also da Costa (1986).

[16] See da Costa (1974): 506; Marconi (1979): 55-6.

the main motivation for a paraconsistent set theory was that we aim at having full, unrestricted Abstraction: there should be more sets around, than are dreamt of in the consistentist's philosophy. Therefore, the fact that such set-theoretic constructions cannot avoid *all* the limitations on set building speaks against them:[17] paraconsistent set theory should be as liberal as possible about sets (apart from the issue of what their admission in the theory entails, concerning their strictly metaphysical status).

11.4 Routley's Dialectical Set Theory, DST

11.4.1 The Basic Ideas

Routley built a paraconsistent-dialectical set theory, **DST**, based on one of his (and Meyer's) relevant paraconsistent logics, namely, **DKQ**. Linguistically, one just adds to ordinary first-order language the membership predicate $\in$ as a primitive symbol. The Abstraction Principle in its unrestricted form now characterizes it:

(Abs) $\exists y \forall x (x \in y \leftrightarrow \alpha[x])$.

(Abs) guarantees full conformity to both Existence and Objectivity, as described in § 3.1: not only, there is a set for any given condition; but also, each set is a full-fledged object capable of unrestricted set membership. But Routley goes even further. When presenting naïve set theory I introduced, as the only restriction on (Abs), that y should not be free in $\alpha[x]$. This has traditionally been taken (from Cantor to Zermelo) as a mere requirement of determinacy on the condition $\alpha[x]$, which would otherwise produce paradoxical "vicious circles" quite independently of the issue of impredicative definitions. However, Routley's theory has no such hindrances against paradox (as is in the spirit of the enterprise, and of the author),[18] so it does not include even this minimal restriction on (Abs). This means that, besides admitting the usual inconsistent sets, such

---

[17] As Marconi has claimed in his evaluation of the da Costa approach, "it seems that one of the goals to be achieved, when we build a paraconsistent set theory, is to avoid any restriction on the formulation of the axioms. Restrictions in a classical system are needed precisely to avoid the antinomic sets, whose dangerous features should be overcome within a paraconsistent framework. Put it otherwise: restrictions of the Comprehension Schema fall foul of the 'spirit' of a paraconsistent system" (Marconi (1979): 306).

[18] See Routley (1979b): 915.

as Russell's, it includes (taking α[x] as $x \notin y$) also the weird self-characterized set of all things that belong to it iff they do not belong to it:

$\exists y \forall x(x \in y \leftrightarrow x \notin y)$.

Therefore, the theory includes non-well-founded sets. Even if we re-introduced the aforementioned minimal restriction on (Abs), we would still have $\{x \mid x \in x\}$. However, this goes beyond the issue of inconsistency and of the (LNC), since, as we have seen, various authors, such as Aczel (1988), independently defend the opportunity of admitting non-well-founded sets. The work on hypersets has shown that the non-well-foundedness of sets is innocent with respect to Russell's paradox.[19]

DST aims at fully including the set-theoretic paradoxes and the reasoning involved in them, so it proves, e.g., that Russell's set both is and is not a member of itself, $R \in R \wedge R \notin R$.[20] A more sensitive issue is how to treat identity in a paraconsistent set theory based on a (depth) relevant logic. The standard formulation of the Extensionality Principle, as we know, is:

(EP) $\forall x(x \in y \leftrightarrow x \in z) \rightarrow y = z$.

It is usually claimed that this is what differentiates purely extensional beings like sets from such "creatures of darkness", rejected by Quine, as properties: a set is completely determined by its elements. The problem with **DST** is that, (EP) notwithstanding, **DKQ** is a fully relevant logic, so $\leftrightarrow$ and $\rightarrow$ should be taken as a (bi)conditional which is invulnerable to any fallacy of relevance. It is, therefore, a strongly *intensional* operator, and it is involved in the principles specifying the identity conditions of set-theoretic entities. Now, if we adopted the standard, Leibnitian treatment of identity, particularly Leibniz's Law:

(LL) $\forall xy(x = y \rightarrow (\alpha[x] \leftrightarrow \alpha[y]))$,[21]

---

[19] For instance, see Barwise and Moss (1996).
[20] See Routley (1979b): 914-5.
[21] This is a schematic formulation. The higher order version, quantifying predicative variables, would give us the so-called Principle of the Indiscernibility of Identicals:

(InId) $\forall xy(x = y \rightarrow \forall P(P(x) \leftrightarrow P(y)))$.

we would face a problem of irrelevance. As Routley acknowledges,[22] by vacuous replacement from (LL) we get $x = y \to (\beta \to \beta)$, which is a case of $\alpha \to (\beta \to \beta)$. This is a fallacy of relevance (which the Routley-Meyer semantics circumvents, as we have seen, by resorting to the three-place accessibility relation on worlds). Therefore, we must "reject the Leibnitzian account of identity".[23] Routley conjectures (with a few provisos) that one may replace (LL) in **DST** with a restricted rule of inference for atomic formulas:

$$\frac{x = y}{x \in z \to y \in z} \quad \text{(ER)}$$

Issues of relevance suggest to be careful also in the definitions of famous sets, such as the universal set, V and the empty set, $\emptyset$. For instance, one cannot define $\emptyset$ as $\{x \mid x \neq x\}$ even though $\forall x(x = x)$ is a theorem of paraconsistent logics, such as **DLQ** and **DKQ** – and, by the way, also of Priest's **LP**, being taken in Priest (1979) as a "self-evident truth".[24] This does not prevent self-identity from failing at impossible worlds; but the issue now is a different one. The point is that such a definition of $\emptyset$ could not entail that $\forall x(\emptyset \subseteq x)$, i.e., that (as is usually admitted) the empty set is included in any set. Such an entailment would be, again a case of irrelevance: one cannot prove $x \neq x \to \alpha$, for any α, in the underlying logic. In addition, one has a dual problem with V. Therefore, it seems that "$\subseteq$ differs significantly from the usual inclusion notion".[25] On the other hand, Routley suggests a different definition, and that something like $\emptyset =_{df} \{x \mid \forall y(x \in y)\}$ may do.

The second order version has been defended in the influential Cartwright (1971). By adding the converse entailment, that is, the Identity of Indiscernibles:

(IdIn)$\forall xy(\forall P(P(x) \leftrightarrow P(y)) \to x = y)$

(which is philosophically much more controversial), we would have the original Leibnitian conception in its full strength.

[22] See Routley (1979b): 920.
[23] *Ibid*: 922.
[24] Priest (1979): 234.
[25] Routley (1979b): 924.

11.4.2 The Pros and Cons of DST

DST has all the advantages one may wish within a theory, which, just as naïve set theory, does not impose limits on (Abs). First, for each set we have its absolute, intuitive complement, that is to say:

$$\forall x \exists y \forall z (z \in y \leftrightarrow z \notin x),$$

whereas in **ZF-ZFC** and relatives we only have relative complements, and in **NF** we have complements of the existing sets, i.e., those allowed by stratified conditions. Even more interesting is the fact that within **DST** one can prove many "Cantorian" things that in the mainstream consistent theories had to be stolen as axioms – for instance, the existence of infinite sets (just as Dedekind did). Furthermore, and strangely enough, the fact that in (Abs) the condition $\alpha[x]$ may include free occurrences of y has a decisive role in the proof of a variant of the (global) Axiom of Choice.[26] This fact is taken by Routley as a sign of the *realist* attitude of a good paraconsistent set theory – in particular, there really are contradictory objects, such as the inconsistent sets, against (LNC3). "There really are" should be taken *cum grano salis*, since we are talking about Routley, and Routley was a Meinongian noneist: he believed that abstract objects do not *exist*, despite their being full-fledged objects. However, this kind of realism need not be taken as Platonism. It has more to do, so to speak, with the refusal of any form of set-theoretic fictionalism. As Woods claimed in his reconstruction of the "dialetheic mini-history of set theory":

> [Consistent axiomatic set theories] cost us realism in mathematics. It stuck us with a nervy postmodernism in mathematical epistemics. Such losses are philosophically painful, and they cannot convincingly be compensated for by naming them "idealism". Better that we stick with *real* sets, rather than stipulations on the *word* "set" in the manner of Russell [...]. The very idea of sets is inconsistent, that is, the *intuitive* idea – the idea that tells us what it is to be a set. This being so, as long as we wish to do set theory, to say with accuracy how sets *actually are*, we will have to honour the intuitive fact of actually inconsistent sets.[27]

---

[26] See Routley (1979b): 924ff.
[27] Woods (2003): 155-6.

Once again, the paraconsistent author can claim she is just helping us to broaden our view: the cumulative hierarchy is now only a consistent and, as it were, relatively "small", fragment of the universe of sets, which includes non-well-founded and inconsistent sets as well. We can also distinguish between two kinds of set-theoretic inconsistency. Some sets are inconsistent because they both have and do not have some property – that is, set-theoretically, some sets x are such that $x \in y \land x \notin y$. But we also have, as Routley labels them, d-inconsistent sets, that is, sets x such that $y \in x \land y \notin x$, i.e., something (which is itself inconsistent) both is and is not a member of them. Russell's set R of non-self-membered sets, for instance, is both inconsistent and d-inconsistent.[28]

DST is probably the paradigm case of a paraconsistent set theory capable of providing as much set-theoretic structure as the professional mathematician or the category theorist may ever need – things you could never dream about in **ZF-ZFC**. By developing category theory from a framework that admits unrestricted Abstraction, one is allowed to introduce the category of all groups or of all categories and to fully investigate their properties (for instance, since the category of all sets is a set, it will be a member of itself, etc.).

I have been talking mainly about things that can be proved in **DST**. Let us turn to things that cannot be proved. As we have already hinted at, many standard proofs of ordinary set-theoretic results break down within a relevant-paraconsistent framework because of the weakness of the underlying logics, and recapturing the involved theorems can be cumbersome. In addition, even if we accept the surrogate characterizations of ∅ and V proposed by Routley, this does not prevent sets in **DST** from having some unavoidable "intensional" flavour. As Dunn has shown, in this kind of relevant set-theoretic framework the structure of the universe of sets is not a Boolean algebra, and there appear to be more than one empty and universal sets. If we add principles that *force* the uniqueness of the universal set and of the empty set, full classical logic with its irrelevance falls out.[29] Doesn't this make sets look more like Quine's "creatures of darkness"? **DST** may turn out to be, in the end, more a theory of *properties* than a theory of sets.

---

[28] See Routley (1979b): 927.
[29] See Dunn (1988).

A more general subject concerning what cannot be proved within a paraconsistent set theory is the aforementioned issue of non-triviality, the paraconsistent counterpart of the usual requirement of consistency proofs for standard theories. Now, we have non-triviality proofs for **DST** (also a proof for an extensional variant of the theory that disposes of the relevant conditional), due to Routley and to Ross Brady, that use the Routley-Meyer semantics with the Routley Star.[30] The point with this kind of result is that the boundaries of what cannot be proved in the non-trivial theory are rather unstable. One may want to be guaranteed, for instance, that garden sets, e.g., finite sets, are not inconsistent: contradictions are not true by default, even for a thorough dialetheist, and nobody so far has produced a positive argument to the effect that finite sets have inconsistent properties. However, Priest (1987) has admitted that "just how far the contradictions in [this kind of paraconsistent set theories] spread, or even how this idea can be formulated precisely, is an open problem",[31] and things have not changed much today.[32] This points to a more general philosophical problem for paraconsistent set theory, that is, how to find the middle way between weakening our logic so much that we cannot produce all the set theory we might want, and fortifying it so much that it delivers triviality.

11.5 Brady's Paraconsistent Set Theory

Ross Brady proposed a paraconsistent set theory which also adopts (Abs) and (ER) as characterized for Routley's theory. It is not based, though, on **DKQ**, but on the logic **DJ^{d}Q**,[33] and it also avoids (LL) and the Leibnitian treatment of identity because of considerations of relevance – in this case, because of Brady's meaning containment semantics I have introduced in § 9.10.2.

An interesting aspect of the approach is that the theory includes *classes*. This may sound strange: as we have seen, von Neumann introduced the set/class distinction specifically as a stratagem to avoid the set-theoretic paradoxes by denying Objectivity, instead of Existence. Brady's goal, though, is to reformulate the distinction in order to both (a) clearly

---

[30] See Brady (1989); Brady and Routley (1989).
[31] Priest (1987): 142.
[32] See Mortensen (1995): 141-6; Woods (2003): 170.
[33] See Brady (2000): 129ff.

discriminate between consistent and inconsistent aggregates, and (b) take the latter more seriously than in the consistent von Neumann-Bernays approaches. (Abs) holds unrestrictedly for classes, and their underlying logic is the weak and relevant **DJ^{d}Q**. Abstraction is limited for sets, whose underlying logic is fully classical: "on the sets we thus have a completely standard logic and ontology".[34]

The innovative part of the theory, therefore, concerns classes. Brady proves its non-triviality, but he also introduces some logical restrictions which may seem to weaken the paraconsistent spirit of the enterprise. In the case of the Russellian class, the biconditional $R \in R \leftrightarrow R \notin R$ can be proved. An explicit contradiction $R \in R \wedge R \notin R$, though, does not follow,[35] because of the weakness of **DJ^{d}Q**: as we know, neither *reductio* nor the (LEM) hold in it. This may be taken as an *inconveniens*: you may recall the attempts to avoid some set-theoretic paradoxes by dropping the (LEM) or the Principle of Bivalence, and the reasons why they do not appear to be satisfying. If we are supposed to adopt a paraconsistent set theory, why should we avoid some celebrated explicit contradictions by weakening the underlying logic beforehand? Analogous considerations may be made on the fact that the validity of Cantor's Theorem is restricted to sets, so that Cantor's paradox is circumvented.[36] Brady's broadly logical framework, on the other hand, is meant to be a universal one, so it has to maintain a certain degree of neutrality on such issues as dialetheism and the existence of true explicit contradictions. In particular, neither the truth-functionality of negation, nor the holding of the (LEM) for it (or, equivalently, negation-completeness), should be taken for granted in the applications: "whether $A \vee \sim A$ holds or not depends on the particular sentence that A represents".[37]

---

[34] Bremer (2005): 144.
[35] See Brady (2000): 131.
[36] See *Ibid*: 131-2.
[37] Brady (1996): 174. See also 175: "Consider, e.g., Russell's Paradox, which leads to the form $R \in R \leftrightarrow \sim R \in R$, where R is Russell's class. There is nothing in the definition of R which supports $R \in R$ being true or false. Any recourse to logical support here, such as $A \vee \sim A$ being the case, would be circular".

11.6 The Inclosure Schema

An interesting philosophical aspect of the paraconsistent approach to set theory, investigated at length in Priest (1995), is the possibility of providing within a paraconsistent set theory (in this case, having **LPQ** as the underlying logic) the common structure of a whole family of logical paradoxes.

As we have seen, according to Priest the paradoxes constituting the fundamental case on behalf of dialetheism are produced by situations "at the limits of thought": situations in which we are dealing with "a totality (of all things expressible, describable, etc.) and an appropriate operation that generates an object that is both within and without the totality".[38] A part of the relevant structure was grasped by Russell,[39] who ascribed the paradoxes to the consideration of a collection, such that taking it as a totality or as *one thing* produces an object, which both is and is not a member of the totality. Russell followed Kant in taking the paradoxes of the limit as a "natural and inevitable illusion" facing the reason that, as it were, misunderstands the cognitive role of the Unconditioned. According to Priest, on the contrary, Hegel got it right: what we are facing at the limit are true contradictions and inconsistent objects. Nevertheless, the business of finding a general structure for the logical paradoxes is an extremely interesting one, also independently of Priest's dialetheic solution.

The proposed representation, called the *Inclosure Schema*, generalizes the Russellian construction as follows. We need two properties, ϕ and ψ, and a function, δ, such that:

(1) $W = \{y \mid \phi[y]\}$ exists, and $\psi[x/W]$
(2) $x \subseteq W$ and $\psi[x] \Rightarrow$
   (2a) $\delta(x) \notin x$
   (2b) $\delta(x) \in W$

Condition (1) includes what I have called Existence in the first part of the book (and it is typically denied by the theory of types and by axiomatic set theories of the Zermelo-Fraenkel family): the set W, which is one of the "illegitimate totalities", exists. Conditions (2a) and (2b) are dubbed by Priest, respectively, Transcendence and Closure. We obtain a

---

[38] Priest (1995): 3-4.
[39] E.g., in Russell (1903): §§ 346-9.

contradiction by considering a subset x of W, such that there is a function δ which is, as Priest calls it, a *diagonaliser* with respect to property ϕ. It is an operation (of which Cantor's diagonalisation is a typical case) "defined systematically to ensure that the result of applying it to any set cannot be identical with any member of that set".[40] Now, according to (2a) the "diagonal" of x is not in x (it "transcends" x), but according to (2b) it has to be in W, which is an omni-inclusive ("closed") totality. By applying (2a) and (2b) to W itself, we have that $\delta(W) \in W \wedge \delta(W) \notin W$: the diagonalising function, taking W itself as an argument, donates us an inconsistent object that both is and is not in W.

Property $\psi[x]$ is involved in Priest's generalization with respect to the Russellian pattern. In order to handle the semantic paradoxes, one has to assume that W has the property at issue. The diagonaliser works only for subsets of W that satisfy $\psi[x]$; this is unimportant for set-theoretic paradoxes, in which $\psi[x]$ is just self-identity. Many paradoxes can now fit the Inclosure Schema. For instance:[41]

	W	$\phi[y]$	$\psi[x]$	$\delta(x)$	Contradiction
Cantor-Russell	The universal set V	y is a set	$x=x$	$\rho_x = \{y \in x | y \notin y\}$	$\rho_V \in V \wedge \rho_V \notin V$
Burali-Forti	The set of all ordinals Ω	y is an ordinal	$x=x$	$log(x)$[42]	$\Omega \in \Omega \wedge \Omega \notin \Omega$
Mirmanoff	The cumulative hierarchy K	y is well-founded	$x=x$	$U\{P(y) | y \in x\}$	$K \in K \wedge K \notin K$
König[43]	The set of definable ordinals D	y is a definable ordinal	x is definable	$\mu_\phi y \notin x$[44]	$(\mu_\phi y \notin D) \in D \wedge (\mu_\phi y \notin D) \notin D$

Priest also proposes his Principle of Uniform Solution, which sounds more or less as: "Same sort of paradox, same sort of solution". Since all

---

[40] Priest (1995): 130.
[41] See *Ibid*: 131, 134, 146; see also Bremer (2005): 146.
[42] $log(x)$ is *the least ordinal greater than all the elements of x*.
[43] König's paradox is a semantic paradox of definability. Something is *definable* if there is a (non-indexical) noun phrase referring to it. Let D be the set of all definable ordinals. English noun-phrases are at most denumerable, but the set of all ordinals Ω is not denumerable for sure. Therefore, some ordinals are not definable. Ω is well-ordered, so there is the least non-definable ordinal, that is, the least ordinal such that it is not in D. But I have just defined it.
[44] μ is a least number descriptor, so μy is just *the least ordinal y, such that....*

the paradoxes of the limit have the same form, "any solution that can handle only some members of the family is bound to appear not to have got to grips with the fundamental issue".[45] Nevertheless, the Schema is neutral with respect to the issue of true contradictions: it is not itself an argument for dialetheism. Some friends of consistency (including J.L. Mackie (1973), and Alonzo Church (1976), who explicitly retrieved Russell's diagnosis of the paradoxes) have advocated the opportunity of providing a uniform solution. By including also the semantic paradoxes, Priest's Schema makes Ramsey's distinction and divide-and-conquer policy look even less appealing, independently of the issue of which is the right strategy.[46] Of course, a consistent solution has to deny some condition in the Schema. Being a dialetheist, Priest leaves matters there, and this *is* a uniform (dis)solution for sure:

> The only satisfactory uniform approach to all these paradoxes is the dialetheic one, which takes the paradoxical contradictions to be exactly what they appear to be. The limits of thought which are the inclosures are truly contradictory objects.[47]

The case for such a conclusion can be made by producing a model for the whole Inclosure Schema. Now, in the technical appendix of the third part of *Beyond the Limits of Thought* Priest provides a model for a paraconsistent set theory based on **LPQ** which contains the Inclosure Schema. More than the theory itself, the interesting part is Priest's exploitation of a "Collapsing Lemma" and of a model-theoretic strategy, originally due to Meyer and Mortensen, whose ontological function is to generate inconsistent models from consistent ones. I shall cope with it in the next Chapter: the strategy is at the core of some inconsistent arithmetics – that is, of some of the strangest, and most interesting, theories in the land of inconsistency.

---

[45] Priest (1995): 166.
[46] It is fair to say that some critics, e.g., Grattan-Guinness (1998), Tennant (1998), and Weir (1999), have claimed that some logical paradoxes either escape the Schema, or have to be tortured considerably in order to fit it.
[47] Priest (1995): 169.

12 Impossible Numbers

> 'You are a slow learner, Winston,' said O'Brien gently.
> 'How can I help it?' he blubbered. 'How can I help seeing what is in front of my eyes? Two and two are four.'
> 'Sometimes, Winston. Sometimes they are five. Sometimes they are three. Sometimes they are all of them at once. You must try harder. It is not easy to become sane.'
>
> George Orwell, *Nineteen Eighty-Four*

12.1 Meyer's and Routley's Relevant Arithmetics, R# and DKA

The first systematically developed paraconsistent arithmetic has been Robert Meyer's relevant arithmetic **R#**, obtained by simply adding to the classical relevant logic **R** Peano axioms corresponding to the (PA1)-(PA7) we met in § 4.1 for (the standard first-order version of) **PA**. The main difference in the interpretation is that the conditional in them is a relevant one.[1] As Meyer has shown, **R#** is sufficiently powerful, that is, it can represent all the recursive functions. More entertainingly, he has also produced a quite short non-triviality proof for it: **R#** is non-trivial in the sense that $0 = 1$ cannot be proved in it; and the proof is obtained via finitary methods that can be formalized within **R#**. Therefore, Gödel's Second Theorem does not apply, in this sense, to the theory. Actually, **R#** is not itself inconsistent, but some of its models also satisfy inconsistent theories (I will come to the model-theoretic issue quite rapidly).

Nevertheless, Routley noticed that **R#** has some weird features: there are discrepancies between the principles regulating addition and those regulating multiplication, and the plain inclusion of the Peano axioms produces some irrelevant entailments, contrary to the spirit of the underlying relevant logic (e.g., any numeric equation entails any theorem).[2] To improve things, Routley proposed a dialectical arithmetic, **DKA**, obtained

---

[1] See Meyer (1976); see also Bremer (2005): 151; Dunn and Restall (2002): 41ff.
[2] See Routley (1979b): 901 and 929.

by adding to the dialectical logic **DKQ** the following "relevant" arithmetic axioms (with $\tau =_{df} 1 = 1$):[3]

(A1) $\forall xy(x = y \wedge \tau \to Succ(x) = Succ(y))$
(A2) $\forall xy(Succ(x) = Succ(y) \wedge \tau \to x = y)$
(A3) $\forall xyz(x = y \wedge y = z \to x = z)$
(A4) $\forall xy(x = y \to y = x)$
(A5) $\forall x(Succ(x) \neq 0)$
(A6) $\forall x(x + 0 = x)$
(A7) $\forall xy(x + Succ(y) = Succ(x + y))$
(A8) $\forall x(x \times 0 = 0)$
(A9) $\forall xy(x \times Succ(y) = (x \times y) + x)$.

The Induction Principle is taken as a rule:

$$\frac{\alpha[x/0], \ \alpha[x] \to \alpha[x/Succ(x)]}{\forall x \alpha[x]} \quad (IMR)$$

DKA is non-trivial, too, not proving such things as $0 = 1$ or $0 \neq 0$ and, also in this case, the non-triviality proof provided by Routley is finitary and formalizable within **DKA**, which therefore circumvents Gödel's Second Theorem.[4]

12.2 n = n + 1

Philosophically, the most interesting aspects of paraconsistent arithmetics do not emerge, in my opinion, via a proof-theoretic approach engaged in finding the "right" axiomatization; but, once again, via a discussion of the underlying semantics and ontology. Therefore, we had better switch into a model-theoretic perspective, and look at the models of inconsistent arithmetics, and their properties. The fundamental issue within philosophy of mathematics probably lies behind the simple question: what are numbers? When paraconsistency extends from the underlying logic of or-

---

[3] It is common within relevant theories to define a sort of "enthymematic" conditional, e.g., $\alpha > \beta =_{df} \alpha \wedge \tau \to \beta$, where τ is logical constant representing a logical or arithmetical truth, or a conjunction of such truths.

[4] See Routley (1979b): 930-4.

dinary mathematics to the set-theoretic framework providing its foundations, mathematics itself becomes paraconsistent in a significant sense: inconsistent *objects*, such as the sets involved in the set-theoretic paradoxes, are admitted. We may expect inconsistent *numbers* to be around, too. This is indeed what happened with paraconsistent arithmetics. Let us have a look.

12.2.1 Natural, Supernatural, and Unnatural Numbers

With overwhelming originality, this story, too, begins with Gödel. You may recall that, in my presentation of Gödel's First Theorem in § 4.2, the unprovability and irrefutability (that is, the unprovability of the negation) of the Gödel sentence γ within **PA** were, so to speak, asymmetric: the unprovability of γ required the consistency of **PA**, whereas the unrefutability required its ω-consistency. Now ω-consistency is stronger than ordinary consistency: any ω-consistent system is consistent, whereas the converse does not hold.[5] Now, one of the most surprising by-products of Gödel's First Theorem, depending upon the asymmetry, is the non-categoricity of **PA**, that is, the existence of the so-called *non-standard* models of arithmetic.

I claimed that the standard model I referred to in § 4.1 as the "intended" interpretation of **PA**, $\mathbb{N}$, includes natural numbers and the operations on them we have learned at the elementary school. It is with respect to this model that it is claimed that γ is *true*, thereby taking the truth *simpliciter* of γ as truth-in-the-intended-interpretation. Since γ is left undecided in **PA**, though, not only the theory **PA** + γ, but also **PA** + $\neg\gamma$, are consistent extensions of **PA**. In particular, **PA** + $\neg\gamma$ is ω-inconsistent, but consistent: it includes as theorems false sentences (namely: $\neg\gamma$) – false, though, only with respect to the standard model $\mathbb{N}$. γ was an unprovable but arithmetically valid sentence; and now we find out that a system including the negation of a valid formula is consistent. One may expect that **PA** + $\neg\gamma$, being consistent, has a model – call it $\mathbb{K}$. Since **PA** + $\neg\gamma$ extends **PA**, we know that $\mathbb{K}$ satisfies the first-order Peano axioms. Since it also includes $\neg\gamma$, which is not satisfied (it is false) in the standard model,

---

[5] See Gödel (1931): 608-10. ω-consistency is a purely arithmetic kind of consistency. Barkley Rosser re-demonstrated the First Theorem without assuming ω-consistency, by exploiting the well-ordering of natural numbers (see Rosser (1936)). We need not enter into the details, though.

though, $\mathbb{K} \neq \mathbb{N}$ – and structurally different. $\mathbb{K}$ is a non-standard model of Peano arithmetic. This is an outcome of Henkin's studies:[6] in the 50s, Henkin provided evidence for the existence of non-standard models falsifying the Gödel sentence.[7]

I think that the philosophical discrepancy revealed by this situation may be summarized as follows. When one establishes the formal language and axiomatic arrangement of **PA**, one has, so to speak, a certain antecedent intuition on what numbers should be, that is, an intuitive model, which one tries to capture by means of the numerals, the syntactic characterizations, etc. However, Gödel's First Theorem entails the existence of non-standard models with strange non-standard numbers, and we cannot force the universal quantifier in the theory to range only on ordinary natural numbers. To be sure, these numbers inhabiting the unexpected models of **PA** behave consistently – it is *consistency* that provides them in the first place: they belong to models of a ω-inconsistent, but consistent theory. In *Gödel, Escher, Bach*, Hofstadter has poetically proposed to call the numbers announced by ¬γ, *supernatural* numbers.[8] The best way to picture them is as infinitely large numbers. Now, the models for paraconsistent arithmetics involve, as it were, a "symmetric" deviation with respect to the story I just told: they include, in a sense, *less* numbers than those admitted in the standard models and, in particular, they include inconsistent numbers, which we may therefore call *subnatural* or, better, *unnatural* numbers.

The perspective of inconsistent arithmetics is therefore involved in a sort of finitism. The underlying intuition would be that there is a finite (albeit hardly imaginable and unknown to us) number of things in the world. Although we cannot specify the number, we know that it must be "a number larger than the number of combinations of fundamental particles in the cosmos, larger than any number that could be sensibly specified in a lifetime"[9] (which should explain why our intuitions on it are rather unconfident). We may notice that such a strict finitism is not unavoidably tied to inconsistency, nonetheless: van Bendegem (1994),

---

[6] See Henkin (1947), (1950).

[7] Actually, non-standard models of arithmetic were built by Skolem since the Thirties: on this, see Kleene (1976).

[8] See Hofstadter (1979): Ch. XIV.

[9] Priest (1994): 338.

(1999) has exploited the properties of paraconsistent models to argue for a greatest number, which is not an inconsistent one.

12.2.2 The Collapsing Lemma

Let us begin with the simplest inconsistent case. Suppose that n is our unnatural-inconsistent number. Let **N** be the *theory of* $\mathbb{N}$, that is, the set of arithmetic sentences true in the standard model $\mathbb{N}$; and let $\mathbf{M}_n$ be the set of sentences true in the paraconsistent model with the inconsistent number n. $\mathbf{M}_n$ can be closed under **LPQ**-consequence[10] (but picking **FDE** as the underlying logic would not make a big difference). $\mathbf{M}_n$ has, as Priest claims, the following enjoyable properties: it is, of course, inconsistent (including, among other things, both its own Gödel sentence and its negation), but provably non-trivial – and its non-triviality proof can be formalized within it. It fully contains **N**, that is, it includes all the sentences true in the standard model. Finally, $\mathbf{M}_n$ includes its own truth predicate. So the inconsistent arithmetic avoids Gödel's First Incompleteness Theorem; it also avoids the Second Theorem, in the sense that its non-triviality can be established within the theory, and Tarski's Theorem, because including its own predicate is not a problem for an inconsistent theory.[11]

This is more than enough to get interested in the inconsistent model of $\mathbf{M}_n$. The model can be obtained by means of an appropriate filter on $\mathbb{N}$, which reduces its cardinality. Before being exploited by Priest, the technique was developed by Meyer and Mortensen. Some of their main results are summarized in a paper appeared in the "Journal of Symbolic logic", in which different finite models are considered.[12] The filter works more or less as follows: let D be the domain of a model $\mathbb{M}$, and $\approx$ an equivalence relation defined on D, which is also a congruence with respect to the denotations of the function symbols of the language. Given the objects $o_1, \ldots, o_n$ belonging to D, $|o_1|, \ldots, |o_n|$ are the corresponding equivalence classes under $\approx$. Now, let $\mathbb{M}^\approx$ be the new model, called the *collapsed model*, whose domain is $D^\approx = \{|o| \mid o \in D\}$. The role of $\mathbb{M}^\approx$ is to provide substitutes for the initial objects, and particularly to identify the members of D in each equivalence class, thereby producing a composite object that "inherits" the properties of its components: the predicates that

---

[10] *Ibid*: 337; see also Priest (1987): 234, (1997): 223-4.
[11] See Priest (1994): 337-8, and Priest (1987): 234-7.
[12] See Meyer and Mortensen (1984).

were true of the initial objects now apply to the substitute. It is here that the Collapsing Lemma, which I mentioned in § 11.6, comes into play:

(CL) Given any formula α which has the truth value v in $\mathbb{M}$, α has the truth value v also in $\mathbb{M}^\approx$.

Therefore, if the original model satisfied some set of formulas, the collapsed model also satisfies it. The proof of (CL) is by induction over the complexity of formulas.[13] The main idea is that, when the initial model $\mathbb{M}$ is collapsed into $\mathbb{M}^\approx$, no sentence loses a truth value – it can only gain them. Of course, when we begin with the model of a standard theory, the only values around are true and false. But in the collapsed model it may be the case that a formula which was initially true only, or false only, becomes both true and false, that is, paradoxical (recall that the set of values in **LP(Q)** is $\{\{1\}, \{0\}, \{1, 0\}\}$). This happens precisely when the collapsing filter produces an inconsistent object: for instance, it may identify in an equivalence class two initial objects, one of which had, whereas the other did not have, the very same property. The procedure works even if among the relevant sentences we have formulas that seem to put constraints on cardinality, such as $\exists xy(x \neq y)$, precisely because *they* can become paradoxical.

In the case of $\mathbf{M}_n$, the trick consists in choosing for $\mathbb{N}$ a filter that (a) given a number $x < n$, puts x and nothing else in the corresponding equivalence class, so that $|x|$ inherits all and only the properties of x; and (b) puts every number $y \geq n$ in a single equivalence class. Consequently, all the true/false equations involving any number smaller than n in the standard model are now true only/false only of the substitute. Because of this, the initial segment in the succession (which is sometimes called the *tail*) behaves as usual: roughly, up to n things work like in ordinary arithmetic. Nevertheless, anything that could be truly/falsely claimed of anything bigger than n is now true/false of the unnatural-inconsistent number. Many things concerning it are therefore paradoxical now (both true and false), and "of course, n is [now] an inconsistent object [...]. In particular, in the model $\mathbf{n} = \mathbf{n} + \mathbf{1}$ is true even though it is also false",[14] so n is the successor of itself. We may observe that the description of the model of $\mathbf{M}_n$ completely piggy-backs on a previous specification of the

---

[13] See Priest (1991); (1994): 346-7; the result was anticipated in Dunn (1979).
[14] Priest (1994): 338. Priest takes **n** as the numeral for *n*.

standard model. Nevertheless, Priest claims that "there are independent (though pedagogically more complex) specifications".[15]

Priest has declared that (CL) is "the ultimate downwards Loewenheim-Skolem Theorem",[16] which is easy to understand. The downward half of the Löwenheim-Skolem Theorem claims that any first-order theory, which has a model with an infinite domain, has a model with a denumerably infinite domain, too.[17] The filter and the Collapsing Lemma allow us to "shrink" even more, since one can reduce a model with a denumerably infinite domain into one of any smaller size. We can have a collapsed model, $\mathbb{M}^{\approx}$, whose domain, $D^{\approx}$, has cardinality k (smaller than that of the initial model), by choosing an appropriate equivalence relation that produces precisely k equivalence classes. Bremer has therefore suggested the following Paraconsistent Löwenheim-Skolem Theorem: "Any mathematical theory presented in first order logic has a *finite* paraconsistent model".[18]

By employing the Meyer-Mortensen technique, various inconsistent models can be produced. For instance, one may want to have a consistent tail up to n, and an inconsistent "cycle" of period p. Beginning again with $\mathbb{N}$, we can collapse it by characterizing the equivalence relation $\approx$ in such a way that $x \approx y$ iff (a) both $x < n$ and $y < n$ and $x = y$, or (b) both $x \geq n$ and $y \geq n$ and $x = y$ mod p. Now the succession behaves thus:

One may also begin with a *non-standard* (consistent) model of arithmetic, $\mathbb{K}$, and collapse it by stipulating that $x \approx y$ iff (a) both x and y are standard numbers and $x = y$, or (b) both x and y are non-standard num-

---

[15] *Ibid*: 338fn.

[16] *Ibid*: 339.

[17] One of the consequences of the downward Theorem is the so-called Skolem paradox. Since it can be expressed in a first-order language, set theory has a model whose domain has the cardinality of the set of natural numbers. However, within set theory we can prove the existence of sets whose cardinality is more than denumerable.

[18] Bremer (2005): 155.

bers. The collapsed model $\mathbb{K}^\approx$ now includes the standard numbers, and also an inconsistent "point at infinity", ∞. The succession behaves thus:[19]

12.3 Turing vs. Wittgenstein

The philosophical question is, why should we like an inconsistent arithmetical theory, such as $\mathbf{M}_n$, better than $\mathbf{N}$? Are inconsistent models *intrinsically* pathological?[20] Paraconsistent authors can certainly vindicate their metamathematical advantages: as we have seen, most of the limitative theorems afflicting the consistent cousin(s) are avoided by an inconsistent theory like $\mathbf{M}_n$. In particular, an inconsistent arithmetic with a finite model is decidable, so if this is the real arithmetic we have an algorithm for arithmetical truths. And given that mathematics is humanly learnable, having a complete and decidable axiomatization of it is good news for sure. I have addressed these subjects in § 4.

How about the concrete *application* of the inconsistent calculus, though? Alan Turing once objected to Wittgenstein that, if arithmetic were inconsistent, then if engineers used it to build bridges, they would fall down.[21] Now, it is true that, as an extreme case of collapsing, we can have a trivial model: just pick an empty tail and period 1. This has that 0 = 1, 1 = 2, and so on (and Russell's proof that he is the Pope may succeed). Nevertheless, the point is that the fork at which paraconsistent arithmetics part from ordinary arithmetic, and impossible numbers begin to play, can be fairly placed, as it were, far enough to avoid the troubles that Turing feared. Standard and paraconsistent mathematicians would behave the same way under ordinary circumstances. They would all count

---

[19] See Priest (1987): 233-4 and, for more detailed explanations, Priest (2002c): 368ff. It has been recently suggested that not *all* these collapsing procedures may work properly: see Paris and Pathmanathan (2006). To be sure, the technique is a novel one, and positive and negative results are rapidly forthcoming.

[20] As explicitly suggested, e.g., by Denyer (1995).

[21] See Wittgenstein (1976).

in the same manner: 0, 1, ... $n, n + 1$, ...; and they would add and multiply following the same rules. To be sure, from n on a paraconsistent mathematician would upset his consistent colleagues by claiming that, yes, he agrees with them that, for any $x > n$, x is different from n: no number, after all, is identical to its successor; but it is also the case that x is the same as n: some number is identical to its successor.[22] This is unimportant when we come to standard mathematical practice, though: n can be so large that it has no physical significance, and no import at all for engineers.

In a sense, not all of this may be unbelievable news. Surprising and (in a broad sense) paradoxical innovations in the history of mathematics led to the detection of new kinds of numbers: from Hyppasus' irrational numbers refuting Pythagorism, to infinitesimals, Cantor's transfinite numbers, and so on. The early reception of such new entities among mathematicians has always been controversial, from the Pythagoreans condemning and expelling Hyppasus, to Kronecker making Cantor's life impossible. A process of rethinking mathematics in order to come to grips with the newly discovered realm has usually followed. This time, though, the discovery would be one of inconsistent objects, so the rethinking is a deep one. Just as with impossible worlds and inconsistent sets, paraconsistentists are asking us to be unprejudiced, broaden our horizons, and explore the new territory.

---

[22] In general, a supporter of inconsistency does not express disagreement with a friend of the (LNC) by *rejecting* something his consistent peer believes. The former can accept all the things the latter accepts, and express his belief in true contradictions by also accepting things inconsistent with them. I will come back to this conversational phenomenon, in its broader form, in § 14.

Part IV.
The Problems

13 Hypercontradictions

> [The liar] both lies and does not lie [...]. For we have a union of opposites, lying and truth, and their immediate contradiction...
>
> G.W.F. Hegel, *Lectures on the History of Philosophy*

13.1 "This sentence is true and false, false only, true only, neither true nor false..."

In § 2.5, we conjectured that any consistent solution of the liar paradox is doomed to face its own revenge liar. If we reject bivalence, we have such things as "This sentence is false or neither true nor false"; If we propose some sort of hierarchy, we get "This sentence is false at its level"; as old paradoxes become manageable, new ones replace them. Therefore, as the dialetheist's story goes, we had better accept the paradoxical arguments as sound, and modify our logic so that their inconsistent conclusions do not detonate. All the same, within the most recent literature the idea that paraconsistency, too, may face its own revenge liar, has come into sight. The strategy is to build an unmanageable strengthened liar (unmanageable, e.g., because it produces the equivalence $1 = 0$ between the truth values, thereby making the distinction between truth and falsity fade) by exploiting some basic paraconsistent semantic and/or set-theoretic notions. Such a paradox would be or originate a *hypercontradiction*.

The first author to talk of "hypercontradictions" has been Priest himself. Let us begin our story with the harmless set of classical truth values:

$V_0 = \{1, 0\}$.

As we saw in § 8.2.1, the semantics of **LP** is obtained by taking the power set of the set of classical values, and subtracting the empty set (no gaps around). Therefore, we have:

$V_1 = P(V_0) - \emptyset = \{\{1\}, \{0\}, \{1, 0\}\}$.

We know that the intuitive reading of the values is: true (only), false (only), both true and false (or paradoxical). Moreover, the initial intuition would be that they are both exhaustive and exclusive: a sentence should get one and *only one* of the three. The designated values are {1} and {1, 0}, and the connectives are taken as operations on V_1.

In a 1984 essay called precisely *Hyper-contradictions*, Priest wondered whether, once V_1 was posited, one could consider sentences taking such impossible values as *both true and false* ({1, 0}) *and true only* ({1}). The answer was in the positive. Furthermore, "if we allow a sentence to take two values which are mutually exclusive, there seems to be no reason why we should not allow them to take an arbitrary number".[1] We may just reiterate the power set operation that made us pass from the standard values to the **LP** values. Therefore, we may have:

$V_2 = P(V_1) - \varnothing$.

And now we are off and running. We can have a whole hierarchy, recursively characterized:

$V_0 = \{1, 0\}$
$V_{n+1} = P(V_n) - \varnothing$.

Which values would be the designated ones? In line with the magnanimous spirit of **LP**, they should be the values in which there is some truth, that is to say, in which 1 appears at some depth in the sets of values.

Once we advance from V_0 to V_1, and we admit paradoxical (both true and false) sentences, the consequence relation changes (recall that **LP**-consequence is decidedly non-classical, whereas all classical tautologies are **LP**-tautologies). All the additional levels in the hierarchy produce no further changes in the consequence relation as defined for **LP** via V_1. Priest has therefore claimed that "hyper-contradictions make no difference",[2] since they do not affect the notion of validity. It is easy to understand the underlying philosophical motivation: once we have admitted (C2)-contradictions, that is, sentences that are both true and false, sentences taking an indefinite number of incompatible truth values look, at least *prima facie*, just like additional contradictions to be admitted on

---

[1] Priest (1984b): 239fn.
[2] *Ibid*: 241.

board in the dialetheic ship. In subsequent writings, Priest reverted to the three values of the original **LP**.

13.2 The Super-Liar

Anthony Everett and Timothy Smiley showed that things are not so straightforward. We know that the typical strengthened liar goes as follows:

(1) (1) is not true.

On the basis of the pristine **LP**-semantics, the evaluation function v assigns to (1) the value $\{1, 0\}$, that is, (1) is both true and false. However, consider the following strengthened liar:

(2) (2) is false only.

This one ascribes to itself precisely only the **LP** non-designated truth value $\{0\}$, that is, simple falsity. Formally, (2) is a sentence λ_s such that (in the semantic notation introduced in § 8.2.1):

(FP$_{\lambda s}$) $\lambda_s \leftrightarrow v(\lambda_s) = \{0\}$.

λ_s turns out to be a troublemaker if the semantics of **LP** is based upon the evaluation *function* v. Let us begin with V_1, that is, with the three original values of **LP**, and reason by cases. Suppose λ_s is *true only*, or *true and false*: $v(\lambda_s) = \{1\}$, or $v(\lambda_s) = \{1, 0\}$. In both cases:

(3) $1 \in v(\lambda_s)$.

The T-schema in the formal notation goes like this (recall that "$1 \in v(\alpha)$" says that α is true):

(T) $1 \in v(\alpha) \leftrightarrow \alpha$.

Now we can apply *modus ponens* to (3) and to the relevant instance of the T-schema, and obtain λ_s, that is, given (FP$_{\lambda s}$):

(4) $v(\lambda_s) = \{0\}$.

Taken together, (3) and (4) tell us that $1 \in \{0\}$, therefore $\{1\} = \{0\}$ and $1 = 0$. Now suppose λ_s takes the last value available: λ_s is *false only*, that is, (4) holds. From (FP$_{\lambda s}$) and (4) we obtain λ_s again. From λ_s and the relevant instance of the T-schema, via *modus ponens*, (3) follows. We have both (3) and (4) again, so $1 \in \{0\}$, $\{1\} = \{0\}$ and $1 = 0$. In all cases, triviality ensues: all sentences are true only, including those that are false only. The Super-Liar argument uses only basic set-theoretic notions, *modus ponens*, replacement of equivalents, and the T-schema. Smiley took the result as showing that "a dialetheist must stop talking about 'the truth-value of A', and stop treating evaluations as functions with truth-values as objects": this because "the dialetheist's problem is that he can never be sure that [his evaluations] are genuine single-valued functions".[3]

Everett (1993) argued that the problem resurfaces even if we admit a transfinite hierarchy of values, such as the one described in the previous §. The extended Priestian apparatus aims at distinguishing a designated value (that is, a value including a 1 at some depth of membership) from a non-designated one. Therefore, it still provides the resources to claim that a sentence is false only, as in (2) or λ_s: so we can produce a liar which is "genuinely monovalent, [but] which cannot be evaluated by Priest's [1984b] semantics as being both true and false unless we identify those two values". This threatened to "destroy the apparatus he requires for distinguishing designated evaluations from non-designated ones".[4]

13.3 Relational Semantics

The debate has made clear that handling semantic functions in a thoroughly inconsistent framework is a sensitive business: one cannot simply *stipulate*, as it were, that a unique value be attached to sentences. In the writings posterior to 1993, Priest and the other paraconsistent authors working on and with **LP** (such as Manuel Bremer, J.C. Beall) bypassed the problem by substituting the evaluation function with an evaluation *relation*, and showed that things get better if we express truth conditions (or truth-in-an-interpretation conditions) in relational terms. For instance, in Priest (1998) we have a semantic relation (be it ρ) between sentences and

---

[3] Smiley (1993): 31-2.
[4] Everett (1993): 40-1.

truth values, and the semantics for extensional connectives goes as follows:

(S¬1) $\neg\alpha \rho 1 \Leftrightarrow \alpha \rho 0$

(S¬2) $\neg\alpha \rho 0 \Leftrightarrow \alpha \rho 1$

(S∧1) $\alpha \wedge \beta \rho 1 \Leftrightarrow \alpha \rho 1$ and $\beta \rho 1$

(S∧2) $\alpha \wedge \beta \rho 0 \Leftrightarrow \alpha \rho 0$ or $\beta \rho 0$

(S∨1) $\alpha \vee \beta \rho 1 \Leftrightarrow \alpha \rho 1$ or $\beta \rho 1$

(S∨2) $\alpha \vee \beta \rho 0 \Leftrightarrow \alpha \rho 1$ and $\beta \rho 1$.[5]

With no special condition on ρ, this provides the semantics of **FDE**. For **LP**, we add the restriction that each formula relates to at least one truth value, and for classical semantics we also require that each formula relates to at most one truth value.

Even if this overcomes the troubles pointed out by Smiley and Everett, it is not granted that all problems are solved. Joachim Bromand has recently developed a couple of hypercontradictions that seem to affect also the relational semantics for **LP**, by making it incapable, once more, of distinguishing the notions *true only* and *false only*. One of the new hypercontradictions exploits the set-theoretic notions underlying the dialetheic semantics. Of course, the Abstraction Principle should hold unrestrictedly in a strongly paraconsistent framework. Therefore, we can obtain a function once again by using set abstracts: for every sentence α, we can abstract the set of its truth values; then we define $v(\alpha) =_{df} \{x \mid \alpha \rho x\}$. Existence and uniqueness are guaranteed by (unrestricted) Abstraction and Extensionality. Now the super-liar just ascribes to itself the singleton $\{0\}$:

(FP$_{\lambda_{sl}}$) $\lambda_{sl} \leftrightarrow v(\lambda_{sl}) = \{x \mid \lambda_{sl} \rho x\} = \{0\}$,

---

[5] See Priest (1998): 412-3; see also Priest (2001): 140-1; (2002c): 308ff.

Therefore it is, again, a sentence claiming of itself that it is false only within the new relational framework. Reasoning by cases gives us $1 = 0$ over again.[6]

If you are beginning to smell a merely technical discussion on minor issues, you may consider the importance of the philosophical outcomes. The possibility of producing some super-liar (with the involved intractable hypercontradiction) within the dialetheic framework may cast doubts on the dialetheic (dis)solution of the semantic paradoxes. Consistent strategies introduce new semantic concepts that cannot be expressed within the language for which the solution of the paradox is proposed, on pain of new contradictions. Now, what about the alleged advantage of the dialetheic way out? A dialetheic semantic theory, such as the one we met in § 10, was claimed to achieve semantic universality: the language/metalanguage distinction was supposed to fade, no meaningful expression being banished from the language for which the semantics was given. Is Tarski's ghost still haunting us now? According to Bromand, "just as in the previous accounts, *LP* [with its relational semantics] therefore fails as a solution to semantic paradoxes because notions crucial for the specification of the theory cannot be represented adequately within it".[7]

The debate on hypercontradictions is lively and rapidly evolving. For instance, Bremer has addressed the set-theoretic variant of Bromand's hypercontradiction by resorting to the properties of paraconsistent arithmetics. Details aside, the solution might look a bit circular, since an inconsistent arithmetic is used to save a logic, such as **LP**, that should itself underlie the arithmetic. Among the logical laws we should do without, in order to make the strategy work, are Contraposition, and (LL), that is, Leibniz's Law or the Substitutivity of Identicals.[8] As for the former move, we have already examined various paraconsistent perspectives admitting a non-contraposible conditional. The dialetheist can claim that the latter move is not quite *ad hoc* either. As we have seen in the preceding part of the book, there are grounds for rejecting or restricting the Leibnitian treatment of identity within paraconsistent set theories and arithmetics. Furthermore, Bremer reminds us that there are reasons for

---

[6] See Bromand (2002): 744-5 and, for a fully formalized version of the argument, Bremer (2005): 193-4.

[7] Bromand (2002): 747.

[8] See Bremer (2005): 195-8.

discarding (LL), quite independently of the issue of paraconsistency, such as those advanced by supporters of relative identity (an idea that can be found in Locke, and whose main contemporary supporter is Peter Geach).[9] According to relative identity theorists, "$x = y$" is not a fully well-formed expression: it does not make sense to say that x is identical to y, *simpliciter*. What does make sense is to say that x and y are the *same P*, where "*P*" varies on a suitably restricted class of predicates, namely, those expressing sortal properties. The truth of such a claim, though, does not entail congruence with respect to all predicates, so x and y can be the same P without being the same Q. Therefore, "On[c]e we see that there is something wrong with substitution of identicals [(LL)] we can reject the criticism that the avoidance of Bromand's hypercontradiction is *ad hoc*".[10]

Greg Littmann and Keith Simmons (2004) proposed further revenge liars that are supposed to affect dialetheism. They insisted particularly on the fact that, whatever semantic and set-theoretic tricks the dialetheist resorts to, she cannot but have, on pain of trivialism, some idea of a reciprocal *exclusion* between some truth values. Put it otherwise, since the dialetheist is not a trivialist, she accepts a notion of *truth that rules out falsity*, and she accepts that some sentences (take "$0 = 1$", for instance) instantiate that notion. Now, call v the value the dialetheist herself assigns to such sentences. We can present the following "introspective liar":

(5) (5) is v.

If the dialetheist replies that (5) is both v and true, "we accept that the dialetheist can say these words, but we cannot see how they can mean them, if the value v really is the value of [$0 = 1$]". That is to say, "if the dialetheist ascribes truth to a sentence that is v, then it seems to us that she has failed to understand what it is for a sentence to be v. Add 'true' to 'v', and you no longer have 'v'".[11]

Littmann and Simmons developed further super-liars by using "graphic values"; that is, by graphically representing the nesting of values

---

[9] See Geach (1962), (1968).

[10] Bremer (2005): 197. An interesting combination of paraconsistency and a non-Leibnitian treatment of identity has been proposed by Tanaka (1998b), in order to explain some dynamical processes of fission and fusion originating problems variously discussed within analytic ontology.

[11] Littmann and Simmons (2004): 324.

in the various strengthened liars. Details aside, the interesting side of their approach is that within it a general problem for dialetheism begins to emerge – a problem that appears to me to be philosophically more fundamental than the one of hypercontradictions, which constitutes but a particular sub-case of it. I will label it as the *exclusion problem*. This will be the subject of the next, and final, Chapter.

14 The Exclusion Problem

> Determinateness is negation posited as affirmative and is the proposition of Spinoza: *omnis determinatio est negatio*. This proposition is infinitely important.
>
> G.W.F. Hegel, *Science of Logic*
>
> I beg to differ, on the contrary
> I agree with every word that you say.
>
> Green Day, *Walking Contradiction*

14.1 Overview

In this Chapter, I will consider the last and most general problem for (strong) paraconsistency. I have labelled it as the exclusion problem, but one may also call it the problem of the self-application of dialetheism.[1] It has been variously recognized within the literature on the issue, but Aristotle had already foreseen it.

Various authors, beginning with Łukasiewicz, have discredited the arguments in defence of the (LNC) put forth in Aristotle's *Metaphysics* as desperately confused. One of the most common criticisms points at the illicit slides between inconsistency and triviality: Aristotle aims at showing that all contradictions are false, but his considerations show, at best, that some contradictions are false, so they do not work, or beg the question, against the dialetheist.[2] Also, the fact that it is possible to find all the formulations of the kinds (LNC1)-(LNC4) in the *Metaphysics* has been taken as a sign that Aristotle did not know exactly what he aimed at defending.

However, it seems that most critics forget what, according to Aristotle, should be a necessary feature of the (LNC) whatever its exact formulation may be. The (LNC) is supposed to express the deeply metaphysical notion of *ontological incompatibility*, or *exclusion*; and, as re-

---

[1] As in Bremer (2005): 185.
[2] See e.g. Łukasiewicz (1910a): Ch. XIII; (1910b): 57.

lated to it, the idea that any sentence can convey a meaningful and informative content only insofar as it excludes something. Because of this, any speaker who aims at producing a meaningful sentence must implicitly (*in actu exercito*, although not *in actu signato*) concede that by her utterance she excludes some content. If by such a sentence she could not rule out anything, her sentence would mean nothing, and consequently it would not even be a *sentence*. In the following, I shall consider the Aristotelian argumentation whose traditional Greek name is ἔλεγχος (*élenchos*), and try to account for such a basic point. I will show that the main literature has somehow recognized an Aristotelian problem as the relevant trouble for dialetheism. Coherent dialetheists are confronted with difficulties in articulating their basic concepts; in excluding rival positions, or expressing disagreement with them; and in conveying through their theories determinate information, or non-trivial content, on the subjects they wish to discuss (e.g., the liar paradoxes). After a discussion of Priest's dialetheic treatment of the notions of *rejection* and *denial*, I will characterize a negation via the primitive and intuitive notion of *ontological incompatibility* or *content exclusion*. And, finally, I will try to provide via such a negation a formulation of the (LNC) which appears to be unquestionable also from the dialetheist's point of view, in the following sense: the dialetheist is committed to accept it without also accepting something inconsistent with *it*, on pain of trivialism – that is to say, on pain of lapsing again into the position according to which everything is the case. Such a result will not constitute a cheap victory for the friends of consistency. We may just learn that different things have been historically conflated under the label of "Law of Non-Contradiction"; that dialetheists rightly attack *some* formulations of the Law, and orthodox logicians and philosophers have been mistaken in assimilating them to the indisputable one.

14.2 Aristotle's informal remarks on ἔλεγχος

The starting point lies in our naïve notions of *petitio principii* – "you cannot buy as a premise of your proof the very sentence you aim at proving" – and *reductio ad absurdum*. *Reductio* is a minimal rule of inference, but it has been clear since the beginning that discussing with a dialetheist and arguing against her theory via *reductio* can be methodologically troublesome. In general, it is difficult to build what Locke called an *argumentum ad hominem* against dialetheists – not the bad *ad hominem*, i.e., the well-known fallacy of attacking the character of the person who holds a view, but the good one. Given a theory or set of beliefs $T = \{\phi_1, \ldots, \phi_n\}$, one can criticize T by drawing from premises the T-theorist endorses some consequence ψ:

where ψ is something the T-theorist has to reject, or a conclusion unwelcome to her. A standard value for ψ is precisely $\psi = \neg\phi_i$, $1 \leq i \leq n$. In a dialogic context, *reductio* is supposed to work as a tool of criticism, leading someone to change her own beliefs, if such beliefs entail both ϕ_i and $\neg\phi_I$ for some i. The rationale for a *reductio* is that since a contradiction cannot but be (simply) false, some part or other of a theory entailing one has to be dropped. But a dialetheist puts herself in the position of keeping both her entire theory, or set of beliefs, and the criticism showing that it entails a contradiction, since, as Priest claims, she "*may* seriously consider accepting the contradiction and [s]he *may* in the end decide to accept it".[3] Priest often reminds us that the dialetheist is not an untouchable: some values for ψ work for her, too, as we shall see (which would not be the case if one were arguing against a trivialist, that is, someone holding that everything is true). However, the fact that the standard value may not work makes discussion and criticism undoubtedly more complicated. The dialetheist may cheerfully swallow the criticism, maintain her entire theory T, including ϕ_i, and accept $\neg\phi_i$, too: she cannot be forced to give up her theory on pain of contradiction. As Timothy Smiley observed, "dialetheism is not simply a theory *about* contradictions; it requires the theorist himself to assert some"[4] (a look at how this happens will be given in the following). In such a case, as Aristotle explicitly admitted, the defender of the (LNC) "might be thought to be assuming what is at issue",[5] i.e., she would commit a *petitio*. That the dialetheist should not contradict herself, because contradictions

---

[3] Priest (1989): 614.

[4] Smiley (1993): 31.

[5] Arst. *Met.* 1006a17. Łukasiewicz (1910a), Ch. XII, (1910b): 55, argues that some of Aristotle's arguments for the (LNC) commit precisely this fallacy.

are (simply) false, is precisely what is at issue. Thus, the defender of the (LNC) cannot use it as a premise of any argument. This is not the kind of refutation that can be expected to obtain.

Nevertheless, another refutation can be obtained. To put it to work, Aristotle argues, it is necessary and sufficient that the dialetheist say *something*:

> Something which is significant both for himself and for another; for this is really necessary, if [s]he really is to say anything. [...] If any one grants this, the demonstration will be possible; for we shall already have something definite.[6]

The dialetheist can aim at saying something only if she implicitly maintains, *in actu exercito*, that her own words cannot mean or entail anything. What she thereby concedes is the determinacy of the meaning of her words. Now, Aristotle does not need words to have a completely determinate meaning, or to be univocal. He admits that vagueness, equivocation and ambiguity are pervasive phenomena of natural language – although he optimistically suggests that one may always clarify meanings by assigning exactly one single different name to each single different thing. Nevertheless, to produce the œlegcoj of the dialetheist much less is required, i.e., a *minimal* form of determinacy. It is only required that, through the sentences she utters in her tentative refutation of the (LNC), the dialetheist excludes *in actu exercito* that something is the case. Therefore, she must rule out the possibility that the words of her own tentative refutation – whatever the exact words may be – mean or entail some such thing.

By such implicit exclusion, Aristotle continues, it is the rival, not the friend of the (LNC), who somehow begs the question: she needs the very principle she wants to challenge for her words to be meaningful. By implicitly admitting that her own words have such a minimally determinate meaning – which is the condition of their meaning anything, i.e., of their being words, not mere noise – the dialetheist's position entails some version of the (LNC). Any meaning, be it a material being, a particular, a set (provided that sets are accepted in the furniture of the world), a Fregean *Sinn*, a concept, a state of affairs or a fact (provided that there are states of affairs or facts), etc., is *something*: a being that, to speak metaphori-

---

[6] 1006a21-25.

cally, refuses to be confused with any other being, or substituted with any other being as the meaning of some linguistic expression. And *that* the world is ontologically minimally determinate, besides being rooted in our deepest intuitions about the world itself, is what the relevant version of the (LNC) states. This line of argumentation has been highlighted by several commentators of Aristotle, and some of them have dubbed it a *transcendental* case for the (LNC).[7] The elenctic approach has been retrieved in various forms by the few who attempted to defend the (LNC) after Aristotle. We can find an indication of the strategy, for instance, in Bradley:

> Ultimate reality is such that it does not contradict itself; here is an absolute criterion. And it is proved absolute by the fact that, either in endeavouring to deny it, or even in attempting to doubt it, we tacitly assume its validity.[8]

It is clear that observations of this kind are quite informal. As such, they are hardly an *argument* for the (LNC). However, they can be worked out in such a way as to spell trouble for the dialetheist. To begin with, Aristotle's remarks on œlegcoj also display at least two virtues. The first one is that, owing precisely to their generality, such remarks do not need to presuppose any particular theory of meaning, or of the content of linguistic expressions. For instance, they do not presuppose the thesis that meanings are mental representations, instead of material objects, or *Sinne*, etc. *A fortiori*, such remarks do not depend on the thesis that the meaning of any particular, syntactically individuated kind of linguistic expression, such as a sentence, is – on a Fregean account – a *Sinn* (the thought it expresses) or a *Bedeutung* (the truth value it enjoys); or – on a Wittgensteinian account – the state of affairs it pictures; etc. They also do not entail the particular characterization of the content of a sentence σ as some set of possible worlds, as preferred, for instance, to the inferential account of the content of σ as a set of consequences (the set of sentences that σ entails, or the set of assertions commitment to which goes with commitment to σ, etc.).

The second virtue is that such elenctic remarks may be used to face different attacks to the (LNC). If it is admitted that the (LNC) can be expressed by different sentences, then also the opposition to it is supposed to take different shapes. Given such a proliferation, it is possible to characterize Aristotle's œlegcoj by recalling what David Wiggins said about his the-

---

[7] See e.g. Irwin (1977-8); Kirwan (1993): 204.
[8] Bradley (1893): 120.

ory of individuation, as based on sortal concepts providing a reply to the Aristotelian *what is it* question:

> An answer to the *what is it* question does both less and more than provide that which counts as evidence for or against an identity. It does less because it may not suggest any immediate test at all. It does more because it provides that which *organizes* the tests or findings.[9]

Similarly, Aristotle's elenctic argumentation does both less and more than providing an indisputable refutation of the rival of the (LNC). It does less because it may not suggest a single, immediate argument. It does more because it *organizes* the justification of various forms of the (LNC), providing a common pattern of defence against the dialetheist's attacks.

In the following §§, some evidence will be given that this is indeed the case. I will begin by pointing out that some of the most serious critiques of dialetheism produced so far (those of Tennant, Grim, Shapiro, Littmann and Simmons, Goldstein, Batens, and others) involve the charge according to which thorough dialetheism ends up as unable to express its own views in a non-trivial way, or exclude its opponents' views. This is exactly the fate of foes of the (LNC), according to Aristotle. The dialetheist is confronted by a dilemma: either to unreservedly assume[10] some version of the (LNC), for this is required by her very trying to express her position as something meaningful; or to be unable to convey informative content, thus becoming, as Aristotle says, "no better than a mere plant",[11] for "if words have no meaning reasoning with other people, and indeed with oneself, has been annihilated".[12]

14.3 Looking for the Exclusion-Expressing Device

We observed that arguing *ad hominem* against the dialetheist can be troublesome. However, the trouble has a dialetheic side, too. A much-quoted passage from Terence Parsons (1990), concerning the expression of dialetheism, provides a good starting point:

---

[9] Wiggins (2001): 60.
[10] "Assume" should be taken *cum grano salis*. In a sense, the thorough dialetheists does *assume* all the main formulations of the (LNC). More on this in the following.
[11] *Met.* 1006a15.
[12] *Met.* 1006b8-9.

Suppose that you say 'β', and Priest replies '¬β'. Under ordinary circumstances you would think he has disagreed with you. But then you remember that Priest is a dialetheist, and it occurs to you that he might very well agree with you after all – since he might think that β and ¬β are *both* true. How can he indicate that he genuinely disagrees with you? The natural choice is for him to say 'β is not true'. However, the truth of this assertion is also consistent with β's being true – for a dialetheist anyway. So [...] Priest has difficulty asserting disagreement with other[s'] view. [Fn.] Priest could indicate genuine disagreement with you if he could assert '¬β', and also say that β *is not a dialetheism*. However, the usual way to say this is to say 'β is not both true and false', i.e. '¬(β is true & β is false).' This, however, gets us nowhere, since if β is the liar sentence, it *is* a dialetheism, yet this statement about it is true. [...] Priest has no means within his symbolism of adequately expressing 'not being a dialetheism'.[13]

14.3.1 Failure at the Level of the "Object Language"

Let us consider Parsons' initial point in its most striking case: the case in which β is exactly (an instance of) the (LNC) in what has been called its syntactic version, i.e., formulation (LNC1): $\beta = \neg(\alpha \wedge \neg\alpha)$. Thus, the dialetheist's reply, "¬β", given Strong Double Negation (which, as we have seen, holds in most paraconsistent logics), amounts to saying: "$\alpha \wedge \neg\alpha$". By stating this latter, the dialetheist would be trying to convey the information that α is a dialetheia. But did she *disagree* with, or *exclude*, the assertion "$\neg(\alpha \wedge \neg\alpha)$"? At least *prima facie*, this does not appear to be the case. We know that, within **LP**, (LNC1) is a logical law; we also know that Routley and Priest have criticized the Brazilian positive-plus systems rejecting (LNC1) as missing the very point of the meaning of negation.

As a result of this, if "α is a non-dialetheia" is expressed by something like (LNC1), *every* sentence is a non-dialetheia, including all dialetheias. At this level, the dialetheist seems to be unable to express disagreement with the statement that all sentences are non-dialetheias. And this sounds like a form of the elenctic argumentation. A dialetheist may want to disagree with such a statement, since it appears to articulate the idea she aims at dismissing. She aims at doing so, but she cannot, for she also has

---

[13] Parsons (1990): 345.

dismissed, so to speak, an *exclusion-expressing device* from the object language.

Some years ago, problems of this kind produced a certain schism between inconsistency-sympathizers. Rescher and Brandom suggested that no (C2c)-contradiction, i.e., no contradiction of the form $T(\ulcorner\alpha\urcorner) \wedge \neg T(\ulcorner\alpha\urcorner)$, is tolerable and, thus, that formulations of the (LNC) of the kind of (LNC2c) should not be called into question. The underlying idea is that inconsistency should not spread into the metalanguage:

> No doubt it is important in the interests of rationality to keep our own claims about a given subject consistent. But this is no reason to insist on consistency in the context of the object of our assertions. It is a key upshot of these deliberations that an inconsistent world can be discussed in perfectly consistent terms.
> An important safety-valve in this regard is provided by distinction between (1) discourse at the level of an object language, or in the present context, ground-level descriptions of the world, and (2) discourse at the theoretical meta-level. [...] Inconsistency can be tolerated in the objects of thought and assertion, while, ultimately, discussion about them *can and should* be consistent at the meta-level of our cognitive commitments...[14]

14.3.2 Failure at the Level of the "Metalanguage"

However, in § 8.2 we have also learned that Priest rejects such a way out: "once we have given up demanding that the object theory be consistent, there is no reason to demand that the metatheory be consistent".[15] Dialetheists surely hold the T-schema, $T(\ulcorner\alpha\urcorner) \leftrightarrow \alpha$, in its unrestricted form, because it is essential for the derivation of various liar paradoxes. Now, in some versions of dialetheism, if α is a dialetheia, then so too is $T(\ulcorner\alpha\urcorner)$. The truth predicate reproduces the truth-status of the sentence it is applied to:

α	$T(\ulcorner\alpha\urcorner)$
1	1
1, 0	1, 0
0	0

---

[14] Rescher and Brandom (1980): 138 and 141.
[15] Priest (1979): 239.

This is Priest's position in *Logic of Paradox*. Actually, he has later allowed the possibility that a claim that a dialetheic sentence is true is itself simply true and not paradoxical (e.g., in *In Contradiction*): so the truth predicate may be a "*partial* consistenciser".[16] However, this does not save the "metatheory" from the possibility of inconsistent behaviour. Priest "tentatively rejects" the Exclusion Principle, i.e., the left-right direction of (Neg2), or the claim that falsity entails untruth:

(Exc) $T(\ulcorner \neg \alpha \urcorner) \to \neg T(\ulcorner \alpha \urcorner)$.

So he cautiously dismisses the idea that any internal (C2b)-contradiction, $T(\ulcorner \alpha \urcorner) \wedge T(\ulcorner \neg \alpha \urcorner)$, entails an external (C2c)-contradiction, $T(\ulcorner \alpha \urcorner) \wedge \neg T(\ulcorner \alpha \urcorner)$: a contradiction that, by the standards of Rescher and Brandom, would be directly asserted in the "metatheory". But he admits that this is exactly what happens if α is a strengthened liar, i.e., a sentence λ_1 such that:

(FP$_{\lambda_1}$) $\lambda_1 \leftrightarrow \neg T(\ulcorner \lambda_1 \urcorner)$.

This is easily understood. Given (FP$_{\lambda_1}$), and the relevant instance of the T-schema, we obtain via replacement:

(1) $T(\ulcorner \lambda_1 \urcorner) \leftrightarrow \neg T(\ulcorner \lambda_1 \urcorner)$,

and from (1), by means of the (LEM) (which, as we know, holds in **LP** and in many other paraconsistent logics) we obtain an explicit (C2c)-contradiction:

(2) $T(\ulcorner \lambda_1 \urcorner) \wedge \neg T(\ulcorner \lambda_1 \urcorner)$.

---

[16] See Priest (1987): 79.

The argument uses neither Contraposition, nor the Exclusion Principle. And (2) is nothing but a contradiction in the "metatheory". Therefore, "untruth and falsity seem to behave in very similar ways",[17] after all.

Now, it turns out that also (LNC2c), i.e., $\neg(T(\ulcorner\alpha\urcorner) \wedge \neg T(\ulcorner\alpha\urcorner))$, is a theorem that holds for any α in *In Contradiction*.[18] Nevertheless, this recasts the trouble. Suppose now "α is a non-dialetheia" is defined as "α is not both true and not true", i.e., precisely $\neg(T(\ulcorner\alpha\urcorner) \wedge \neg T(\ulcorner\alpha\urcorner))$. As Stewart Shapiro has stressed, "the locution 'not both true and not true' *does not make any distinctions at all*. For the dialetheist and the non-dialetheist alike, every sentence is not both true and not true, including any that may be both true and not true".[19] Again, the dialetheist seems to be deprived of an exclusion-expressing device.

14.3.3 Inconsistency All the Way Up

Once again, let us not overlook the following crucial philosophical point. The thorough dialetheist has independent theoretical reasons to refuse to submit to the requirement that her own semantic "metatheory" be consistent. As we know, it is of the essence of dialetheic semantics to reject Tarskian and hierarchical approaches that aim at solving the problem caused by the various versions of the liar paradox by appeal to a rigid object language/metalanguage separation. A weak form of dialetheism, such as Rescher and Brandom's, from this viewpoint looks like a useless concession. The dialetheic "metatheory" is nothing but a fragment of a unique theory. And once it has been accepted that inconsistency can move up to the "metalanguage", there seem to be good reasons to push the possibility of it all the way up, so that it spreads into any level of discourse (supposing it is

---

[17] Priest (1993): 40. "Dialetheists hold that truth and falsity overlap. They need not, I suppose, hold that truth and untruth overlap (that is, that a contradiction of a certain kind is true). But if their rationale for dialetheism includes the semantic paradoxes, and in particular the liar paradox, then exactly the same considerations will lead them to this position. The sentence λ: λ is untrue, would appear to be both true and untrue." (*Ibid*). In *Doubt Truth to be a Liar* (Priest (2006): 85) it is claimed that the dialetheist can accept (Exc) or the whole (Neg2) and give a homophonic dialetheic semantics. On the other hand, in the second edition of *In Contradiction* (Exc) is, again, dismissed; it is also claimed that λ_1 is the *only* denizen of the twisted area in which truth and untruth overlap (see Priest (1987): 294).

[18] See Priest (1987): 71-2.

[19] Shapiro (2004): 343, my italics.

still reasonable to talk about "levels", which at this point may be doubtful). Priest has accepted this. As we saw in § 8.2, the reason why the Disjunctive Syllogism cannot be enthymematically valid is that there can be no "metalinguistic" predicate, and no sentential operator, that *forces* consistency. When we say "α behaves consistently", what we say may itself be inconsistent, therefore, it cannot rule out, as it were, *a priori*, the inconsistency of α. This Priest calls "one of the hard facts of paraconsistency".[20]

And it makes the dialetheist's life even harder. If the expression "α behaves consistently", i.e., "α is a non-dialetheia", can itself be inconsistent, as Parsons maintained in the quote given above, the dialetheist may have no means of excluding the possibility that α is a dialetheia. It comes as no surprise, then, that a paraconsistent logician like Batens refused "global paraconsistency" and Priest's idea that no sentential operator can force consistency. One cannot push the possibility of inconsistency all the way up: if inconsistencies could occur at the "highest level" of a theory, then that simply would not be a *theory* anymore. Batens also claimed that "to whatever extent one might disagree with some of Popper's views, I cannot see how one could disagree with his basic insight that only those theories are informative that 'forbid' something".[21] Such a point appears to be related, again, to the elenctic idea that a sentence can be informative only insofar as it rules out something.

Priest and Routley carefully replied to Batens' argument, and the reply is worth listening:

> The argument fails, and does so at the first step. There is no reason to suppose that for a sentence to have a determinate and non-trivial content it must exclude anything. Consider '2 + 2 = 4' and 'Perth is in Australia'. If paraconsistency is right, neither of these assertions *logically* excludes its negation, or anything else. Yet each has a different determinate but non-trivial content. This is so because each carries information the other *does not include*. So the second implies that Perth is somewhere, that Perth is in either Australia or Indonesia etc. whilst the first does not.[22]

Does this get the point of the elenctic strategy? The strategy is based on the idea that "Perth is in Australia" can be a sentence with a determi-

---

[20] See Priest (1989): 623-4.
[21] Batens (1980): 227. See also Batens (1990).
[22] Priest and Routley (1989b): 513.

nate content only insofar as it excludes something *simpliciter*, i.e., only if it entails the exclusion that some state of affairs holds, or that something is the case. If someone says "Perth is in Australia", she cannot be taken as seriously speaking of the geographical position of a city if she admits (possibly, *in actu exercito*) that her utterance is compatible with the holding of any state of affairs, such as Perth being in Norway, on the dark side of the moon, in both places simultaneously, in both and neither, etc. Correspondingly, if by her critique of the (LNC) the dialetheist could not exclude the content the (LNC) expresses, or that what the (LNC) says is the case, then how could dialetheism be taken seriously? As Grim's remark goes:

> Dialetheism is conceived in denial of the LNC [...]. But how is the dialetheist to express [her] own position in a way that makes it clear what [s]he is opposing? The threat here is that what the dialetheist seeks to claim regarding the LNC will become as ineffable as rival positions [s]he condemns on the basis of inexpressibility.[23]

This brings us to the more general issue of theories not conveying a determinate and non-trivial content. For instance, J.C. Beall claimed that dialetheism is obviously understandable and capable of conveying information: after all, the supporter of the (LNC) criticizes dialetheism and argues against it, so she must have understood it. She also has been able to explain dialetheism to her audience; otherwise, they would not understand what she is criticizing.[24] The problem, though, may be slightly more complicated. Consider D, the set of sentences in which thorough, i.e., self-contradictory, dialetheism consists. Consider some consistent subset of D (let us name it D_1). We certainly understand D_1. However, troubles arise as soon as the logical consequences of D_1 are tracked down. Littmann and Simmons claimed that such consequences simply *take back* what has been said in some other (consistent and comprehensible) subset of D.[25] And D ends up as non-informative. Consistent subsets of D are meaningful; and it is also possible to understand what the theory as a whole *is* (after all, it is exactly D). But according to Littmann and Simmons it is difficult to understand is what D as a whole *tells* one about language, or about the world.

---

[23] Grim (2004): 61.
[24] See Beall (2001b).
[25] See Littmann and Simmons (2004): 319.

Can the dialetheist reply by constructing what might be called a *restricted* dialetheist theory? For instance, she may somehow limit the possibility of true contradictions to the field of semantics, and even adopt a classical approach for the fragment of her theory that is free of semantic predicates. But this does not seem to work, for the following two reasons. First, the initial aim of dialetheism was to provide a standardized solution to a very wide range of paradoxes. Therefore, restricting dialetheism to some previously delimited semantic fragment does not appear (contrary to Priest's purposes) to be a "uniform solution" at all, but rather a withdrawal. Second, the problem of exclusion and of conveying informative content can now be reformulated exactly for the restricted semantic fragment. According to Littmann and Simmons, one may have problems in understanding the dialetheic treatment of the (strengthened) liar, λ_1. Even if the dialetheist rejects (Exc) or (Neg2), according to her λ_1 is both true and untrue, that is, as we have seen:

(2) $T(\ulcorner \lambda_1 \urcorner) \wedge \neg T(\ulcorner \lambda_1 \urcorner)$.

This is supposed to be a distinctive claim of D, of the dialetheist's theory. But the negation of (2),

(Non2) $\neg(T\ulcorner \lambda_1 \urcorner \wedge \neg T\ulcorner \lambda_1 \urcorner)$,

is also in D, since it holds for *any* sentence. Therefore, {(2), (Non2)} is a subset of D. Now, Littmann and Simmons claim, since it is an inconsistent subset, each sentence can at best (or at worst?) be understood separately. However, exactly what does the subset {(2), (Non2)} as a whole tell us about the Liar?

> We are faced with two claims which do not seem to supplement one another: rather each seems to take back what the other says. We are presented with an evaluation of [λ_1] that is no clearer than [λ_1]. [...] But the real question is whether or not we understand what the theory is telling us about [λ_1]. And the answer to that, we think, is in the negative: we do not understand what is being said of [λ_1] when it is said that [λ_1] is and isn't both true and [untrue], any

more than we understand what is being said of someone when it is said that he both shaves and does not shave himself.[26]

14.4 The Pragmatic Way Out

14.4.1 Rejection-Consistency and the Rationality Principle

Dialetheists have provided several, and sometimes quite convincing, replies to criticisms of this kind. I will now focus on t pragmatics, for, according to Priest's (i.e., the most influential) version of dialetheism, all these troubles with ruling out things can be solved by turning entirely into the realm of pragmatics. In order to help the dialetheist rule out something, he has provided an interesting treatment of the notion of *rejection*. It is time to recover our two pragmatic operators, $\vdash_x$ and $\dashv_x$.

The dialetheist adopts the following Rationality Principle:

(RP) If you have good evidence for (the truth of) α, you ought to accept α.[27]

Belief, acceptance, and assertion have a *point*: when we believe and assert, what we aim at is believing and asserting what is the case or, equivalently, the truth. Therefore, the dialetheist will accept and, sometimes, assert both α and $\neg\alpha$ if she has evidence that both α and $\neg\alpha$ are true. Consequently, we will sometimes have not only (C1)- and (C2)-contradictions, but, when x is a dialetheist, (C4a)-contradictions, too:

(C4a) $\vdash_x \alpha \wedge \vdash_x \neg\alpha$.

Now, in § 2.4.1 we have seen that the Frege-Geach equivalence between rejection and acceptance of negation:

(Acc) $\dashv_x \alpha \leftrightarrow \vdash_x \neg\alpha$,

has been challenged by the gappers. Priest has also claimed that accepting $\neg\alpha$ is different from rejecting α: a dialetheist can do the former

---

[26] Littmann and Simmons (2004): 319.
[27] See Priest (2006): 109.

and not the letter – exactly when she thinks that α is paradoxical.[28] The classical equivalence (Acc) gives sentential negation a double foundation in the concepts of *disagreement* and *incompatibility*, but such a fusion, Priest argues, is a confusion. He may even concede that the assertion of ¬α amounts to a denial of α in ordinary circumstances: (Acc) can be maintained as a defeasible principle,[29] thereby doing justice to the intuition that rejection and negation should have something to do with one another. However, in special circumstances this natural assumption breaks down, and negation and denial come apart. A denial/rejection of α becomes a non-derivative mental or linguistic act, in that it is directly aimed at α (or at the content of α, or at the proposition expressed by α, etc).

Given that (Acc) can fail, the fact that a dialetheist instantiates (C4a)-cases of contradiction (that is, she accepts both a sentence and its negation) does not entail that she also instantiates (C4b)-cases:

(C4b) $\vdash_x α \land \dashv_x α$.

She can accept both α and ¬α but she does not need to accept and reject α. Actually, she *cannot* even do it, that is, Priest considers acceptance and rejection as reciprocally *incompatible*, even though α and ¬α are not:

> Someone who rejects A cannot simultaneously accept it any more than a person can simultaneously catch a bus and miss it, or win a game of chess and lose it. If a person is asked whether or not A, he can of course say 'Yes and no'. However this does not show that he both accepts and rejects A. It means that he accepts both A and its negation. Moreover a person can alternate between accepting and rejecting a claim. He can also be undecided as to which to do. But do both he can not.[30]

It seems we have found a way at last for the dialetheist to rule out something and to express this. Although the dialetheist cannot rule out α by simply saying "¬α", she can *reject* α.[31] It also seems we have a version of the (LNC) that Priest, too, accepts, namely:

---

[28] See e.g. Priest (2006): 104.
[29] As suggested, e.g., by Tappenden (1999), Mares (2000).
[30] Priest (1989): 61; see also Priest (1987): 98-9.
[31] I will not deal here with another option some claim to be available to the dialetheist in order to express disagreement: a dialetheist can disagree with respect to a given sen-

(RC) **Not** $\vdash_x \alpha \wedge \dashv_x \alpha$

(keep your eye on the boldfaced **Not**). We can call (RC) *rejection-consistency*, borrowing the terminology used within the treatment of rejection operator(s) in formal logic. Actually, to say that (RC) is a version of the Law which Priest *accepts* is a bit misleading. As we have seen, he accepts them all – or, at least, he accepts all the traditional formulations. $\neg(\alpha \wedge \neg\alpha)$ is a logical truth in most paraconsistent logics (a notable exception being, as we know, the positive-plus systems), and the necessitation of the Law is a logical truth in their modal extensions. As we have seen, $\neg(T(\ulcorner\alpha\urcorner) \wedge \neg T(\ulcorner\alpha\urcorner))$ is also endorsed in *In Contradiction*. As Routley says, "despite the correctness of contradictions, *Aristotle's principle of non-contradiction is correct, both in syntactical and semantical formulations. For Aristotle's syntactical principle $\sim(A \;\&\; \sim A)$ is a theorem, hence valid, hence true*".[32] Now, given (RP), the rational dialetheists *accepts* logical truths.

However, dialetheists often embrace pragmatic counterparts of the fundamental logical laws or rules of inference that fail in some way or other within their favourite paraconsistent logics. As it is all too clear, the reason is that in the dialetheic framework the pragmatic operator(s) for rejection/denial take over the *exclusive* features traditionally ascribed to negation.[33] Of course, assertion is meaningful only insofar as not every-

---

tence α by asserting: "$\alpha \to \phi$", where ϕ is something particularly repugnant – typically, $\phi =$ "Everything is true", $\forall x T(x)$ (see, e.g., Priest (1996): 644-5). Then, ϕ expresses the trivial position, and trivialism is unacceptable if anything is, even by the dialetheist's standards. One may find such a way out odd, nevertheless. First, as Priest admits, $\alpha \to \phi$ is still *logically* compatible with α, at least given the trivial model of **LP**. As a consequence, if uttered by a trivialist "$\alpha \to \phi$" would not express disagreement yet – and nothing, indeed, would (see Priest (1996): 644; (2006): 107). More importantly, a dialetheist living in Perth may want to disagree on "Perth is in Norway" on the basis of the simply empirical fact that she *knows* where she lives; but it seems strange that she is thereby committed to something like "If Perth is in Norway, then everything is true". Can't we have any slightly gentler form of disagreement?

[32] Routley (1979a): 312.

[33] Besides rejection-consistency or rejection-soundness (Brady (2004): 45; Mares (2000): 504; Goodship (1996): 153), we have: $\vdash_x(\alpha \vee \beta) \wedge \dashv_x\alpha \Rightarrow \vdash_x\beta$, the pragmatic correlate of the Disjunctive Syllogism (Priest (1989): 618; Mares (2000): 508-9); the so-called rejection by detachment: $\vdash_x(\alpha \to \beta) \wedge \dashv_x\beta \Rightarrow \dashv_x\alpha$, the correlate of *modus tollens* (Brady

thing could be asserted: "an intelligible assertion must rule something out (it must determine something and *omnis determinatio est negatio*".[34] And this seems to be mandatory, if dialetheism is to be able to rule out something and express its views. As Grim has observed:

> The retention of pragmatic exclusion between assertion and denial seems a necessary foothold against the charge of dialetheic inability to either champion or contest any position. But retention of that foothold is peculiar as well. It is unclear, to begin with, why the argument should stop at this point. If dialetheism has so much going for it, why stop it short of assertion and denial? It is also unclear that exclusion can be restricted to the pragmatics of assertion and denial alone.[35]

Now, it seems to me that exclusion had better not be restricted to the pragmatics of acceptance/assertion and rejection/denial, since it is a semantic and ontological notion. A denial is supposed to convey some content but, as Grim also notes, "any content that inherits the exclusionary characteristics that Priest recognizes for denial will thereby have precisely the exclusionary characteristics he refuses to recognize for negation".[36] In order to appreciate this point, let us turn to the boldfaced **Not** in (RC). At one time, various authors (including me)[37] thought that, by saying things like "it is *im*possible jointly to accept and reject the same thing"; or "acceptance and rejection are mutually *in*compatible"; or "someone who rejects A can*not* simultaneously accept it",[38] Priest was asserting, thus accepting, the negation of something, that is, he was asserting a pragmatic version of the (LNC):

(LNC4b) $\neg(\vdash_x \alpha \land \dashv_x \alpha)$.

(2004): 45, Mares (2000): 507); etc. When embedded into a logic, such principles work as axioms or rules expressing within the object language of the formal system how formulas are accepted or rejected as theorems of the system itself. The intuitive link with pragmatics is that provability and disprovability are to a formal system the analogue, in a context of demonstration, of what acceptance and rejection are to a believer.

[34] Rescher ahd Brandom (1980): 25.
[35] Grim (2004): 62.
[36] *Ibid.*
[37] Berto (2006b): 291.
[38] Priest (1987): 103, (1989): 618, my italics.

For instance, Batens has claimed that such a negation (*"im-"*, *"in-"*, *"not"*) has to be taken as an exclusive, non-paraconsistent one. Otherwise, by asserting her paraconsistent negation of any (C4b)-contradiction, the dialetheist would not have managed to rule out the possibility that someone who rejects something accepts it simultaneously.[39]

However, Priest has recently clarified that the logical form of (RC), despite its **Not**, is not manifested by (LNC4b). Ordinary language "not" is ambiguous (at the very least) between a content modifier and a force operator/speech act indicator – as Horn says in his *Natural History of Negation*, "not" is *pragmatically* ambiguous. And acts of denial *may well* be performed by asserting negations. Only an inspection of the context and of the intentions of the utterer can help us disambiguate her claims (so no surprise that someone gets it wrong sometimes). Priest explained that "when I said [...] that one cannot accept and reject something, I was denying the claim that one can do this".[40] Therefore, Priest commits himself to something: he rejects/denies that anyone (any rational agent x) can both accept and reject the same thing. Supposing acceptance and rejection *are* exclusive, therefore, Priest cannot accept (and, since he is a sincere man, assert) $\vdash_x \alpha \wedge \dashv_x \alpha$ for any rational agent x and sentence α, i.e., any (C4b)-contradiction. Of course, the fact that dialetheists countenance contradictions of various kinds does not commit them to countenancing them all. Once again, dialetheists are not trivialists; they do not believe that *everything* is the case:

> The paraconsistentist is by no means committed to the view that all contradictions (or pairs of contraries) are realizable. In particular, the pair $\dashv_x A$ and $\vdash_x A$ would not seem to be so.[41]

The general incompatibility of acceptance and rejection plays a pivotal role in Priest's strategy.[42] Besides providing a tool for ruling things

---

[39] "Talking classically, I am able to say all Priest is able to say, but *not conversely*. [...] The reason for the trouble lies with paraconsistent negation. [...] If [the negation used by Priest] were a paraconsistent negation, then the meaning of 'incompatible' would be took weak for Priest's argument. The least a sufficiently strong sense of 'incompatibility' should warrant is that incompatibles cannot possibly be true together" Batens (1990): 215 and 220.

[40] Priest (2006): 107.

[41] Priest (1989): 618.

out and expressing disagreement, such incompatibility is essential to the dialetheist also for rational reasons. Suppose we embrace an extreme form of dialetheism, in which acceptance and rejection are psychologically compatible. The dialetheist holds that when we are dealing with a sentence α for which we have good evidence that it is paradoxical or a dialetheia, i.e., both true and false, we should accept it on the basis of the Rationality Principle (RP). Of course, α is also false, but this is irrelevant: since some truths are false, if we accept all truths we will have to accept some falsehoods. Therefore, we should not criticize an argument that has α as its conclusion: the whole point of the dialetheist's strategy concerning logical paradoxes is precisely that they should be accepted as sound proofs of their inconsistent conclusions. If we *could* also reject the dialetheia we have accepted (for instance, because it is false anyway), we would have to criticize the argument, too: there must be something bad in an argument that takes us to a rejectable conclusion.[43] Therefore, in addition to accepting the paradoxical arguments as sound, the dialetheist would have to do what anyone else does, i.e., condemn them and find some questionable premise or inferential step in them. Then, these questionable premises or inferential steps might turn out to be acceptable, too… It would seem that we have lost contact with rationality *tout-court*: argumentation could not get any grip on the assessment of acceptances, rejections, beliefs and disbeliefs.

14.4.2 Some Inconvenientia

For several reasons, therefore, the dialetheist had better maintain that, even though truth and falsity can be compatible, at least acceptance and rejection are incompatible. I have no knock-down argument against Priest's treatment of rejection. However, it seems to me that a handful of problems faces this pragmatic way out.

First, if incompatibilities have to be evaluated, so to speak, one by one and each one on its own merits, which are the particular demerits of the

---

[42] Actually, the incompatibility may hold only for the psychological states. Therefore, as anticipated at the beginning of the book, the treatment of the psychological states and that of the corresponding illocutionary acts can come apart in some circumstances. More on this in the following.

[43] As pointed out by Sainsbury (1997).

(C4b)-schema? To say that it would spell trouble for the dialetheist because of the above considerations seems rather circular. Priest does not appear to provide many independent arguments for the incompatibility of acceptance and rejection, except maybe by claiming that "characteristically, the behaviour patterns that go with doing X and refusing to do X cannot be displayed simultaneously".[44] However, behaviour patterns usually do not help us in conceptual subjects. Mental acceptance, rejection, and simultaneous acceptance and rejection (if available), may entail no determinate behaviour pattern at all. The fact that someone simultaneously accepts and rejects α may lead to no practical consequences, if α expresses something quite abstract and theoretical: exactly which behavioural outcomes would be necessarily entailed by the simultaneous acceptance and rejection of the strengthened liar? Furthermore, behaviour and psychology may also come apart, in that linguistic acts and the corresponding mental states can split. Priest himself has sometimes admitted that one can act in such a way as to express acceptance and rejection of the same thing at the same time. It may turn out that assertion and denial, as broadly linguistic or expressive acts, are not exclusive. Priest's example is: I can deny over the phone that I went to the Whiskey-a-Go-Go, and simultaneously assert it to someone watching, with a wink. If this is not a simple case of equivocation, we had better restrict real incompatibility to mental states.[45] This is why Priest usually claims that he prefers to run the whole discourse in terms of the psychological states only, not in terms of assertion and denial.

Mares (2000) calls rejection-consistency (RC) a "principle of coherence",[46] and takes this incompatibility as simply constitutive of the notion: "by virtue of the nature of rejection, it is a necessary condition on a sentence's being rejected that it also *not* be accepted".[47] Is there any such nature? In § 1.3.2, we saw how both Husserl and Łukasiewicz argued against pragmatic and psychological versions of the (LNC) because of the weakness of their warrants. Such principles are, at most, inductive theses

---

[44] Priest (1987): 99.

[45] See Priest (1993): 36.

[46] The relevant formulation in Mares' paper is to the effect that the "acceptance box" and "rejection box", i.e., the mental boxes containing accepted and rejected sentences, are disjoint: nothing can be in both.

[47] Mares (2000): 504. Is that italicized *not* also a rejection on Mares' side?

based upon considerations of empirical psychology. They do not discover any *nature* or *essence* at all.[48]

Can the dialetheist reply that the mutual exclusiveness of mental acceptance and rejection is ascertainable by introspection, and link this to the infallibility of some First Person Authority? She may claim that our knowledge of our own mental states is infallible or incorrigible, at least in this respect: the fact that acceptance and rejection cannot co-occur simultaneously in the same mind, and with respect to the same sentence, is self-intimating. However, I doubt it. Against commentators who maintain that the Aristotelian (LNC) is a psychological law to be established by introspection, Priest observes that "the unsatisfactoriness of trying to establish psychological laws in this way hardly needs to be laboured".[49] We observed in § 8 that nothing is ruled out *on logical grounds alone* in the mainstream dialetheic framework: given any set of sentences, some **LP** model satisfies it (eventually, the trivial model). Something can be discarded *a posteriori*, and we do have evidence that some contradictions do not hold – that the world is not trivial. But according to Priest there is no infallible (exterior or interior) observation at all:

> We know, then, that the world is not trivial, since we can see that this is so. [...] There is something, then, about the world, that fails to obtain. These considerations, like all *a posteriori* considerations, are defeasible. Observation is a fallible matter, and what appears to be the case may not, in fact, be so.[50]

To the epistemic difficulties raised by Łukasiewicz and Husserl, the dialetheist may simply answer: "*c'est la vie*, and you cannot do any better". She may rest on a reliabilist account of perception and experience.[51] If everything obtained, she would see many inconsistent states of affairs that she just does not see, and she relies on perception.

---

[48] We saw that Łukasiewicz directly blamed Aristotle for claiming that nobody can believe a contradiction in the sense of accepting both a sentence and its negation, $\vdash_x \alpha \wedge \vdash_x \neg \alpha$; but analogue considerations may hold against Priest's thesis that nobody can both accept and reject a sentence, $\vdash_x \alpha \wedge \dashv_x \alpha$.

[49] Priest (2006): 10.

[50] *Ibid*: 62.

[51] "We [...] take the inference from the statement of perception to a statement about the world to be a reasonable default inference" (*Ibid*: 64).

Another case in which the dialetheist has to bite the bullet may be the following. We saw that the typical dialetheist can, and does, straightforwardly accept all the classical formulations of the (LNC): (LNC1)-formulations are theorems of **LP**, and (LNC2)-formulations are endorsed in *In Contradiction*. Given that acceptance and rejection are incompatible, she cannot reject them. But isn't she supposed to be able to *somehow* rule out the position advocated by supporters of the Law? The same holds for the very idea that truth is consistent, which is usually associated with the (LNC) itself. It seems that the dialetheist may want to reject at least some formulation of the idea:

> If, for example, I am in a discussion with someone who claims that the truth is consistent, it is natural for me to mark my rejection of the view by uttering 'it is not', thereby denying it.[52]

Denying *what*, exactly? A natural way to express the idea that truth is consistent is, once again, via some formulation of the (LNC): for all α, it is not the case that α is both true and untrue; or: for all α, it is not the case that both α and $\neg\alpha$ are true. We have seen that these are taken as logical truths in the mainstream dialetheic position, and given the Rationality Principle one ought to accept truths; so the rational dialetheist accepts them. If this is what is meant by the idea that truth is consistent, then the dialetheist accepts that truth is consistent and, again, since acceptance and rejection are exclusive she cannot reject it. It seems that, even though the dialetheist can disagree via rejection, and express this via denial, she what she cannot disagree with are the main claims of the supporters of consistency and of the (LNC) in their standard formulations. Quite so, the thorough dialetheist replies: she accepts that truth is consistent, provided she is allowed to *add*, for the sake of completeness, that it is also inconsistent; she accepts that all contradictions are false, provided she can add that some are also true; *et cetera*.[53] Hence comes the understandable frustration of the orthodox logician: as she advances α with all her dedication, the dialetheist answers: "I completely agree with you. And $\neg\alpha$, too" (recall the paraconsistent mathematician we met in § 12.3, who agrees with her conservative colleagues that all numbers are distinct from their successors; and also adds that some number is not distinct

---

[52] *Ibid*: 105.
[53] See e.g. Priest (1987): 294.

from its successor). Wouldn't it be nice to find *at least one* formulation of the (LNC) which the dialetheist is forced not to accept? Can the very notion of rejection help the dialetheist? I think not. Force operators cannot be embedded in a sentence, which makes it difficult to express through them something aiming at having general validity.[54] And, as Priest notices, in this context it would be simply mistaken to formulate the (LNC) by saying that one ought to deny/reject (that) $\alpha \wedge \neg\alpha$, for all α: "the claim that two sentences are contradictories concerns their truth-relations: it has nothing, of itself, to do with rationality of obligation".[55]

The difficulty of finding a clear formulation of the (LNC) that the dialetheist can reject might be taken as just another *inconveniens*. However, the sharp distinction between force and content in this context seems to raise yet another problem. On the one hand, restricting exclusion to pragmatics facilitates the dialetheist in the following sense: the pragmatic incompatibility does not allow us to rebuild any hypercontradiction by using denial/rejection. No super-liar can be formulated via the notion of denial, which, "being a force operator, has no interaction with the content of what is uttered":[56] since we are dealing with an illocutionary act, not with a connective, no extended paradox is expected.

On the other hand, exactly this feature of denial may spell troubles. Suppose we accept (RC), with its **Not** understood in terms of denial: then we know that Priest rejects something and commits himself on that. We may ask, once again: has he thereby managed to rule out that he *also* accepts that very thing? Couldn't he both accept and reject the same thing, that is, instantiate both mental states simultaneously? (RC) rules out the possibility that Priest also accepts that, for some x and α, $\vdash_x \alpha \wedge \dashv_x \alpha$, only on the *presupposition* that acceptance and rejection *are* incompatible. This is something we can say ("talking consistently", so to speak) by claiming that the very content of Priest's rejection does not hold, or is false. But it won't work in a thoroughly dialetheic framework: in such a framework, to say that some content does not hold, or that it is false, does

---

[54] Therefore, even though the Frege-Geach argument supporting (Acc) fails, it may still work as a generally reliable *test* for aspects of the use of "not" that belong to content, and aspects of it that belong to performance: if "not" makes sense when the sentence having it as its main operator is embedded in a larger sentence, it is likely that we are dealing with a content-modifying negation, not with a denial negation (see, e.g., Tappenden (1999): 277-9).

[55] Priest (2006): 78.

[56] *Ibid*: 108.

not rule out that it also holds, or that it is true. The possibility of ruling something out via rejection, and expressing this via denial, seems to *presuppose some content exclusion*, i.e., that some states of affairs in the (mental or so-called external) world are reciprocally *incompatible* or that the holding of one rules out the holding of the other. In this case, it presupposes that the situation in which some x accepts something rules out the situation in which that very x rejects that very thing. However, (RC) opens with a force operator for denial. And "$\dashv$, being a force operator has no interaction with the content of what is uttered". It is exactly this feature that spares dialetheism the trouble of facing a super-liar formulated in terms of denial. Furthermore, according to Priest there is no operator on content that can mimic the force operator for denial.[57] Because of this, the information we get from (RC) is that Priest is committed to something: he rejects that acceptance and rejection are compatible. This is sufficient to rule out some content, only on the presupposition that acceptance and rejection are incompatible, i.e., that one rules out the other.

We saw that the classical account of negation runs together two different ideas: the one of disagreement and the one of incompatibility, or exclusion. Now, Priest has made two moves with respect to the standard account.

(a) He has dissociated the assertion of a negation, $\neg$, as a content operator, from denial, as a force operator expressing rejection, $\dashv_x$. Rejection is now a *sui generis* intentional state, largely independent from the acceptance/assertion of any negation. The equivalence (Acc) is nothing pertaining to logic: it holds as a defeasible principle, and it can be defeated in unusual circumstances – those in which gappers and glutters begin to play. This first move might be not at issue, at least insofar as it can be questioned independently from dialetheism. However, it may lead to troubles when it is combined with the following second move.

(b) Priest has also deprived ordinary negation of the capacity to express content exclusion, or incompatibility. As Priest and Routley claim, "we [as dialetheists] cannot use content-exclusion as a way of defining the sense, or content, of negation. But then there are plenty of other ways of doing this, for example, through a semantic account".[58] Now, of course they *can* give a semantic account of negation – such as the one of **LP** or

---

[57] *Ibid.*
[58] Priest and Routley (1989d): 513.

FDE, or via the Routley Star. But this is, by the admission we have just heard them making, not strong enough to support content exclusion.

The speech act of denial, in the dialetheist's mouth, aims at expressing her commitment to the failure of the conditions that would have to obtain for the rejected/denied content to obtain. Now it seems that rejection can rule out something only if excludes acceptance, and *that* acceptance and rejection are exclusive is an incompatibility between *contents* (particularly, mental states). As a consequence of this, it seems to me that we *still* need some exclusion-expressing device that works on *content*, for we want to make the point that *some things in the world* (be it the so-called external world, or the world of our mental states) rule out each other – not just that we commit ourselves on rejecting something, or that we are in a certain mental state. Otherwise, expressing our rejection of something would not be very different from uttering: "I dislike ξ", with no hint whatsoever on what, exactly, ξ is:

> Given that the dialetheist can deny certain claims [...] what is the information that he conveys by his denial? If we accept his denial, what precisely is that we have accepted? If we learn that he is right, what precisely is it that we have learned?[59]

The situation would be somehow analogue to the one of those radical moral non-cognitivists who expel any content from the notion of good except for that of subjective appraisal. Holding an action as good does not describe it in any way – so that to claim that action A is good would be nothing else but uttering something like: "Hurrah for A!".[60]

My upshot is that the dialetheist can indeed express disagreement, and rule out something, only if she has a notion of content exclusion which is not reducible to any force operator. And I think she does have one, indeed. To focus on this will be the task of the subsequent §§.

14.5 The Notion of Material Incompatibility

It seems that we all, even as dialetheists, have an intuition of content exclusion. We may search for an operator (arguably, a negation) that allows

---

[59] Grim (2004): 62.
[60] See Geach (1960).

us to capture and express such intuition. How about starting from *the very notion of* exclusion or incompatibility in order to obtain it?

14.5.1 The Intuition of Exclusion

However, we had better avoid explicitly employing the concepts of *truth* and *falsity* to characterize such an operator. The dialetheist casts doubts on *their* being exclusive by pointing out that some truth-bearers, notably, the liars, fall under both concepts simultaneously – and I have taken the dialetheist quite seriously throughout this book.

> Understanding negation involves a sensitivity to incompatibility, but this notion does not have to be specified [by direct reference to truth and falsity]. For instance, one might suggest that the basic notion of incompatibility in directly semantic terms consists in the fact that incompatible sentences must have opposite truth values, which makes true contradictions conjunctions of incompatibles. However, one might prefer to avoid an account of understanding which involved attributing such semantic notions to speakers, for example on the grounds that the account would not be neutral with respect to realist and intuitionist preconceptions[61]

… And dialetheic preconceptions, too. This entails that we have to advance very carefully. We should refrain from expressing exclusion via the traditional concept of *contrariness*, since, as we have seen, in most accounts such a concept typically depends upon those of truth and falsity. Defining α and β as contraries iff $\alpha \wedge \beta$ is logically false, as Huw Price observed, "clearly depends on our knowing that truth and falsity are incompatible", so that "if we do not have a sense of *that*, the truth tables for negation give us no sense of the connection between negation and incompatibility".[62] On the other hand, one may take the intuitive notion of exclusion as a *primitive* basis for the definition of a negation:

> The apprehension of incompatibility [is] an ability more primitive than the use of negation. The negation operator is being explained as initially a means of *registering* (publicly or privately) a perceived incompatibility. […] For present purposes, what matters is that *incompatibility be a very basic feature*

---

[61] Sainsbury (1997): 224.
[62] Price (1990): 226.

of a speaker's (or proto-speaker's) experience of the world, so that negation can plausibly be explained in terms of incompatibility.[63]

We may begin with the *single* basic assumption that ordinary speakers and rational agents have some acquaintance with incompatibility: they can recognize it in the world, and in their commerce with the world. I shall talk of *material exclusion* or, equivalently, of *material incompatibility*. It may be explained in terms of concepts, properties, states of affairs, propositions, or worlds, depending on one's metaphysical preferences – and we want to be as neutral as possible not only on logical, but also on metaphysical issues. For instance, we may view it as the relation that holds between a couple of properties, P_1 and P_2, iff, by having P_1, an object has dismissed any chance of simultaneously having P_2. Or we may also claim that material incompatibility holds between two concepts C_1 and C_2, iff the very instantiating C_1 by a puts a bar on the possibility that a also instantiates C_2. Or we may say that it holds between two states of affairs s_1 and s_2, iff the holding of s_1 (in world w, at time t) precludes the possibility that s_2 also holds (in world w, at time t).

Put it any way you like, material exclusion has to do with *content*, not mere performance: it is rooted in our experience of the world, rather than in pragmatics.[64] It has been dubbed *material* to stress the fact that it is not a merely *logical*, in the sense of *formal*, notion: it is based on the material content of the involved concepts, or properties, etc. Neil Tennant calls such concepts *antonyms*, and observes that:

> Here the antonyms A and B are so simple and primitive that there cannot be any question of their 'dialetheically' holding simultaneously. Such antonyms A and B are antonymic not on the basis of their logical form, but on the basis of their primitive non-logical contents. The tension between them – their mutual exclusivity – is a matter of deep metaphysical necessity.[65]

---

[63] *Ibid*: 226-8, my italics.

[64] Material exclusion appears to be inescapably modal, though (which, admittedly, may make it unpalatable at least to the unshakably extensionalist Quinean): it does not hold between two merely different properties, like *being circular* and *being red*, which can be instantiated by the same object, even though sometimes they are not. It holds between two properties, such that an object instantiating one of them has lost any opportunity of simultaneously instantiating the other, like *being circular* and *being square*.

[65] Tennant (2004): 362.

Tennant's examples are: phenomenological colour incompatibilities, such as *being (solidly) Red* and *being (solidly) Green*; concepts that express our categorization of physical objects in space and time, such as *x being here right now* and *x being way over there right now*, for a suitably small *x*. Other cases provided by Patrick Grim are *x being less than two inches long* and *x being more than three feet long*.[66] But we may also take Priest's above *x's catching the bus* and *x's missing the bus*.

14.5.2 Whither Formalization?

One may wonder, don't we need some axiomatic or broadly formal characterization? Exactly which logical and inferential properties does a negation expressing material exclusion have? My instinctive answer would be: pretty much the ones you like, *provided* you stick to the fundamental intuition. What we are dealing with is one of the most basic insights we can appeal to. It may therefore be susceptible to different logical characterizations – I would call it a *determinable* concept: something open to different further determinations. For instance, a feasible formal account may adapt (by avoiding direct reference to truth and truth conditions) the idea developed by Michael Dunn (1996), that "one can define negation in terms of one primitive relation of incompatibility [...] in a metaphysical framework".[67] Dunn has in mind a notion initially developed within quantum logic: the Birkoff-von Neumann-Goldblatt definition of *ortho negation*. What makes it attractive is that it uses precisely a relation of incompatibility (usually called "orthogonality", or simply "perp").[68] A first attempt would be to frame it in terms of properties: take an ordered couple $<S, \perp>$, where S is a set of properties, and $\perp$ is our binary relation of material exclusion,[69] defined on S. Then we have something like the following NOT:

(3a) NOT $P_1(x) =_{df} \exists P_2(P_2(x) \wedge P_1 \perp P_2)$.

---

[66] Grim (2004): 63.
[67] Dunn (1996): 9.
[68] See Birkoff and von Neumann (1936); Goldblatt (1974).
[69] I stick to Dunn's notation even though it may be a little bit confusing: logicians know "$\perp$" mainly as the *falsum*, a 0-adic logical constant, not as a symbol for a binary relation.

To say that something is NOT P_1 is to say that it has some property, P_2, which is materially exclusive with respect to P_1. A quite simple fact of ordinary language is mirrored by the partial indeterminacy in the information conveyed by an expression containing NOT. When you declare "The car is red", this is not the logically weakest, or less informative, sentence incompatible with the sentence "The car is blue" (Assuming, for the sake of the argument, that *red* and *blue* are exclusive; we shall say something on the opportunity of choosing our incompatibilities carefully in the following). The weakest sentence incompatible with "The car is blue" is "The car is NOT blue", which, given (3a), merely says that the car has some property incompatible with that of being blue, not specifying which one. "The car is red" specifically says which other, incompatible colour the car has.

However, we may prefer to talk in terms of states of affairs, or maybe facts. We want to adhere to a conception of NOT as a sentential operator, therefore, in the left-hand side of (3a) NOT is supposed to attach to the whole sentence (or open formula), not to the predicate. It is also supposed to work for relations of any ariety, not only for properties. So let us identify facts with *propositions* (that which is expressed by a sentence) and embed our operator in a little bit of algebra. Since the following account is phrased in terms of propositions as facts, and operations on them, various classical issues may rise, e.g., on the metaphysical status of negative and general facts. I will pass on for the sake of simplicity, but we may recall that so-called realist dialetheism appears to be committed, for instance, to *negative* facts, via the development of a dialetheic correspondence theory of truth.[70]

We may think of a structure $<U, \angle, V, \perp>$, where U is a set of facts or propositions; $\angle$ and $\perp$ are binary relations defined on U; and V is a unary operation on subsets of U. $\angle$ is to be thought of as a pre-order, i.e., reflexivity and transitivity hold, and "$p \angle q$" can be read as "The proposition p entails the proposition q". Given a set of propositions $P \subseteq U$, VP is the (possibly infinitary) disjunction of all the propositions in P. A proposition may have one or more incompatible peers: it need not exclude only one other, but it may rule out a whole assortment of alternatives (Grim (2004), for instance, talks about the *exclusionary class* of a given property). If we have set abstracts, the exclusionary class of a given proposition p is the set $E = \{x \mid x \perp p\}$. Then, NOT-p is nothing but VE. If E has

---

[70] See Beall (2000); Priest (1987): 299-302; (2006): 51-4.

a finite cardinality, i.e., the set of propositions incompatible with p is a finite one, then NOT-p is nothing but an ordinary disjunction: $q_1 \vee \ldots \vee q_n$, where $q_1, \ldots, q_n$ are all the members of E. If, on the other hand, we make the metaphysical assumption of an infinity of propositions incompatible with p, (as one might expect with *the car being blue*, given the continuum of colours), NOT-p turns out to be an infinitary disjunction. If one has problems with infinitary disjunctions, we cannot avoid quantifying on facts/propositions:

(3b) NOT-$p =_{df} \exists x(x \wedge x \perp p)$.

In both cases, it is clear in which sense NOT-p is the logically weakest among the n incompatibles: it is entailed by any q_i, $1 \leq i \leq n$, such that $q_i \perp p$. The point may also be expressed via the following equivalence:

(4) $x \angle$ NOT-p iff $x \perp p$.

Putting NOT-p for x, and by detachment, we get:

(5) NOT-$p \perp p$,

NOT-p is incompatible with p. The right-left direction of (4), then, tells us that NOT-p is the weakest incompatible, i.e., it is entailed by any incompatible proposition. It is clear that such an account is the heir of the one traditionally made between *contraries* and *contradictories*, which as we know was usually defined by reference to truth and falsity.[71]

Any formal characterization of NOT, though, is likely to make at least one logician unhappy. For instance, it may be natural to assume that $\perp$ is symmetric, that is, if $p \perp q$, then $q \perp p$.[72] But if in the algebraic framework NOT is stipulated as an operation of period two, i.e.:

---

[71] Variations on the theme of the characterization of negation via incompatibility and on negation as minimal incompatible, both from a classical and a constructivist point of view, can be found in Brandom (1985), (1994): 381ff.; Peacocke (1987); and Lance (1988). It is fair to say that one may take issue both with the existence and the uniqueness of a minimal incompatible: for instance, Wright (1993) raises doubts about uniqueness, and Tappenden (1999) about existence.

[72] Though even this is not so straightforward: see the discussion in Dunn (1999): 13-4.

(6) NOT-NOT-$p = p$,

this is likely to be rejected by an intuitionist, though not by many paraconsistent logicians. The intuitionist may also object to the fact that NOT has been defined using other operators, which goes against the independence of logical constants in a constructivist framework. Or, if we make the *prima facie* natural assumption that:

(7) If $p \angle q$ and $x \perp q$, then $x \perp p$,

we can easily get Contraposition,[73] and such a result would be an unwelcome one, as we now know, for several paraconsistentists.

It is certainly true that more work is required to show that a basic characterization of NOT via material exclusion yields something with the features of negation. But the point is that, as we have learned at length throughout this book, different philosophical parties (classicists, intuitionists, paraconsistentists, etc.) have opposed views on what negation is, whereas the aim here is to provide an intuitive depiction on which all parties can, or, better, have to, agree. By claiming that various systematizations are possible, as anticipated, we want to formulate the point in such a way as to maintain all the neutrality available, both on logical and on metaphysical issues. Let us inspect further.

14.5.3 The Niceties of NOT

Technical details aside, it seems that our operator has some nice features.

(1) First, is not explicitly defined via the concept *truth*. To quote Price again:

> Where *P* signals a state of affairs of a certain kind – whether an intention to act, or the obtaining of some condition in the world – [NOT-]*P* signifies [a] corresponding incompatible state. [...] It is the beginning of an answer to the questions with which we began; and of an answer which does not depend on the notions of truth and falsity.[74]

---

[73] See Dunn (1997): 10.
[74] Price (1990): 228.

Of course, this may not prevent truth from jumping in again at some point. We have been forced to admit that, given some (albeit debatable) metaphysical assumptions, we may need propositional or predicate quantification to spell out the details of NOT. And such quantification is interdefinable with truth.[75] Nevertheless, what NOT is explicitly referred to is the concept *exclusion*, whose primitiveness is now clear: it is entailed, for instance, by our experience of the world as agents, facing choices between performing some action or other – something we think non-linguistic animals as well do every day. To face a choice is to perceive an incompatibility. But it may also be entailed by the simple and basic capacity to recognize the boundary (even a blurred one) between something and something else, between an object and another one. Exclusion is such a basic feature that "without some fundamental grasp of precisely that notion to begin with it seems quite possible that it cannot later be specified […]. If exclusion is not understood to begin with, what possible exposition could we rely on to nail it?".[76] Furthermore, due to such a connection with perception and action in the real world, this framework of intuitions supporting NOT shares much with Priest's basic insight laying behind his preference for rejection: the insight that exclusion has to be linked to the concrete realm of action, pragmatics and behaviour – not to the mere association of a sentence with a truth value. What we add to Priest's point is that marking disagreements makes sense only insofar as, to speak metaphorically, it is primarily *things in the world* that can "disagree". Pragmatics is now rooted again in content.

(2) Secondly, NOT has a strong pre-theoretical appeal as an exclusion-expressing tool. I could hardly improve Price's description of what a conversation between me and you would be if we had no means to exclude (via negation, rejection, falsity, or whatever) the possibility of Fred's being simultaneously in the kitchen and in the garden:

Me: 'Fred is in the kitchen.' (Sets off for kitchen.)
You: 'Wait! Fred is in the garden.'
Me: 'I see. But he is in the kitchen, so I'll go there.' (Sets off.)
You: 'You lack understanding. The kitchen is Fred-free'.
Me: 'Is it really? But Fred's in it, and that's the important thing.' (Leaves for kitchen.)[77]

---

[75] I am indebted to Graham Priest (in private communication) for this remark.
[76] Grim (2004): 70.
[77] Price (1990): 224.

A simple: "Look, Fred is NOT in the kitchen" (that is to say: "Fred is somewhere else – in the garden – and his being there *excludes* his being in the kitchen"), would definitely make things easier.

(3) Finally, and more importantly, I claim that paraconsistent logicians and dialetheists *do* grasp the notion of exclusion. Dialetheists ask us to stop using "not" or "true" as exclusion-expressing devices, because "not-α" is insufficient by itself to rule out α, and "α is true" is insufficient by itself to rule out that α is also false. Priest and Routley have declared that dialetheists cannot define negation via content-exclusion, and we have examined at length Priest's preference for rejection and denial. But the dialetheists' account of acceptance and rejection shows that they do believe in the impossibility that particular couples of facts, or states of affairs, simultaneously obtain; or, equivalently, that they assume that some properties materially exclude some others: x's simultaneously catching and missing the bus, for instance; and, of course, x's simultaneously accepting and rejecting the same α.

14.5.4 Consistency and Fallibility

The whole story suggests that the notion of material exclusion is somehow inescapable. However, I have not considered a main objection yet. A dialetheist may obviously contest the *irreflexivity* of $\bot$: a property (proposition, state of affairs, etc.), *can* be incompatible with itself! Typically: *true* and *untrue* are incompatible properties: "polar opposites", by the dialetheist's own account.[78] But some very nasty paradoxical constructions deliver something, i.e., the strengthened liar, which is both true and untrue. Therefore, truth seems to be incompatible with itself.

Priest has sometimes replied to the charge of inexpressibility that it is just a misunderstanding: the dialetheist can rule out that α is the case, with *these very words*. What she cannot *guarantee*, or *force*, is the consistency of any concept, or property, etc. When the dialetheist claims that α is not true, she cannot ensure that the very words she utters behave consistently. However, as we have seen in § 8.4.2, nothing can force consistency. So the same holds for the orthodox logician and, indeed, for *anyone*:

---

[78] See Priest (2006): 110.

Once the matter is put this way, it is clear that a classical logician cannot do this either. Maybe they would like to; but that does not mean they succeed. Maybe they intend to; but intentions are not guaranteed fulfilment. Indeed, it may be logically impossible to fulfil them.[79]

As we have seen, the notion of logical possibility is indeed somehow empty (or, if we want, radically omni-inclusive) in such a dialetheic framework as the one of **LP**: given any α, there is a model (the trivial one) both for α and for anything else. From the dialetheist's point of view, we may say that any contradiction, therefore, any claim, is certainly out there in the logical space: "there is no logical guarantee against a person being a trivialist".[80] However, is it rational, or just *feasible*, to embrace any claim as a consequence of this?

> One cannot choose between this and that if one believes that this and that are the same thing, which the trivialist does. Of course, the trivialist believes that this and that are distinct too. But, as before, for the trivialist, two things being distinct does not rule out their being identical.[81]

The only properly working argument against the trivialist, according to Priest, is a "transcendental argument" based on the phenomenology of choice: the trivialist's behaviour, particularly her choice-based acting, shows that she does operate purposefully, accepting something and rejecting something else, and sees her activities as making a real difference in the empirical world. We can infer from this that she does not actually believe everything. But transcendental arguments of this kind are usually taken as a commitment to some form of anti-realism or verificationism, for they aim at establishing categorical conclusion on the world from features of our consciousness. And this is *not* what Priest takes his argument to bring about:

> This does not show that trivialism is untrue. As far as the above considerations go, it is quite possible that everything is the case; but not for me – or for any other persons. [...] The argument I have deployed provides a transcen-

---

[79] *Ibid*: 106-7.
[80] *Ibid*.
[81] Priest (2000c): 194.

dental deduction from certain features of consciousness to the impossibility of being a trivialist.[82]

We may notice that this is not a transcendental argument in the traditional sense, for it does not take us from the impossibility of a trivial viewpoint to the *falsity* of trivialism – only to the much more modest result of the impossibility of *believing* in it. According to Kroon, therefore, "Priest's way of putting his own transcendental argument is likely to signify a commitment to some kind of realism", and to a realist dialetheism.[83]

Regardless of the transcendental-phenomenological argument Priest uses to criticize the trivialist's position, it is not controversial that the paradigmatic dialetheist is not a trivialist. Priest has clarified that there are good reasons to reject some contradictions, or even most of them. The dialetheist does not believe that anything is compatible with anything, or that all states of affairs obtain, or that anything can be anything (else?). We have also seen that from the dialetheist's point of view any assumption of incompatibility is rationally retractable, for instance, on the basis of further evidence. But then, Priest's claim that nothing can force consistency rests on a superposition of metaphysics and epistemology: it is simply a claim of general *fallibility*, and one on which almost anyone can agree (with perhaps the exception of the few surviving epistemic foundationalists). We can come to believe that some properties, or concepts, or states of affairs, are incompatible, and then find out that they are not. The standard strategy is simply to retract our previous assumption that they were. Given two properties P_1 and P_2, the question whether they are exclusive can involve broadly empirical matters, difficult analyses of our conceptual toolkit and/or of our use of ordinary language expressions. Some cases may be easy to resolve; but others may produce battles of intuitions: are *young* and *old* actually exclusive? *Blue* and *red*? *True* and *false*? *Circular* and *square*? I claimed that material exclusion is based on

---

[82] Priest (2006): 70-1.

[83] "The reason why even the compulsory rejection of trivialism is compatible with the truth of trivialism is that there is a conceptual gap between the content of our best beliefs and the way the world really is [...]. For all our clever attempts to insulate contradictions and provide reasons for thinking that the space of true contradictions is in fact very sparse, the world may yet fool us. [...] As a realist, Priest must grant that internal constraints on what we must believe are not constraints on what the world is like independently of our best theories and beliefs" (Kroon (2004): 251-2).

the content of facts, concepts, or properties, but how do we know what the content of a concept is, or which are the actual fields of application of a property? This is the kind of disquisition one should avoid when dealing with the claim that there are true contradictions, or that a sentence can be both true and false. We must keep in mind that the characterization of $\bot$ does not entail special commitments on the specific properties, or concepts, or states of affairs, between which it holds. If this sounds disappointing, you may recall that such merely formalistic descriptions are expected when dealing with purely metaphysical notions: they often leave our epistemic troubles just where they are.

On the other hand, the rights of NOT are *not* just the rights of a stipulative definition. Prior has taught us that a stipulative operator like *tonk* can spell troubles in the face of its plain definition. Nevertheless, NOT appears to be much more than a stipulation: it appeals to our intuition of exclusion, which the dialetheist shares – even though she disagrees on what rules out what in some of our most basic conceptual tools. While Dummett made a plea for a logical foundation for metaphysics, we are doing the opposite: we look for a metaphysical foundation for logic. Our sense of exclusion, I have claimed, comes from our having to do with mutually exclusive colour ascriptions, spatial locations, actions, and, of course, mental states. I began with the couples of properties, or concepts, or states of affairs, the dialetheist herself assumes as materially exclusive (acceptance and rejection, or x's catching the bus and x's missing the bus, etc.). These have been taken as instances of a primitive, intuitive notion of exclusion, $\bot$. Then, I characterized via $\bot$ a sentential operator, NOT, which works as an exclusion-expressing device. It seems that there is no point with the dialetheist refusing the procedure now: NOT does exactly the job that rejection is supposed to do in the dialetheic framework, but that it may not do, *unless* a material incompatibility holds between acceptance and rejection themselves. To put it another way: NOT should work even in a theoretical situation in which nothing is ruled out on logical grounds alone, because it is not merely logically, i.e., formally, but metaphysically ("materially") founded. The dialetheist may have a vacuous notion of logical, formal incompatibility (at least in the sense that logic alone cannot rule out the trivial world – a world in which every atomic sentence is true and false and, therefore, every sentence is true and false). However, she does have a notion of material incompatibility.

14.6 The ἀδύνατον

Now for the final step: express our (LNC) via NOT. That we are dealing with a fundamental intuition on contents explains why Aristotle never discussed the question of the undeniable truth of the (LNC) in his *Organon*, that is, in his writings on the subject of logic. He undertook the issue in his *Metaphysics*, because he thought it to be an ontological subject, not to be solved by mere logical, in the sense of *formal*, tools.[84] Now, take one of Aristotle's traditional formulations of the (LNC) in the *Metaphysics*, and just put in it our NOT. The formulation can be simply taken as a definition of ἀδύνατον, "the impossible":

(Ad) For the same thing to hold good and NOT hold good simultaneously of the same thing and in the same respect is impossible [ἀδύνατον].[85]

'Αδύνατον is that which has no chance, no power (δύναμις) to be. "P_1 does NOT hold goof of x" should be a short form for "to x belongs some property P_2, which is materially incompatible with P_1". This does not seem to be questionable by the dialetheist anymore, provided she has understood NOT – and to understand NOT is to understand exclusion (what the dialetheist does, as we have seen). "Not questionable", as should be clear by now, does not mean only that the dialetheist is forced to accept (Ad) – she may well do it, given the principle (RP) and the consequent fact that she accepts all the traditional formulations of the (LNC). It means that she cannot *also* coherently accept claims inconsistent with (Ad).

None of these remarks should count as an easy discharge of dialetheism (if it could be easily discharged, one should not write a big book on it). After characterizing a negation, which is very similar to the one proposed here, Grim observes:

---

[84] "[Aristotle] argues that no rational person can fail to accept the LNC. The ability to speak demands the ability to identify and name objects and this implies being able to recognize the boundary between an object and its background – the line (possibly a blurred one) between what is the object and what is *not* the object. From his ability to speak about things, we can transcendentally deduce that an individual must acknowledge that what is a particular object is separated by a boundary from what is *not* that object, that what is that object cannot be what is *not* that object" (Goldstein (2004): 308).

[85] *Met.* 1005b18-21.

One option for the dialetheist is to concede a minor battle and hold out for victory in a larger way. The victory for the LNC outlined above applies only to a particular form of the LNC phrased in terms of that sense of contradiction. Any defeat for dialetheism is thus only a very limited defeat.[86]

If the dialetheist refuses to subscribe to the characterization of NOT via the intuitive notion of exclusion, she actually seems to end up as unable to express the exclusion of any position (is she trying to exclude exclusion?). A dialetheism without the (LNC) stated in terms of NOT looks very much like a trivialism. Such a (LNC), to use Aristotle's words, is "a principle which every one must have who knows anything about being".[87] And the exclusionary NOT promises to offer an exclusion-expressing tool, and the prospect of a discussion, which may avoid ending into a hard conflict of intuitions between foes and friends of consistency.

---

[86] Grim (2004): 68. See also 69-71: "The outline above uses various forms of negation, including the English 'not', prominently and repeatedly in trying to get the idea across. If these forms of negation can be understood a particular way, it seems inevitable that [NOT] can be understood a particular way. Given a dialetheic interpretation of all the various forms of negation in the outline, then, one might well end up with a dialetheic interpretation of [NOT]. The result could be that every claim made above is allowed but without the concept of exclusion that is their main intent [...]. All I can say is that those forms of dialetheism seem less interesting to me: I don't see how the prospect of impasse is then to be avoided, and such forms don't seem to me to promise any deeper understanding of notions as central to our conceptual toolkit as is the notion of contradiction".

[87] *Met.* 1005b 14-15.

References

Aczel, P. (1988), *Non Well-Founded Sets*, Stanford: CLSI Lecture Notes 14.
Agazzi, E. (1978), *Non-contradiction et existence en mathématique*, "Logique et Analyse", 21: 459-81.
Agazzi, E. (1990), *Can Knowledge Be Acquired through Contradiction?*, "Studies in Soviet Thought", 39: 205-8.
Anderson, A.R. (1963), *Some Open Problems Concerning the System E of Entailment*, "Acta Philosophica Fennica", 16: 7-18.
Anderson, A.R. and N.D. Belnap (1959), *A Simple Treatment of Truth Functions*, "Journal of Symbolic Logic", 24: 301-12.
Anderson, A.R. and N.D. Belnap (1975), *Entailment. The Logic of Relevance and Necessity*, vol. I, Princeton, N.J.: Princeton University Press.
Anderson, A.R., N.D. Belnap and J.M. Dunn (1982), *Entailment. The Logic of Relevance and Necessity*, vol. II, Princeton, N.J.: Princeton University Press.
Anderson, C.A. (1984), *General Intensional Logic*, in Gabbay and Guenthner (1983-1989), vol. II: 355-86.
Anscombe, G.E.M. (1956), *Aristotle and the Sea Battle*, "Mind", 64: 1-15.
Aristotle, *The Complete Works* (ed. by J. Barnes), Princeton: Princeton University Press.
Aristotle, *The Basic Works* (ed. by R. McKeon), New York: Random House.
Armour-Garb, B. (2004), *Diagnosing Dialetheism*, in Priest, Beall, Armour-Garb (2004): 113-25.
Armour-Garb, B. and J.C. Beall (2003), *Minimalism and the Dialetheic Challenge*, "Australasian Journal of Philosophy", 81: 383-401.
Arruda, A.I. (1967), *Sur certaines Hiérarchies de calculs propositionnels*, "Comptes Rendus de l'Académie des Sciences de Paris": 641-4.
Arruda, A.I. (1989), *Aspects of the Historical Development of Paraconsistent Logic*, in Priest, Routley and Norman (1989): 99-130.
Arruda, A.I. and D. Batens (1982), *Russell's Set versus the Universal Set in Paraconsistent Set Theories*, "Logique et Analyse", 25: 121-36.
Asenjo, F.G. (1965), *Dialectic Logic*, "Logique et Analyse", 8: 321-6.
Asenjo, F.G. (1966), *A Calculus of Antinomies*, "Notre Dame Journal of Formal Logic", 16: 103-5.
Asenjo, F.G. (1971), *Logica dialectica*, "Teorema", 1: 7-13.
Asenjo, F.G. (1989), *Toward an Antinomic Mathematics*, in Priest, Routley and Norman (1989): 394-414.
Asenjo, F.G. and J. Tamburino (1975), *Logic of Antinomies*, "Notre Dame Journal of Formal Logic", 16: 272-8.
Ayer, A. (1984), *Philosophy in the 20th Century*, London: Allen & Unwin.
Barcan-Marcus, R. (1983), *Rationality and Believing the Impossible*, "Journal of Philosophy", 80: 321-38.

Bar-Hillel, Y. (1957), *New Light on the Liar*, "Analysis", 18: 1-6.
Bartley, W.W. III (1984) (ed.), *The Retreat to Commitment*, LaSalle, Ill.: Open Court.
Barwise, J. and J. Etchemendy (1987), *The Liar: an Essay on Truth and Circularity*, Oxford: Oxford University Press.
Barwise, J. and L. Moss (1996), *Vicious Circles*, Stanford, Calif.: CSLI Publications.
Barwise, J. and J. Perry (1983), *Situations and Attitudes*, Cambridge, Mass.: MIT Press.
Batens, D. (1980), *Paraconsistent Extensional Propositional Logic*, "Logique et Analyse", 90: 196-234.
Batens, D. (1986), *Dialectical Dynamics Within Formal Logics*, "Logique et Analyse", 114: 161-73.
Batens, D. (1989), *Dynamic Dialectical Logics*, in Priest, Routley and Norman (1989): 187-217.
Batens, D. (1990), *Against Global Paraconsistency*, "Studies in Soviet Thought", 39: 209-29.
Batens, D. (2000), *A Survey of Inconsistency-Adaptive Logics*, in Batens, Mortensen, Priest and van Bendegem (2000): 49-73.
Batens, D. (2001), *A Dynamic Characterization of the Pure Logic of Relevant Implication*, "Journal of Philosophical Logic", 30: 267-80.
Batens, D. and K. De Clercq (2004), *A Rich Paraconsistent Extension of Full Positive Logic*, "Logique et Analyse", 185-8: 227-57.
Batens, D., C. Mortensen, G. Priest and J.-P- van Bendegem, (2000) (eds.), *Frontiers of Paraconsistent Logic*, Baldock: Research Studies Press.
Beall, J.C. (2000a), *Is the Observable World Consistent?*, "Australasian Journal of Philosophy", 78: 113-8.
Beall, J.C. (2000b), *On Truthmakers for Negative Truths*, "Australasian Journal of Philosophy", 78: 264-8.
Beall, J.C. (2001a), *Dialetheism and the Probability of Contradictions*, "Australasian Journal of Philosophy", 79: 114-8.
Beall, J.C (2001b), *Speaking of Paradox*, unpublished MS.
Beall, J.C. (2001c), *Is Yablo's Paradox Non-Circular?*, "Analysis", 61: 176-87.
Beall, J.C (2004a), *At the Intersection of Truth and Falsity*, in Priest, Beall and Armour-Garb (2004): 1-19.
Beall, J.C. (2004b), *True and False – As If*, in Priest, Beall and Armour-Grab (2004): 197-216.
Beall, J.C. and G. Restall (2000), *Logical Pluralism*, "Australasian Journal of Philosophy", 78: 475-93.
Beall, J.C and M. Colyvan (2001), *Looking for Contradictions*, "Australasian Journal of Philosophy", 79: 564-9.
Beall, J.C. and B. van Fraassen (2003), *Possibilities and Paradox. An Introduction to Modal and Many-Valued Logic*, Oxford: Oxford University Press.

Bell, J. (1981), *Category Theory and the Foundations of Mathematics*, "British Journal for the Philosophy of Science", 32: 349-58.
Belnap, N.D. (1962), *Tink, Tonk and Plink*, "Analysis", 22: 130-4.
Belnap, N.D. (1977), *A Useful Four-Valued Logic*, in Dunn and Epstein (1977): 8-37.
Benardete, J.A. (1989), *Metaphysics*, Oxford: Oxford University Press.
Bendegem, J.-P. van (1994), *Strict Finitism as a Viable Alternative in the Foundations of Mathematics*, "Logique et Analyse", 37: 23-40.
Bendegem, J.-P. van (1999), *Why the Largest Number Imaginable Is Still a Finite Number*, "Logique et Analyse", 165-6: 107-26.
Benthem, J. van (1978), *Four Paradoxes*, "Journal of Philosophical Logic", 7: 49-72.
Benthem, J. van (1979), *What is Dialectical Logic?*, "Erkenntnis", 14: 333-47.
Benthem, J. van and K. Doets, (1983), *Higher Order Logic*, in Gabbay and Guenthner (1983-1989), vol. I: 275-330.
Bernays, P. and A. Fraenkel (1958), *Axiomatic Set Theory*, Amsterdam: North-Holland.
Béziau, J.Y. (1998), *Idempotent Full Paraconsistent Negations are Not Algebraizable*, "Notre Dame Journal of Formal Logic", 39: 135-9.
Berto, F. (2004), *Il primo teorema di Gödel e l'indeterminabilità del riferimento*, "Epistemologia", 27(2004): 29-54.
Berto, F. (2005), *Che cos'è la dialettica hegeliana? Un'interpretazione analitica del metodo*, Padova: Il Poligrafo.
Berto, F. (2006a), *Characterizing Negation to Face Dialetheism*, "Logique et Analyse", 195: 241-63..
Berto, F. (2006b), *Meaning, Metaphysics and Contradiction*, "American Philosophical Quarterly", 43: 283-97.
Berto, F. (2006c), *Teorie dell'assurdo. I rivali del Principio di Non-Contraddizione*, Roma: Carocci.
Berto, F. (2007), *Hegel's Dialectics as a Semantic Theory. An Analytic Reading*, "European Journal of Philosophy", forthcoming.
Berto, F. (2008), *Is Dialetheism an Idealism?*, "Dialectica", forthcoming.
Birkoff, G. and J. von Neumann (1936), *The Logic of Quantum Mechanics*, "Annals of Mathematics", 37: 823-43.
Blackburn, S. and K. Simmons (1993) (eds.), *Truth*, Oxford: Oxford University Press.
Bloesch, A. (1993), *A Tableau Style Proof System for Two Paraconsistent Logics*, "Notre Dame Journal of Formal Logic": 295-301.
Bobenrieth, A. (1998), *Five Philosophical Problems Related to Paraconsistent Logic*, "Logique et Analyse", 161: 21-30.
Bochvar, D.A. (1939), *Ob odnom tréhznacnom isčislenii I égo priménénii k analiza paradoxov klassičéskogo rässirénnogo funkcional'nogo isčisléniá*, "Matématičeskij sbornik", 4: 287-308.

Boolos, G.S., J.P. Burgess and R. Jeffrey (2002), *Computability and Logic. Fourth Edition*, Cambridge: Cambridge University Press.
Bradley, F.H. (1893), *Appearance and Reality*, Oxford: Clarendon Press.
Brady, R.T. (1984), *Depth Relevance of Some Paraconsistent Logics*, "Studia Logica", 43: 63-74.
Brady, R.T. (1989), *The Non-Triviality of Dialectical Set Theory*, in Priest, Routley and Norman (1989): 437-71.
Brady, R.T. (1996), *Relevant Implication and the Case for a Weaker Logic*, "Journal of Philosophical Logic", 25: 151-83.
Brady, R.T. (2000), *Entailment, Negation and Paradox Resolution*, in Batens, Mortensen, Priest and van Bendegem (2000): 113-36.
Brady, R.T. (2004), *On the Formalization of the Law of Non-Contradiction*, in Priest, Beall and Armour-Garb (2004): 41-7.
Brady, R.T. (2006), *Normalized Natural Deduction Systems for Some Relevant Logics I: the Logic DW*, "Journal of Symbolic Logic", 71: 35-66.
Brady, R.T. and R. Routley (1989), The Non-Triviality of Extensional Dialectical Set Theory, in Priest, Routley and Norman (1989): 415-36.
Brandom, R.B. (1985), *Varieties of Understanding*, in Rescher (1985): 27-51.
Brandom, R.B. (1994), *Making it Explicit*, Cambridge, Mass.: Harvard University Press.
Bremer, M. (1999), *Can Contradictions Be Asserted?*, "Logic and Logical Philosophy", 7: 167-77.
Bremer, M. (2005), *An Introduction to Paraconsistent Logics*, Frankfurt a.M.: Peter Lang.
Bremer, M., and D. Cohnitz (2004), *Information and Information Flow. An Introduction*, Frankfurt a.M.: Peter Lang.
Brody, B. (1967), *Logical Terms, Glossary of*, in Paul Edwards (ed.), *The Encyclopaedia of Philosophy*, Macmillan and Free Press.
Bromand, J. (2002), *Why Paraconsistent Logic Can Only Tell Half the Truth*, "Mind", 111: 741-9.
Brouwer, J.L.E. (1975), *Collected Works* (ed. by A. Heyting), Vol. I, Amsterdam: North-Holland.
Brown, B. (1999a), *Yes, Virginia, there Really Are Paraconsistent Logics*, "Journal of Philosophical Logic": 489-500.
Brown, B. (1999b), *Old Quantum Theory: a Paraconsistent Approach*, "Proceedings of the Philosophy of Science Association", 2: 397-441.
Brown, B. (2000), *Simple Natural Deduction for Weekly Aggregative Paraconsistent Logics*, in Batens, Mortensen, Priest and van Bendegem (2000): 137-48.
Brown, B. (2004), *Knowledge and Non-Contradiction*, in Priest, Beall and Armour-Garb (2004): 127-55.
Brown, B. and G. Priest (2004), *Chunk and Permeate, a Paraconsistent Inference Strategy. Part I: The Infinitesimal Calculus*, "Journal of Philosophical Logic", 33: 379-88.

Brown, B. and P.E. Schotch (1999), *Logic and Aggregation*, "Journal of Philosophical Logic", 28: 265-87.
Bull, R.A. and K. Segerberg (1984), *Basic Modal Logic*, in Gabbay and Guenthner (1983-1989), vol. II: 1-88.
Bunder, M.W. (1984), *Some Definitions of Negation Leading to Paraconsistent Logics*, "Studia logica", 43: 75-8.
Bunder, M.W. (1989), *Some Results in Some Subsystems and in an Extension of C_n*, "The Journal of Non-Classical Logic", 6: 45-56.
Burali-Forti, C. (1897), *Una questione sui numeri transfiniti*, "Rendiconti del circolo matematico di Palermo", 11: 154-64.
Burge, T. (1979), *Semantical Paradox*, "Journal of Philosophy", 76: 169-98.
Cantor, G. (1895), *Beiträge zur Begründung der transfiniten Mengenlehre*, "Mathematische Annalen", 46: 481-512, tr. in *Contributions to the Founding of the Theory of Transfinite Numbers*, New York: Dover 1955.
Cantor, G. (1899), *Cantor an Dedekind*, in van Heijenoort (1967): 113-7.
Cantor, G. (1932), *Gesammelte Abhandlungen mathematischen und philosophischen Inhalts* (ed. by E. Zermelo), Berlin: Springer.
Carnap, R. (1937), *The Logical Syntax of Language*, London: Routledge & Kegan Paul.
Carnap, R. (1947), *Meaning and Necessity*, Chicago: University of Chicago Press.
Carnielli, W.A., M. Coniglio and I. D'Ottaviano (2002) (eds.), *Paraconsistency. The Logical Way to the Inconsistent*, New York-Basel.
Carnielli, W.A. and L.P. de Alcantara (1984), *Paraconsistent Algebras*, "Studia logica", 43: 79-88.
Carnielli, W.A. and J. Marcos (2002), *A Taxonomy of C-Systems*, in Carnielli, Coniglio and D'Ottaviano (2002): 1-94.
Cartwright, R. (1971), *Identity and Substitutivity*, in Munitz (1971): 119-34.
Chellas, B.F. (1980), *Modal Logic: an Introduction*, Cambridge: Cambridge University Press.
Chihara, C. (1984), *Priest, the Liar, and Gödel*, "Journal of Philosophical Logic", 13: 117-24.
Church, A. (1939), *Review of* Bochvar (1939), "Journal of symbolic Logic", 4: 98-9.
Church, A. (1956), I*ntroduction to Mathematical Logic*, Princeton, N.J.: Princeton University Press.
Church, A. (1976), *Comparison of Russell's Resolution of the Semantical Antinomies with that of Tarski*, "Journal of Symbolic Logic", 41: 747-60.
Cohen, L.J. (1986), *The Dialogue of Reason: an Analysis of Analytical Philosophy*, Oxford: Clarendon Press.
Cole, P., and J.L. Morgan (1975) (eds.), *Syntax and Semantics, vol. III: Speech Acts*, New York: Academic Press.
Copi, I.M. and C. Cohen (1994), *Introduction to Logic*, Prentice Hall, Englewood Cliffs, N.J.: Prentice Hall.

Copeland, B.J. (1979), *On When a Semantics is Not a Semantics: Some Reasons for Disliking the Routley-Meyer Semantics for Relevance Logic*, "Journal of Philosophical Logic", 8: 399-413.

Copeland, B.J. (1980), *The Trouble Anderson and Belnap Have with Relevance*, "Philosophical Studies", 37: 325-34.

Copeland, B.J. (1986=, *What is a Semantics for Classical Negation?*, "Mind", 95: 478-90.

Costa, H.A. (2005), *Non-Adjunctive Inference and Classical Modalities*, "Journal of Philosophical Logic", 34: 581-605.

Da Costa, N.C.A. (1974), *On the Theory of Inconsistent Formal Systems*, "Notre Dame Journal of Formal Logic", 15: 497-510.

Da Costa, N.C.A. (1975), *Remarks on Jaskowki's Discussive Logic*, "Reports on Mathematical Logic": 7-16.

Da Costa, N.C.A. (1977), *A Semantical Analysis of the Calculi C_n*, "Notre Dame Journal of Formal Logic", 18: 621-30.

Da Costa, N.C.A. (1982), *The Philosophical Import of Paraconsistent Logic*, "The Journal of Non-Classical Logic", 1: 1-19.

Da Costa, N.C.A. (1986), *On Paraconsistent Set Theory*, "Logique et Analyse", 29: 361-71.

Da Costa, N.C.A. and E.H. Alves (1976), *Une sémantique pour le calcul C1*, "Comptes Rendus de l'Académie des Sciences de Paris": 729-31.

Da Costa, N.C.A, and L. Dubikajtis (1968), *Sur la logique discorsive de Jaskowski*, "Bulletin of the Polish Academy of Sciences", 16: 551-7.

Da Costa, N.C.A. and D. Marconi (1989), *An Overview of Paraconsistent Logic in the 80s*, "The Journal of Non-Classical Logic", 6: 5-23.

Da Costa, N.C.A. and R.G. Wolf (1980), *Studies in Paraconsistent Logic I: the Dialectical Principle of the Unity of Opposites*, "Philosophia", 9: 187-217.

Dalen, D. van (1986), *Intuitionistic Logic*, in Gabbay and Guenthner (1983-1989), vol. III: 225-339.

Davidson, D. (1984), *Inquiries into Truth and Interpretation*, Oxford: Oxford University Press.

Davis, M. (1965) (ed.), *The Undecidable*, New York: The Raven Press.

De Morgan, A. (1846), *On the Syllogism: I: On the Structure of the Syllogism*, in *On The Syllogism and Other Logical Writings* (ed. by P. Heath), London: Routledge & Kegan Paul: 1966.

Dennett, D. (1987), *The Intentional Stance*, Cambridge, Mass.: MIT Press.

Denyer, N. (1989), *Dialetheism and Trivialization*, "Mind", 98: 261-3.

Denyer, N. (1995), *Priest's Paraconsistent Arithmetic*, "Mind", 104: 567-75.

Detlefsen, M. (1979), *On Interpreting Gödel's Second Theorem*, "Journal of Philosophical Logic", 8: 297-313.

Diaz, M.R. (1981), *Topics in the Logic of Relevance*, München: Philosophia Verlag.

Dummett, M. (1973a), *The Justification of Deduction*, "Proceedings of the British Academy", 59.

Dummett, M. (1973b), *Frege. Philosophy of Language*, London: Duckworth.
Dummett, M. (1977), *Elements of Intuitionism*, Oxford: Clarendon Press.
Dummett, M. (1978), *Truth and Other Enigmas*, London: Duckworth.
Dummett, M. (1991), *The Logical Basis of Metaphysics*, London: Duckworth.
Dunn, J.M. (1976), *Intuitive Semantics for First-Degree Entailment and 'Coupled Trees'*, "Philosophical Studies", 29: 149-68.
Dunn, J.M. (1979), *A Theorem in 3-Valued Model Theory with Connections to Number Theory, Type Theory and Relevant Logic*, "Studia Logica", 38: 149-69.
Dunn, J.M. (1986), *Relevance Logic and Entailment*, in Gabbay and Guenthner (1983-1989), vol. III: 117-224.
Dunn, J.M. (1988), *The Impossibility of Certain Second-Order Non-Classical Logics with Extensionality*, in D.F. Austin (1988) (ed.), *Philosophical Analysis*, Dordrecht: Kluwer: 261-79.
Dunn, J.M. (1996), *Generalized Ortho Negation*, in Wansing (1996): 3-26
Dunn, J.M and G. Epstein (1977) (eds.), *Modern Uses of Multiple-Valued Logics*, Dordrecht: Reidel.
Dunn, J.M. and G. Restall (2002), *Relevance Logic*, in Gabbay and Guenthner (2002): 1-128.
Epstein, R.L. (2005), *Paraconsistent Logics with Simple Semantics*, "Logique et Analyse", 189-92: 71-86.
Esser, O. (2003), *A Strong Model of Paraconsistent Logic*, "Notre Dame Journal of Formal Logic", 44: 149-56.
Evans, G. (1978), *Can There Be Vague Objects?*, "Analysis", 38: 208.
Everett, A. (1993), *A Note on Priest's "Hypercontradictions"*, "Logique et Analyse", 141-2: 39-43.
Everett, A. (1994), *Absorbing Dialetheia?*, "Mind", 103: 413-9.
Everett, A. (1996), *A Dilemma for Priest's Dialethism?*, "Australasian Journal of Philosophy", 74: 657-68.
Feferman, S. (1977), *Categorical Foundations and Foundations of Category Theory*, in Hintikka and Butts (1977): 149-69.
Feferman, S. (1983), *Kurt Gödel: Conviction and Caution*, in Shanker (1988).
Fidel, M. (1977), *The Decidability of the Calculi C_n*, "Reports on Mathematical Logic", 8: 31-40.
Field, H. (2002), *Saving the Truth Schema from Paradox*, "Journal of Philosophical Logic", 31: 1-27.
Fine, K. (1974), *Models for Entailment*, "Journal of Philosophical Logic", 3: 347-72.
Fine, K. (1975), *Vagueness, Truth and Logic*, "Synthese", 30: 265-300.
Fitch, F.B. (1946), *Self-Reference in Philosophy*, "Mind", 55: 64-73.
Fitch, F.B. (1953), *Self-referential Relations*, in *Actes du Xième Congrès International de Philosophie*, 14, Amsterdam: North-Holland.
Foley, R. (1979), *Justified Inconsistent Beliefs*, "American Philosophical Quarterly", 16: 247-57.

Foley, R. (1986), *Is it Possible to have Contradictory Beliefs?*, "Midwest Studies in Philosophy", 10: 327-55.
Fraassen, B. van (1966), *Singular Terms, Truth-Value Gaps, and Free Logic*, "Journal of Philosophy", 63: 481-95.
Fraassen, B. van (1968), *Presuppositions, Implication and Self Reference*, "Journal of Philosophy", 65: 136-51.
Fraenkel, A., Y. Bar Hillel and A. Levy (1973), *Foundations of Set Theory*, Amsterdam: North-Holland.
Frege, G. (1879), *Begriffsschrift, eine der arithmetischen nachgebildete Formelsprache des reinen Denkens*, Halle: Nebert, tr. in van Heijenoort (1967): 1-82.
Frege, G. (1884), *Die Grundlagen der Arithmetik. Eine logisch-mathematische Untersuchung über den Begriff der Zahl*, Brealsu: Koebner.
Frege, G. (1903), *Grundgesetze der Arithmetik. Begriffsschriftlich abgeleitet*, voll. I-II, Jena: Pohle.
Frege, G. (1976), *Wissenschaftlicher Briefwechsel* (ed. by G. Gabriel and H. Hermes), Hamburg: Meiner.
Gabbay, D. and F. Guenthner (1983-1989) (eds.), *Handbook of Philosophical Logic*, voll. I-IV, Dordrecht: Kluwer.
Gabbay, D. and F. Guenthner (2002) (eds.), *Handbook of Philosophical Logic. 2nd Edition*, vol. VI, Dordrecht: Kluwer.
Gabbay, D. and H. Wansing (1999) (eds.), *What is negation?*, Dordrecht: Kluwer.
Garson, J.W. (1984), *Quantification in Modal Logic*, in Gabbay e Guenthner (1983-1989), vol. II: 249-308.
Geach, P.T. (1962), *Reference and Generality. An Examination of Some Medieval and Modern Theories*, Ithaca, N.Y.: Cornell University Press.
Geach, P.T. (1965), *Assertion*, "Philosophical Review", 74: 449-65.
Geach, P.T. (1968), *Identity*, "Review of Metaphysics", 21: 3-12.
Goble, L. (2000), *An Incomplete Relevant Modal Logic*, "Journal of Philosophical Logic", 29: 103-19.
Goble, L. (2006), *Paraconsistent Modal Logic*, "Logique et Analyse", 193: 3-29.
Gödel, K. (1930), *Die Vollständigkeit der Axiome des logischen Funktionenkalküls*, "Monatshefte für Mathematik und Physik", 37: 349-60, tr. *The completeness of the Axioms of the Functional Calculus of Logic*, in van Heijenoort (1967): 582-91.
Gödel, K. (1931), *Über formal unentscheidbare Sätze der* Principia mathematica *und verwandter Systeme I*, "Monatshefte für Mathematik und Physik", 38: 173-98, tr. *On Formally Undecidable Propositions of* Principia mathematica *and Related Systems I*, in van Heijenoort (1967): 596-617.
Gödel, K. (1944), *Russell's Mathematical Logic*, in P.A. Schlipp (1944) (ed.), *The Philosophy of Bertrand Russell*, Evanston, Ill.: Northwestern University Press: 125-53.

Gödel, K. (1947), *What is Cantor's Continuum Problem?*, "American Mathematical Monthly", 54: 515-25.
Goldblatt, R.I. (1974), *Semantic Analysis of Orthologic*, "Journal of Philosophical Logic", 3: 19-35.
Goldstein, L. (2004), *The Barber, Russell's Paradox, Catch-22, God and More: A Defence of a Wittgensteinian Conception of Contradiction*, in Priest, Beall and Armour-Garb (2004): 295-313.
Goodman, N. (1978), *Ways of Worldmaking*, Indianapolis: Hackett.
Goodship, L. (1996), *On Dialethism*, "Australasian Journal of Philosophy", 74: 153-61.
Grant, J. (1975), *Inconsistent and Incomplete Logics*, "Mathematics Magazine", 48: 154-9.
Grant, J. (1978), *Classifications for Inconsistent Theories*, "Notre Dame Journal of Formal Logic", 19: 435-44.
Grattan-Guinness, I. (1998), *Structural Similarity or Structuralism*, "Mind", 107: 823-34.
Grice, H.P. (1975), *Logic and Conversation*, in Cole e Morgan (1975): 41-58.
Grim, P. (1991), *The Incomplete Universe: Totality, Knowledge and Truth*, Bradford Books, MIT Press.
Grim, P. (2004), *What is a Contradiction?*, in Priest, Beall and Armour-Garb (2004): 49-72.
Guccione, S. (1980), *A Semantics for a Dialectical Logic*, "Logique et Analyse", 23: 461-70.
Haack, S. (1978), *Philosophy of Logics*, Cambridge: Cambridge University Press.
Hahn, L.E. and P.A. Schlipp (1986), *The Philosophy of W.V. Quine*, La Salle, Ill.: Open Court.
Hallett, M. (1984), *Cantorian Set Theory and Limitation of Size*, Oxford: Clarendon Press.
Halmos, P.R. (1960), *Naïve Set Theory*, Princeton, N.J.: van Nostrand.
Havas, K. (1990), *Dialectic and Inconsistency in Knowledge Acquisition*, "Studies in Soviet Thought", 39: 189-98.
Hegel, G.W.F. (1830), *Enzyklopädie der der philosophischen Wissenschaften in Grundrisse. Mit Erläuterungen und Zusätzen versehen von L. von Henning, K.L. Michelet und L. Boumann*, in *Werke in zwanzig Bände*, hrg. von E. Moldenhauer und K.M. Michel, Bände 8-10, Frankfurt a.M.: Suhrkamp, 1970; tr. *The Encyclopaedia Logic (with the Zusätze)*, Indianaplois: Hackett, 1991.
Hegel, G.W.F. (1831), *Wissenschaft der Logik*, vols. 11 (1978) and 12 (1981) of *Gesammelte Werke*, in Verbindung mit der Deutschen Forschungsgemeinschaft, hrg. von der Rheinisch-Westfälischen Akademie der Wissenschaften, Hamburg: Meiner, 1968ff; tr. *Hegel's Science of Logic*, New York: Humanity Books, 1969.
Heijenoort, J. van (1967) (ed.), *From Frege to Gödel. A Source Book in Mathematical Logic*, Harvard: Harvard University Press.
Heintz, J. (1979), *Reference and Inference in Fiction*, "Poetics", 8: 85-99.

Henkin, L. (1947), *The Completeness of Formal Systems* (tesi di dottorato), Princeton.
Henkin, L. (1949), *The Completeness of the First Order Functional Calculus*, "Journal of Symbolic Logic", 14: 159-66.
Henkin, L. (1950), *Completeness in the Theory of Types*, "Journal of Symbolic Logic", 15: 81-91.
Herzberger, H.G. (1970), *Paradoxes of Grounding in Semantics*, "Journal of Philosophy", 67: 145-67.
Heyting, A. (1956), *Intuitionism*, Amsterdam: North-Holland.
Hilbert, D. (1904), *Über die Grundlegung der Logik und der Arithmetik*, in *Verhandlungen des Dritten Internationalen Mathematiker-Kongress in Heidelberg vom 8. bis 13. August 1904*, Leipzig: Teubner: 174-185, tr. *On the Foundations of Logic and Arithmetic*, in van Heijenoort (1967): 129-38.
Hilbert, D. (1925), *Über das Unendliche*, "Jahresbericht der Deutschen Mathematiker-Vereinigung", 36, s. I: 201-15, tr. *On the Infinite*, in van Heijenoort (1967): 367-92.
Hintikka, J. (1969), *Models for Modalities*, Dordrecht: Reidel.
Hintikka, J. (1975), *Impossible Possible Worlds Vindicated*, "Journal of Philosophical Logic", 4: 475-84.
Hintikka, J. and R. Butts (1977) (eds.), *Logic, Foundations of Mathematics and Computability Theory*, Dordrecht: Reidel.
Hirsch, E. (1982), *The Concept of Identity*, Oxford: Oxford University Press.
Hodges, W. (1983), *Elementary Predicate Logic*, in Gabbay and Guenthner (1983-1989), vol. I: 1-132.
Hofstadter, D. (1979), *Gödel, Escher, Bach: an Eternal Golden Braid*, New York: Basic Books.
Horn, L.R. (1989), *A Natural History of Negation*, Chicago: University of Chicago Press.
Horwich, P. (1990), *Truth*, Oxford: Blackwell; 2nd edition, Oxford: Oxford University Press, 1998.
Howe, M. (1970), *Introduction to Human Memory*, New York: Harper & Row.
Hughes, G.E. and M.J. Cresswell (1968), *An Introduction to Modal Logic*, London: Methuen.
Hughes, G.E. and M.J. Cresswell (1984), *A Companion to Modal Logic*, London: Methuen.
Husserl, E. (1900), *Logische Untersuchungen*, vol. I, Halle, tr. *Logical Investigations*, London, 1970.
Hyde, D. (1997), *From Heaps and Gaps to Heaps of Gluts*, "Mind", 106: 640-60.
Irwin, T. (1977), *Aristotle's Discovery of Metahpysics*, "Review of Metaphysics", 31: 210-29.
Jaskowski, S. (1948), *Rachunek zdan dla systemow dedukcyjnych sprzecznychi*, "Studia Societatis Scientiarum Torunensis", sectio A, I, 5: 55-77, tr. *Propositional Calculus for Contradictory Deductive Systems*, "Studia Logica", 24:

143-57, and *O konjunkcji dyskusyjnej w rachunku zdan dla systmów dedukcyjnych sprzecznych*, "Studia Societatis Scientiarum Torunensis", sectio A, I, 8: 171-2.
Johansson, I. (1936), *Der Minimalkalkül, ein reduzierter intuitionistischer Formalismus*, "Compositio mathematica", 4: 119-36.
Kahane, H. (1995), *Logic and Contemporary Rhetoric*, Belmont, CA: Wadsworth.
Kalish, D., R. Montague and G. Mar (1980), *Logic: Techniques of Formal Reasoning*, New York: Harcourt Brace Jovanovich.
Kant, I. (1781), *Kritik der reinen Vernunft*, vols. 3 and 4 (1911) of *Gesammelte Schriften*, hrg. von der Königlich Preu☐ischen Akademie der Wissenschaften, Berlin: de Gruyter & Co., 1969; tr. *Critique of Pure Reason*, New York: Palgrave Macmillan, 2003.
Katzley, A. (1975), *Human Memory: Structures and Processes*, San Francisco: W.H. Freeman.
Keefe, R. (2000), *Theories of Vagueness*, Cambridge: Cambridge University Press.
Keefe, R. and P. Smith (1996) (eds.), *Vagueness. A reader*, Cambridge, Mass.: MIT Press.
Kielkopf, C.F. (1975), *Adjunction and Paradoxical Derivations*, "Analysis", 35: 127-9.
Kirkham, R.L. (1992), *Theories of Truth. A Critical Introduction*, Cambridge, Mass.: MIT Press.
Kirwan, C. (1978), *Logic and Argument*, London: Duckworth.
Kirwan, C. (1993), *Aristotle: Metaphysics, Books Γ, Δ, E*, 2nd edition, Oxford: Oxford University Press.
Kleene, S. (1952), *Introduction to Metamathematics*, Amsterdam: North-Holland.
Kleene, S. (1976), *The work of Kurt Gödel*, "Journal of Symbolic Logic", 41: 761-78.
Kotas, J. (1971), *On the Algebras of Classes of Formulae of Jaskowski's Discussive System*, "Studia Logica": 81-90.
Kotas, J. (1975), *Discussive Sentential Calculus of Jaskowski*, "Studia Logica": 149-68.
Kotas, J. and N.C.A. da Costa (1979), *A New Formulation of Discussive Logic*, "Studia Logica", 38: 429-44.
Kripke, S. (1959), *A Completeness Theorem in Modal Logic*, "Journal of Symbolic Logic", 24: 1-14.
Kripke, S (1963a), *Semantical Analysis of Intuitionistic Logic I*, in J. Crossley and M. Dummett (1963) (eds.) *Formal Systems and Recursive Functions*, Amsterdam: North-Holland: 92-129.
Kripke, S. (1963b), *Semantical Analysis of Modal Logic I: Normal Modal Propositional Calculi*, "Zeitschrift für mathematische Logik und Grundlagen der Mathematik", 9: 67-96.

Kripke, S. (1965), *Semantical Analysis of Modal Logic II: Non-normal Modal Propositional Calculi*, in Addison et al. (1965) (eds.) *The Theory of Models*, Amsterdam: North-Holland: 206-20.

Kripke, S. (1971), *Identity and Necessity*, in Munitz (1971) (ed.): 135-64.

Kripke, S. (1972), *Naming and Necessity*, in D. Davidson and G. Harman (1972) (eds.), *Semantics of Natural Language*, Dordrecht: Reidel: 253-355, repr. Oxford: Basil Blackwell, 1980.

Kripke, S. (1975), *Outline of a Theory of Truth*, "Journal of Philosophy", 72: 690-716, repr. in Martin (1984): 53-81.

Kroon, F. (2004), *Realism and Dialetheism*, in Priest, Beall and Armour-Garb (2004): 245-63.

Kyburg, H.E. (1997), *The Rule of Adjunction and Reasonable Inference*, "Journal of Philosophy", 94: 109-25.

Lakatos, I. (1970), *Falsification and the Methodology of Scientific Research Programs*, Ch. 1 of ID. *Collected Papers* (ed. by J. Worrall and G. Currie), Cambridge: Cambridge University Press, 1978.

Lance, M. (1988), *Normative inferential Vocabulary: The Explicitation of Social Linguistic Practice*, Ph.D. thesis, University of Pittsburgh.

Lear, J. (1977), *Sets and Semantics*, "Journal of Philosophy", 74: 86-102.

Leblanc, H. (1973) (ed.), *Truth, Syntax and Semantics*, Amsterdam: North-Holland.

Lemmon, E.J. (1965), *Beginning Logic*, London: Th. Nelson & Sons and van Nostrand Reinhold.

Lenzen, W. (1996), *Necessary Conditions for Negation Operators*, in Wansing (1996): 37-58.

Levin, G.D. (1982), *Dialectics and the Paradoxes of Set Theory*, "Soviet Studies in Philosophy", 24: 26-45.

Lewis, C.I. (1917), *The Issues Concerning Material Implication*, "Journal of Philosophy, Psychology, and Scientific Methods", 14: 350-6.

Lewis, C.I. and C.H. Langford (1932), *Symbolic Logic*, New York and London: The Century Co.

Lewis, D. (1973), *Counterfactuals*, Cambridge, Mass.: Harvard University Press.

Lewis, D. (1978), *Truth in Fiction*, "American Philosophical Quarterly", 15: 37-46, repr. in Lewis (1983): 261-75.

Lewis, D. (1982), *Logic for Equivocators*, "Nous", 16: 431-41, repr. in Lewis (1998): 97-110.

Lewis, D. (1983), *Philosophical Papers*, vol. I, Oxford: Oxford University Press.

Lewis, D. (1986), *On the Plurality of Worlds*, Oxford: Basil Blackwell.

Lewis, D. (1998), *Papers in Philosophical Logic*, Cambridge: Cambridge University Press.

Lewis, D. (2004), *Letters to Beall and Priest*, in Priest, Beall and Armour-Garb (2004): 176-7.

Lindsay, P. and D. Norman (1977), *Human Information Processing*, New York: Academic Press.
Littmann, G. and K. Simmons (2004), *A Critique of Dialetheism*, in Priest, Beall and Armour-Garb (2004): 314-35.
Łukasiewicz, J. (1910a), *O zasadzie sprzeczności u Arystotelesa,* Cracow: Studium Krytyczne.
Łukasiewicz, J. (1910b), *Über den Satz des Widerspruchs bei Aristoteles*, "Bulletin International de l'Académie des Sciences de Cracovie": 15-38. tr. *Aristotle on the Law of Contradiction*, in J, Barnes, M. Schofield and R Sorabij (1979) (eds.), *Articles on Aristotle, iii. Metaphysics*, London: Duckworth, 1979.
Łukasiewicz, J. (1957), *Aristotle's Syllogistic from the Standpoint of Modern Formal Logic*, 2nd edition, Oxford: Clarendon Press.
Mackie, J.L. (1973), *Truth, Probability, and Paradox*, Oxford: Oxford University Press.
Marconi, D. (1979) (ed.), *La formalizzazione della dialettica. Hegel, Marx e la logica contemporanea*, Torino: Rosenberg & Sellier.
Marconi, D. (1981), *Types of Non-Scotian Logic*, "Logique et Analyse", 96: 407-14.
Marconi, D. (1984), *Wittgenstein on Contradiction and the Philosophy of Paraconsistent Logic*, "History of Philosophy Quarterly", 1: 333-52.
Marconi, D. (1997), *Lexical Competence*, Cambridge, Mass.: MIT Press.
Mares, E. (1996), *Relevant Logic and the Theory of Information*, "Synthese", 109: 345-60.
Mares, E. (2000), *Even Dialetheists Should Hate Contradictions*, "Australasian Journal of Philosophy", 78: 503-16.
Mares, E. (2004), *Semantic Dialetheism*, in Priest, Beall and Armour-Garb (2004): 264-75.
Mares, E. and R. Goldblatt (2006), *An Alternative Semantics for Quantified Relevant Logic*, "Journal of Symbolic Logic", 71: 163-87.
Martin, G. (1955), *Kant's Metaphysics and Theory of Science*, Manchester: Manchester University Press.
Martin, R.M. (1967), *Towards a Solution to the Liar Paradox*, "Philosophical Review", 76: 279-311.
Martin, R.M. (1984) (ed.), *Recent Essays on Truth and the Liar Paradox*, Oxford: Oxford University Press.
Mayberry, J. (1977), *On the Consistency Problem for Set Theory: an Essay on the Cantorian Foundations of Classical Mathematics (I)*, "British Journal for the Philosophy of Science", 28: 1-34.
McTaggart, J.M.E. (1922), *Studies in the Hegelean Dialectic*, 2nd edition, Cambridge: Cambridge University Press.
Meheus, J. (2002) (ed.), *Inconsistency in Science*, Dordrecht; Kluwer.
Meheus, J. (2003), *Paraconsistent Compatibility*, "Logique et Analyse", 183-4: 251-87.

Meyer, R.K. (1974), *Entailment is Not Strict Implication*, "Austraslasian Journal of Philosophy", 52: 212-31.
Meyer, R.K. (1976), *Relevant Arithmetic*, "Bulletin of the Section of Logic of the Polish Academy of Science": 133-7.
Meyer, R.K. and E.P. Martin (1986), *Logic on the Australian Plan*, "Journal of Philosophical Logic", 15: 305-32.
Meyer, R.K. and C. Mortensen (1984), *Inconsistent Models for Relevant Arithmetic*, "Journal of Symbolic Logic", 49: 917-29.
Meyer, R.K., R. Routley and J.M. Dunn (1979), *Curry's Paradox*, "Analysis", 39: 124-8.
Miró Quesada, F. (1989), *Paraconsistent Logic: Some Philosophical Issues*, in Priest, Routley and Norman (1989): 627-52.
Moore, A.W. (1985), *Set Theory, Skolem's Paradox, and the* Tractatus, "Analysis", 45: 13-20.
Moore, A.W. (1990), *The Infinite*, London: Routledge & Kegan Paul.
Mortensen, C. (1980), *Every Quotient Algebra for C_1 is Trivial*, "Notre Dame Journal of Formal Logic", 21: 694-700.
Mortensen, C. (1983), *The Validity of the Disjunctive Syllogism is Not so Easily Proved*, "Notre Dame Journal of Formal Logic", 24: 35-40.
Mortensen, C. (1989), *Anything is Possible*, "Erkenntnis", 30: 319-37.
Mortensen, C. (1995), *Inconsistent Mathematics*, Dordrecht: Kluwer.
Munitz, M.K. (1971) (ed.), *Identity and Individuation*, New York: New York University Press.
Nagel, E. and J.R. Newman (1958), *Gödel's Proof*, New York: New York University Press.
Neumann, J. von (1925), *Eine Axiomatisierung der Mengenlehre*, "Journal für die reine und angewandte Mathematik", 154: 219-40, tr. *An Axiomatization of Set Theory*, in van Heijenoort (1967): 393-413.
Norman, J. and R. Sylvan (1989) (eds.), *Directions in Relevant Logic*, Dordrecht: Kluwer.
Orwell, G. (1949), *Nineteen Eighty-Four,* New York: Everyman's Library.
Paoli, F. (2003), *Quine and Slater on Paraconsistency and Deviance*, "Journal of Philosophical Logic", 32: 531-48.
Paris, J.B. and N. Pathmanathan (2006), *A Note on Priest's Finite Inconsistent Arithmetics*, "Journal of Philosophical Logic", 35: 529-37.
Parsons, C. (1974), *Informal Axiomatization, Formalization, and the Concept of Truth*, "Synthese", 27: 27-47.
Parsons, T. (1984), *Assertion, Denial and the Liar Paradox*, "Journal of Philosophical Logic", 13: 137-52.
Parsons, T. (1990), *True Contradictions*, "Canadian Journal of Philosophy", 20: 335-54.
Parsons, T. and P. Woodruff (1995), *Worldly Indeterminacy of Identity*, "Proceedings of the Aristotelian Society", 95: 171-91.

Pasniczek, J. (1998), *Beyond Consistent and Complete Possibile Worlds*, "Logique et Analyse", 161: 121-34.

Peacocke, C. (1987), *Understanding Logical Constants: a realist's Account*, "Proceedings of the British Academy", 73: 153-200.

Pitcher, G. (1964) (ed.), *Truth*, Englewood Cliffs, N.J.: Prentice Hall.

Pizzi, C. (1987), *Dalla logica della rilevanza alla logica condizionale*, Roma: Euroma.

Plantinga, A. (1974), *The Nature of Necessity*, Oxford: Oxford University Press.

Popper, K.R. (1969), *Conjectures and Refutations*, London: Routledge and Kegan Paul.

Potter, M. (2004), *Set Theory and its Philosophy*, Oxford: Oxford University Press.

Prawitz, D. (1965), *Natural Deduction. A Proof-Theoretical Study*, Uppsala: Almqvist & Wilksell.

Price, H. (1990), *Why 'Not'?*, "Mind", 99: 221-38.

Priest, G. (1978), *Classical Logic aufgehoben* in Priest, Routley and Norman (1989): 131-48.

Priest, G. (1979), *The Logic of Paradox*, "Journal of Philosophical Logic", 8: 219-41.

Priest, G. (1982), *To Be and Not To Be: Dialectical Tense Logic*, "Studia Logica", 41: 249-68.

Priest, G. (1984a), *Logic of Paradox Revisited*, "Journal of Philosophical Logic", 13: 153-79.

Priest, G. (1984b), *Hyper-contradictions*, "Logique et Analyse", 107: 237-43.

Priest, G. (1987), *In Contradiction: A Study of the Transconsistent*, Dordrecht: Martinus Nijhoff, 2nd expanded edition, Oxford: Oxford University Press, 2006.

Priest, G. (1989), *Reductio ad Absurdum et Modus Tollendo Ponens* in Priest, Routley and Norman (1989): 613-26.

Priest, G. (1990), *Boolean Negation and All That*, "Journal of Philosophical Logic", 19: 201-15.

Priest, G. (1991), *Minimally Inconsistent LP*, "Studia Logica", 50: 321-31.

Priest, G. (1992), *What is a Non-Normal World?*, "Logique et Analyse", 35: 291-302.

Priest, G. (1993a), *Can Contradictions Be True? II*, "Proceedings of the Aristotelian Society", suppl. vol. 67: 35-54.

Priest, G. (1993b), *Another Disguise of the Same Fundamental Problems: Barwise and Etchemendy on the Liar*, "Australasian Journal of Philosophy", 71: 60-9.

Priest, G. (1994), *Is Arithmetic Consistent?*, "Mind", 103: 337-49.

Priest, G. (1995), *Beyond the Limits of Thought*, Cambridge: Cambridge University Press, 2nd extended edition, Oxford: Oxford University Press, 2002.

Priest, G. (1996), *Everett's Trilogy*, "Mind", 105: 631-47.

Priest, G. (1997a), *Inconsistent Models of Arithmetic – Part I: Finite Models*, "Journal of Philosophical Logic", 26: 223-35.

Priest, G. (1997b), *Yablo's Paradox*, "Analysis", 57: 236-42.
Priest, G. (1997c), *Sylvan's Box*, "notre Dame Journal of Formal Logic", 38: 573-82.
Priest, G. (1998), *What is so Bad about Contradictions?*, "Journal of Philosophy", 8: 410-26.
Priest, G. (1999a), *Perceiving Contradictions*, "Australasian Journal of Philosophy", 77: 439-46.
Priest, G. (1999b), *Could Everything Be True?*, "Australasian Journal of Philosophy", 78: 189-95.
Priest, G. (1999c), *What Not? A Defence of a Dialetheic Account of Negation*, in Gabbay and Wansing (1999): 101-20.
Priest, G. (2000a), *Truth and Contradiction*, "Philosophical Quarterly", 50: 305-19.
Priest, G. (2000b), *Motivations for Paraconsistency: the Slippery Slope from Classical Logic to Dialetheism*, in Batens, Mortensen, Priest and van Bendegem (2000).
Priest, G. (2000c), *Could Everything be True?*, "Australasian Journal of Philosophy", 78: 189-95.
Priest, G. (2001), *An Introduction to Non-Classical Logic*, Cambridge: Cambridge University Press.
Priest, G.. (2002a), *Rational Dilemmas*, "Analysis", 62: 11-6.
Priest, G. (2002b), *Inconsistency and the Empirical Sciences*, in Meheus (2002): 119-28.
Priest. G. (2002c), *Paraconsistent Logic*, in Gabbay and Guenthner (2002): 287-393.
Priest, G.. (2006), *Doubt Truth to Be a Liar*, Oxford: Oxford University Press.
Priest, G., J.C. Beall and B. Armour-Garb (2004) (eds.), *The Law of Non-Contradiction. New Philosophical Essays*, Oxford: Oxford University Press.
Priest, G. and R. Routley (1982), *Lessons from Pseudo-Scotus*, "Philosophical Studies", 42: 189-99.
Priest, G. and R. Routley (1989a), *An Outline of the History of (Logical) Dialectic*, in Priest, Routley and Norman (1989): 76-98.
Priest, G. and R. Routley (1989b), *Applications of Paraconsistent Logics*, in Priest, Routley and Norman (1989): 367-93.
Priest, G. and R. Routley (1989c), *Systems of Paraconsistent Logics*, in Priest, Routley and Norman (1989): 151-86.
Priest, G. and R. Routley. (1989d), *The Philosophical Significance and Inevitability of Paraconsistency*, in Priest, Routley and Norman (1989): 483-539.
Priest, G., R. Routley and J. Norman (1989) (eds.), *Paraconsistent Logic. Essays on the Inconsistent*, München: Philosophia Verlag.
Prior, A.N. (1962), *Formal Logic*, Oxford: Oxford University Press.
Prior, A.N. (1967), *Negation*, in Paul Edwards (1967) (ed.), *The Encyclopedia of Philosophy*, Macmillan and Free Press.

Quine, W.V.O. (1937), *New Foundations for Mathematical Logic*, "American Mathematical Monthly", 44, repr. with additions in Quine (1953): 80-101

Quine, W.V.O. (1951). *Mathematical Logic*, New York: Harper & Row.

Quine, W.V.O. (1953), *From a Logical Point of View*, Cambridge, Mass.: Harvard University Press.

Quine, W.V.O. (1960), *Word and Object*, Cambridge, Mass.: MIT Press.

Quine, W.V.O. (1963), *Set Theory and its Logic*, Cambridge, Mass.: Harvard University Press.

Quine, W.V.O. (1970), *Philosophy of Logic*, Englewood Cliffs, N.J.: Prentice Hall.

Quine, W.V.O. (1976), *The Ways of Paradox and Other Essays*, Cambridge: Harvard University Press.

Raggio, A.R. (1968), *Propositional Sequence-Calculi for Inconsistent Systems*, "Notre Dame Journal of Formal Logic": 359-66.

Raju, P.T. (1954), *The Principle of Four-Cornered Negation in Indian Philosophy*, "Review of Metaphysics", 7: 694-713.

Ramsey, F.P. (1931), *The Foundations of Mathematics and Other Logical Essays*, London: Routledge & Kegan Paul.

Read, S. (1988), *Relevant Logic*, London: Basil Blackwell.

Reid, T. (1974). *An Inquiry into the Human Mind: On the Principle of Common Sense*, ed. D.R. Brooks, University Park, Penn.: Pennsylvania State University Press, 1997.

Rescher, N. (1973), *The Coherence Theory of Truth*, Oxford: Oxford University Press.

Rescher, N. (1979), *Non-standard Possible Worlds*, in Marconi (1979): 354-416.

Rescher, N. (1985) (ed.), *Reason and Rationality in Natural Science*, Lanham: University Press of America.

Rescher, N. (2001), *Paradoxes: Their Roots, Range, and Resolution*, Chicago: Open Court.

Rescher, N. and R.B. Brandom (1980), *The Logic of Inconsistency: a Study in Non-standard Possible Worlds Semantics and Ontology*, Oxford: Basil Blackwell.

Rescher, N. and R. Manor (1970), *On Inference from Inconsistent Premisses*, "Theory and Decision", 1: 179-217.

Resnik, M.D. (1974), *On The Philosophical Significance of Consistency Proofs*, "Journal of Symbolic Logic", 3: 133-47.

Resnik, M.D. (2004), *Revising Logic*, in Priest, Beall and Armour-Garb (2004): 178-93.

Restall, G. (1992), *A Note on Naive Set Theory in LP*, "Notre Dame Journal of Formal Logic", 33: 422-32.

Restall, G. (1995), *Four-Valued Semantics for Relevant Logics (and Some of Their Rivals)*, "Journal of Philosophical Logic", 24: 139-60.

Restall, G. (1997), *Ways Things Can't Be*, "Notre Dame Journal of Formal Logic", 39: 583-596.

Restall, G. (1999), *Negation in Relevant Logics (How I stopped worrying and learned to love the Routley Star)*, in Gabbay nd Wansing (1999): 53-76.
Restall, G. (2002), *Paraconsistency Everywhere*, "Notre Dame Journal of Formal Logic", 43: 147-56.
Restall, G. (2004a), *Laws of Non-Contradiction, Laws of the Excluded Middle, and Logics*, in Priest, Beall and Armour-Garb (2004): 73-84.
Restall, G. (2004b), *Assertion, Denial, Accepting, Rejecting, Symmetry, Paradox, and All That* (paper presented at the annual meeting of the Australasian Association of Philosophy), http://consequently.org/writing/assertiondenialparadox.
Riche, J. (1998), *Finitization Procedures and Finite Model Property*, "Logique et Analyse", 161-3: 155-66.
Rogerson, S. and G. Restall (2004), *Routes to Triviality*, "Journal of Philosophical Logic", 33: 421-36.
Rosser, B. (1936), *Extension of Some Theorems of Gödel and Church*, "Journal of Symbolic Logic", 1: 87-91.
Rosser, B. (1942), *The Burali-Forti Paradox*, "Journal of Symbolic Logic", 7: 251-76.
Routley, R. (1977), *Ultralogic as Universal?*, "Relevance Logic Newsletter", 2: 50-90 and 138-75, repr. as the Appendix of Routley (1979b).
Routley, R. (1979a), *Dialectical Logic, Semantics and Metamathematics*, "Erkenntnis", 14: 301-31.
Routley, R. (1979b), *Exploring Meinong's Jungle and Beyond. An Investigation of Noneism and the Theory of Items*, Canberra: Dipartimental Monograph, Australian National University.
Routley, R. and R.K. Meyer (1973), *The Semantics of Entailment I*, in Leblanc (1973): 194-243.
Routley, R. and R.K. Meyer (1976), *Dialectical Logic, Classical Logic and the Consistency of the World*, "Studies in Soviet Thought", 16: 1-25.
Routley, R., R.K. Meyer, V. Plumwood and R. Brady (1982) (eds.), *Relevant Logics and Their Rivals I*, Atascadero, Ca.: Ridgeview.
Routley, R., and G. Priest (1992), *Simplified Semantics for Basic Relevant Logics*, "Journal of Philosophical Logic", 21: 217-32.
Routley, R. and V. Routley (1972), *Semantics for First Degree Entailment*, "Nous", 6: 335-59.
Routley, R. and V. Routley. (1985), *Negation and Contradiction*, "'Revista Colombiana de Matemáticas", 19: 201-31.
Russell, B. (1903), *The Principles of Mathematics*, Cambridge: Cambridge University Press.
Russell, B. (1905), *On Some Difficulties in the Theory of Transfinite Numbers and Order Types*, "Proceedings of the London Mathematical Society", 4: 29-53.
Russell, B. (1908), *Mathematical Logic as Based on the Theory of Types*, "American Journal of Mathematics", 30: 222-62, repr. in van Heijenoort (1967): 150-82.

Russell, B. (1910), *The Theory of Logical Types*, in *The Collected Papers of Bertrand Russell*, vol. 6, London-New York: Routledge 1992: 4-31.
Russell, B. (1919), *Introduction to Mathematical Philosophy*, London: Allen & Unwin.
Russell, B. (1923), *Vagueness*, "Australasian Journal of Philosophy and Psychology", 1: 84-92, now in Keefe and Smith (1996): 61-8.
Russell, B. (1940), *An Inquiry into Meaning and Truth*, London: Allen & Unwin.
Russell, B. (1959), *My Philosophical Development*, London: Allen & Unwin.
Russell, B. and A.N. Whitehead (1910-25), *Principia mathematica*, Cambridge: Cambridge University Press.
Ryle, G. (1950), *Heterologicality*, "Analysis", 11: 61-9.
Sainsbury, R.M. (1991), *Logical Forms*, Oxford: Blackwell.
Sainsbury, R.M. (1995), *Paradoxes*, Cambridge: Cambridge University Press.
Sainsbury, R.M. (1997), *Can Rational Dialetheism Be Refuted By Considerations about Negation and Denial?*, "Protosociology", 10: 215-28.
Sainsbury, R.M. (2004), *Option Negation and Dialetheias*, in Priest, Beall and Armour-Garb (2004): 85-92.
Sacks, M. (1989), *The World We Found. The Limits of Ontological Talk*, LaSalle, Ill: Open Court.
Saka, P. (2001), *Exploding the Myth of Paraconsistent Logics*, Unpublished MS.
Schiffer, S. (1998), *Two Issues of Vagueness*, "The Monist", 81/2: 193-214.
Schlick, M. (1930), *Die Wende der Philosophie*, "Erkenntnis", 1: 4-11.
Schlick, M. (1932), *Positivismus und Realismus*, "Erkenntnis", 3: 1-31.
Schotch, P.K. (1993), *Paraconsistent Logic: the View from the Right*, "Philosophy of Science Association", 2: 421-9.
Schotch, P.K. and R.E. Jennings (1980), *Inference and Necessity*, "Journal of Philosophical Logic", 9: 327-40.
Schotch, P.K. and R.E. Jennings (1989), *On Detonating*, in Priest, Routley and Norman (1989): 306-27.
Schurz, G. (1991), *Relevant Deduction*, "Erkenntnis", 35: 391-437.
Shanker, S.G. (1988) (ed.), *Gödel's Theorem in Focus*, London: Croom Helm.
Shapiro, S. (2002), *Incompleteness and Inconsistency*, "Mind", 111: 817-32.
Shapiro, S. (2004), *Simple Truth, Contradiction, and Consistency*, in Priest, Beall and Armour-Garb (2004): 336-54.
Skolem, T. (1962), *Abstract Set Theory*, Notre Dame: Notre Dame University Press.
Slaney, J.K. (1984), *A Metacompleteness Theorem for Contraction-Free Relevant Logics*, "Studia Logica", 43: 159-68.
Slaney, J.K. (1989), *RWX is not Curry Paraconsistent*, in Priest, Routley and Norman (1989): 472-81.
Slater, B.H. (1995), *Paraconsistent Logics?*, "Journal of Philosophical Logic", 24: 451-4.
Smiley, T. (1993), *Can Contradictions Be True? I*, "Proceedings of the Aristotelian Society", 67: 17-34.

Smullyan, R. (1988), *Forever undecided. A puzzle guide to Gödel*, Oxford: Oxford University Press.
Smullyan, R. (1992), *Gödel's Incompleteness Theorems*, Oxford: Oxford University Press.
Sorensen, R. (2001), *Vagueness and Contradiction*, Oxford: Oxford University Press.
Sorensen, R. (2003), *A Brief History of the Paradox: Philosophy and the Labyrinths of the Mind*, Oxford: Oxford University Press.
Sosa, E. (1999), *Existential Relativity*, "Midwest Studies in Philosophy", 23: 132-43.
Stalnaker, R.C. (1968), *A Theory of Conditionals*, in N. Rescher (ed.), *Studies in Logical Theory*, (suppl. monograph to "American philosophical Quarterly"): 98-112.
Stalnaker, R.C. (1984), *Inquiry*, Cambridge, Mass.: MIT Press.
Stephenson, G.H. (1975), *Entailment, Negation and Disjunctive Syllogism*, "Philosophical Studies", 27: 377-87.
Strawson, P.F. (1952), *Introduction to Logical Theory*, New York: Wiley & Sons.
Strawson, P.F. (1959), *Individuals. An Essay in Descriptive Metaphysics*, London: Methuen.
Strawson, P.F. (1997), E*ntity & Identity and Other Essays*, Oxford: Oxford University Press.
Sundholm, G. (1983), *Systems of Deduction*, in Gabbay and Guenthner (1983-1989), vol. I: 133-88.
Suppes, P. (1960), *Axiomatic Set Theory*, Princeton, N.J.: van Nostrand.
Szabo, M.E. (1969) (ed.), *The Collected Papers of G. Gentzen*, Amsterdam: North-Holland.
Talmy, L. (2000), *Toward a Cognitive Semantics*, Cambridge, Mass.: MIT Press.
Tanaka, K. (1998a), *What Does Paraconsistency Do? The Case of Belief Revision*, in T. Childers (ed.), *The Logica Yearbook 1997*, Prague.
Tanaka, K. (1998b), *To be Something and Something Else: Dialetheic Tense Logic*, "Logique et Analyse", 161-3: 189-202.
Tanaka, K. (2005)*, The AGM Theory and Inconsistent Belief Change*, "Logique et Analyse", 189-92: 113-50.
Tappenden, J. (1993), *The Liar and Sorites Paradox: Toward a Unified Treatment*, "Journal of Philosophy", 90: 551-77.
Tappenden, J. (1999), *Negation, Denial and Language Change in Philosophical Logic*, in Gabbay and Wansing (1999): 261-98.
Tarski, A. (1936), *O ugruntowaniu naukowej semantyki*, "Przeglad Filozoficzny", 39: 50-7.
Tarski, A. (1944), *The Semantic Conception of Truth*, "Philosophy and Phenomenological Research", 4: 341-76, repr. in Blackburn and Simmons (1993): 115-43.

Tarski, A. (1956), *Logic, Semantics, Metamathematics. Papers from 1923 to 1938*, Oxford: Oxford University Press.
Tennant, N. (1982), *Proof and Paradox*, "Dialectica", 36: 265-85.
Tennant, N. (1984), *Perfect Validity, Entailment and Paraconsistency*, "Studia Logica", 43: 181-200.
Tennant, N. (1998), *Critical Notice of* Beyond the Limits of Thought, "Philosophical Books", 39: 20-38.
Tennant, N. (2004), *An Anti-Realist Critique of Dialetheism*, in Priest, Beall e Armour-Garb (2004): 355-84.
Troelstra, A.S. and D. van Dalen (1988), *Constructivism in Mathematics: An Introduction*, Amsterdam: North-Holland.
Tye, M. (1990), *Vague Objects*, "Mind", 99: 535-57.
Unger, P. (1979a), *There Are No Ordinary Things*, "Synthese", 41: 117-54.
Unger, P. (1979b), *Why There Are No People*, "Midwest Studies in Philosophy", 4: 177-222.
Urbas, I. (1989), *Paraconsistency and the C-systems of da Costa*, "Notre Dame Journal of Formal Logic", 30:583-97.
Urbas, I. (1990), *Paraconsistency*, "Studies in Soviet Thought", 39: 343-54.
Urchs, M. (1995), *Discursive Logic: Towards a Logic of Rational Discourse*, "Studia logica", 54: 231-49.
Urchs, M. (2002), *On the Role of Adjunction in Para(in)consistent Logic*, in Carnielli, Coniglio and D'Ottaviano (2002): 487-500.
Urquhart, A. (1984), *The Undecidability of Entailment and Relevant Implication*, "Journal of Symbolic Logic", 49: 1059-73.
Urquhart, A. (1986), *Many-Valued Logic*, in Gabbay and Guenthner (1983-1989), vol. III: 71-116.
Vanackere, G. (1999), *Minimizing Ambiguity and Paraconsistency*, "Logique et Analyse", 165-6: 139-60.
Varzi, A.C. (1997), *Inconsistency without Contradiction*, "Notre Dame Journal of Formal Logic", 38: 621-39.
Varzi, A.C. (1999), *An Essay in Universal Semantics*, Dordrecht: Kluwer.
Varzi, A.C. (2000), *Supervaluationism and Paraconsistency*, in Batens, Mortensen, Priest and van Bendegem (2000): 279-97.
Varzi, A.C. (2001), *Parole, oggetti, eventi e altri argomenti di metafisica*, Roma: Carocci.
Varzi, A.C. (2004), *Conjunction and Contradiction*, in Priest, Beall and Armour-Garb (2004): 93-110.
Visser, A. (1984), *Four-Valued Semantics and the Liar*, "Journal of Philosophical Logic", 12: 181-212.
Wang, H. (1950), *A Formal System of Logic*, "Journal of Symbolic Logic", 15: 35-32.
Wang, H. (1986), *Quine's Logical Ideas in Historical Perspective*, in Hahn and Schlipp (2004): 623-43.

Wansing, H. (1996) (ed.), *Negation. A Notion in Focus*, Berlin-New York: De Gruyter.
Weingartner, P. (1990), *Antinomies and Paradoxes and their Solutions*, "Studies in Soviet Thought", 39: 313-32.
Weir, A. (1999), *Review of* Beyond the Limits of Thought, "Philosophical Quarterly", 49: 122-5.
Wiggins, D. (1980), *Sameness and Substance*, Oxford: Basil Blackwell.
Wiggins, D. (2001), *Sameness and Substance Renewed*, Cambridge: Cambridge University Press.
Williams, J.N. (1982), *Believing the Self-Contradictory*, "American Philosophical Quarterly", 19: 279-86.
Williamson, T. (1992), *Vagueness and Ignorance*, "Proceedings of the Aristotelian Society", 66: 145-62; repr. in Keefe and Smith (1996): 265-80.
Wittgenstein, L. (1921), *Logisch-philosophische Abhandlung*, "Annalen der Naturphilosophie", 14, rev. ed. *Tractatus logico-philosophicus*, London: Routledge & Kegan Paul, 1922, repr. New York: Barnes & Noble.
Wittgenstein, L. (1953), *Philosophische Untersuchungen*, Oxford: Basil Blackwell.
Wittgenstein, L. (1956), *Bemerkungen über die Grundlagen der Mathematik*, Oxford: Basil Blackwell.
Wittgenstein, L. (1976), *Lectures on the Foundations of Mathematics*, Ithaca, N.Y.: Cornell University Press.
Woods, J. (1989), *The Relevance of Relevant Logic*, in Norman and Sylvan (1989): 77-86.
Woods, J. (2003), *Paradox and Paraconsistency. Conflict Resolution in the Abstract Sciences*, Cambridge: Cambridge University Press.
Wright, C. (1980), *Wittgenstein's Philosophy of Mathematics*, London: Duckworth.
Wright, C. (1993), *On an Argument on Behalf of Classical Negation*, "Mind", 102: 123-31.
Yablo, S. (1985), *Truth and Reflection*, "Journal of Philosophical Logic", 14: 297-349.
Yablo, S. (1993), *Paradox Without Self-Reference*, "Analysis", 53: 251-2.
Yagisawa, T. (1988), *Beyond Possible Worlds*, "Philosophical Studies", 53: 175-204.
Zeleny, J. (1990), *On Dialectical Inconsistency*, "Studies in Soviet Thought", 39: 199-203.
Zermelo, E. (1908), *Untersuchungen über die Grundlagen der Mengenlehre I*, "Mathematische Annalen", 65: 261-81, tr. *Investigations on the Foundations of Set Theory I*, in van Heijenoort (1967): 199-215.

www.ingramcontent.com/pod-product-compliance
Ingram Content Group UK Ltd.
Pitfield, Milton Keynes, MK11 3LW, UK
UKHW021316180426
11947UKWH00015B/1272